New Technologies for Electrochemical Applications

New Technologies for Electrochemical Applications

Edited by
Mu. Naushad, Saravanan Rajendran, and
Abdullah M. Al-Enizi

CRC Press is an imprint of the
Taylor & Francis Group, an **informa** business

CRC Press
Taylor & Francis Group
6000 Broken Sound Parkway NW, Suite 300
Boca Raton, FL 33487-2742

© 2020 by Taylor & Francis Group, LLC
CRC Press is an imprint of Taylor & Francis Group, an Informa business

No claim to original U.S. Government works

International Standard Book Number-13: 978-0-367-19067-5 (Hardback)

This book contains information obtained from authentic and highly regarded sources. Reasonable efforts have been made to publish reliable data and information, but the author and publisher cannot assume responsibility for the validity of all materials or the consequences of their use. The authors and publishers have attempted to trace the copyright holders of all material reproduced in this publication and apologize to copyright holders if permission to publish in this form has not been obtained. If any copyright material has not been acknowledged, please write and let us know so we may rectify in any future reprint.

Except as permitted under U.S. Copyright Law, no part of this book may be reprinted, reproduced, transmitted, or utilized in any form by any electronic, mechanical, or other means, now known or hereafter invented, including photocopying, microfilming, and recording, or in any information storage or retrieval system, without written permission from the publishers.

For permission to photocopy or use material electronically from this work, please access www.copyright.com (http://www.copyright.com/) or contact the Copyright Clearance Center, Inc. (CCC), 222 Rosewood Drive, Danvers, MA 01923, 978-750-8400. CCC is a not-for-profit organization that provides licenses and registration for a variety of users. For organizations that have been granted a photocopy license by the CCC, a separate system of payment has been arranged.

Trademark Notice: Product or corporate names may be trademarks or registered trademarks, and are used only for identification and explanation without intent to infringe.

Visit the Taylor & Francis Web site at
http://www.taylorandfrancis.com

and the CRC Press Web site at
http://www.crcpress.com

Contents

Preface vii
Acknowledgments ix
Editors xi
Contributors xiii

Chapter 1 Electrochemistry: Different Materials and Applications: An Overview 1

    D. Durgalakshmi, Saravanan Rajendran, and Abdullah M. Al-Enizi

Chapter 2 Controlled Electrochemical Deposition for Materials Synthesis 25

    T. Sivaranjani, T. A. Revathy, and A. Stephen

Chapter 3 Photovoltaic Energy Generation System: Material, Device, and Fabrication 49

    Ananthakumar Soosaimanickam and Moorthy Babu Sridharan

Chapter 4 Role of Advanced Materials in Electrochemical Supercapacitors 63

    Joyita Banerjee, Kingshuk Dutta, and M. Abdul Kader

Chapter 5 Microbial Electrochemical Technologies for Fuel Cell Devices 83

    S. V. Sheen Mers, K. Sathish-Kumar, L. A. Sánchez-Olmos, M. Sánchez-Cardenas, and Felipe Caballero-Briones

Chapter 6 Photoelectrochemical Process for Hydrogen Production 105

    S. Devi and V. Tharmaraj

Chapter 7 Recent Trends in Chemiresistive Gas Sensing Materials 121

    Baskaran Ganesh Kumar, J. Nimita Jebaranjitham, and Saravanan Rajendran

Chapter 8 Role of Innovative Material in Electrochemical Glucose Sensors 135

    R. Suresh, Claudio Sandoval, Eimmy Ramirez, R. V. Mangalaraja, and Jorge Yáñez

Chapter 9 Electrocatalysts for Wastewater Treatment 153

    Prasenjit Bhunia, Kingshuk Dutta, and M. Abdul Kader

Chapter 10 Development and Characterization of A-MWCNTs/Tetra Functional Epoxy Coatings for Corrosion and Prevention of Mild Steel 169

    D. Duraibabu, A. Gnanaprakasam, and S. Ananda Kumar

Chapter 11 Electrochemical Oxidation Reaction for the Treatment of Textile Dyes 181
Eswaran Prabakaran and Kriveshini Pillay

Chapter 12 Materials Involved in Electrocoagulation Process for Industrial Effluents 193
Carlos Navas-Cárdenas, Herman Murillo, Maibelin Rosales, Cesar Ron, and Florinella Muñoz

Chapter 13 Role of Organic Materials in Electrochemical Applications 217
S. Ganesan

Chapter 14 Advanced Conducting Polymers for Electrochemical Applications 233
R. Suresh, R. V. Mangalaraja, Paola Santander, and Jorge Yáñez

Chapter 15 Electrochemical Studies for Biomedical Applications 251
Rajesh Parsanathan

Index 267

Preface

A big thanks to Faraday, who gave us the beautiful laws of electrochemistry, and to so many electrochemists who have made a contribution to enrich the foundations of electrochemistry, on which we are laying a small pebble as a stepping stone to new technological applications using the basics of the field. Electrochemistry is still moving from its basic principles to new innovations. One recent notable advancement in the field was by Prof. Allen Brad, the father of modern electrochemistry, with his works on electrochemiluminescence. Great technological advances will occur when there is an interdisciplinary work among the branches of science and technology. Buckminster Fuller, the inventor of fullerene carbon nanostructures, said, "You never change things by fighting the existing reality. To change something, build a new model that makes the existing model obsolete." And hence, new changes related to affordable costs, easy handling, and almost zero pollution to the environment are greatly needed in the present scenario. The branches of electrochemistry have guided its development from metal extraction, batteries, fuel cells, and supercapacitors to photovoltaic cells.

In this book, the chapters are organized to give readers an overview of the current field of electrochemical applications. Some of the contents of the first chapter give the historical timeline of the advancements in the field of electrochemistry; most of the work in the 19th century focused on the synthesis of metals and their alloys by the electrochemical deposition method. The second chapter is devoted to the controlled electrochemical deposition method for material synthesis. With a view to covering the most recent technologies and applications of electrochemistry, the topics of the electrochemical application of materials in the fields of (i) photovoltaic energy, (ii) electrochemical supercapacitors, (iii) microbial fuel cells, (iv) hydrogen generation, (v) electrocatalyst and electrocoagulation for wastewater treatments, and (vi) biosensors and gas sensors are emphasized in different chapters. To support the subject of current material advancement in electrochemical applications, chapters are devoted to the topics of organic material and conducting polymers for electrochemical applications.

Although there have been various advances in the fields of science and technology, the alarms related to global warming push us to think about materials with energy-related and environmental applications. This may be the fastest-growing field for the future of electrochemistry. While moving in this direction, one must not forget the words of Dr. Martin Luther King Jr., the youngest man to receive the Nobel Peace Prize: "But today our very survival depends on our ability to stay awake, to adjust to new ideas, to remain vigilant and to face the challenge of change. We must work passionately and indefatigably to bridge the gulf between our scientific progress and our moral progress."

Acknowledgments

Dr. Saravanan Rajendran would like to express his sincere thanks to Prof. Francisco Gracia (DIQBT, University of Chile), Prof. Lorena Cornejo Ponce (Department of Mechanical Engineering, Universidad de Tarapacá), and Prof. Rodrigo Palma (Director, SERC) for their constant encouragement and valuable support to complete the task. Further thanks is extended to financial support from the Government of Chile (CONICYT-FONDECYT-Project No. 11170414), SERC (CONICYT/FONDAP/15110019), and Faculty of Engineering, Department of Mechanical Engineering, Universidad de Tarapacá, Arica, Chile.

Dr. Mu. Naushad and Dr. Abdullah M. Al-Enizi extend their appreciation to the Deanship of Scientific Research at King Saud University, Saudi Arabia, for their support.

Editors

Dr. Mu. Naushad is as an associate professor in the Department of Chemistry, College of Science, King Saud University (KSU), Riyadh, Kingdom of Saudi Arabia. He earned his MSc and PhD in analytical chemistry from Aligarh Muslim University, Aligarh, India, in 2002 and 2007, respectively. He has vast research experience in the fields of analytical chemistry, materials chemistry, and environmental science. He holds several US patents, and has authored more than 280 articles in international journals of repute, 20 book chapters, and several books published by renowned international publishers. He has more than 9100 citations with a Google Scholar h-index of >56. He has successfully run several research projects funded by the National Plan for Science and Technology (NPST) and King Abdulaziz City for Science and Technology (KACST), Kingdom of Saudi Arabia. He is the editor/editorial member of several reputed journals including *Scientific Reports* (Nature); *Process Safety and Environmental Protection* (Elsevier); *Journal of Water Process Engineering* (Elsevier); and *International Journal of Environmental Research and Public Health* (MDPI). He is also the associate editor for *Environmental Chemistry Letters* (Springer) and *Desalination and Water Treatment* (Taylor & Francis). He has been awarded the Scientist of the Year Award (2015) from the National Environmental Science Academy, Delhi, India; Scientific Research Quality Award (2019), King Saud University, Saudi Arabia; and Almarai Award (2017), Saudi Arabia.

Dr. Saravanan Rajendran earned his PhD in physics-material science in 2013 from the Department of Nuclear Physics, University of Madras, Chennai, India. He was awarded the University Research Fellowship (URF) during the years 2009–2011 by the University of Madras. After working as an assistant professor at Dhanalakshmi College of Engineering, Chennai, India, in 2013–2014, he was awarded the SERC and CONICYT-FONDECYT post-doctoral fellowship, University of Chile, Santiago, in 2014–2017. He worked in the research group of Professor John Irvine, School of Chemistry, University of St. Andrews, UK, as a postdoctoral research fellow within the framework of a EPSRC-Global Challenges Research Fund for the removal of blue-green algae and their toxins (2017–2018). Currently, he is a research associate in the Faculty of Engineering, Department of Mechanical Engineering, University of Tarapacá, Arica, Chile. He is also a postdoctoral research fellow at SERC, University of Chile, Santiago, Chile. He is associate editor for the *International Journal of Environmental Science and Technology* (Springer). His research interests focus on the areas of nanostructured functional materials, photophysics, surface chemistry, and nanocatalysts for renewable energy and wastewater purification. He has authored papers in several international peer-reviewed journals, five book chapters, and three books with renowned international publishers.

Dr. Abdullah M. Al-Enizi earned his PhD in 2013 jointly from King Saud University (Kingdom of Saudi Arabia) and University of Texas at Austin, and then joined the Department of Chemistry of King Saud University as an assistant professor. His research interests are polymeric materials, porous nanomaterials, catalysis, and electrochemistry. He has authored more than 62 publications of high impact and has three US patents. His Scopus citation is 2500, and he is a leading researcher in advanced polymers and hybrid nanomaterials. He is also a life member of international scientific societies.

Contributors

M. Abdul Kader
School for Advanced Research in Polymers (SARP)
Central Institute of Plastics Engineering and Technology (CIPET)
Chennai, India

S. Ananda Kumar
Department of Chemistry
Anna University
Chennai, India

Joyita Banerjee
Department of Chemical Engineering
University of Pittsburgh
Pittsburgh, Pennsylvania

Prasenjit Bhunia
Department of Chemical Engineering
Indian Institute of Technology
Kharagpur, India

Felipe Caballero-Briones
Materials and Technologies for Energy, Health and the Environment (GESMAT)
Instituto Politécnico Nacional, CICATA
Altamira, México

S. Devi
Department of Inorganic Chemistry
University of Madras
Chennai, India

Kingshuk Dutta
School for Advanced Research in Polymers (SARP)
Central Institute of Plastics Engineering and Technology (CIPET)
Chennai, India

D. Duraibabu
Department of Chemistry
Anna University
Chennai, India

D. Durgalakshmi
Department of Medical Physics
Anna University
Chennai, India

A. Gnanaprakasam
Department of Chemistry
Anna University
Chennai, India

S. Ganesan
Department of Chemistry
SRM Institute of Science and Technology
Tamil Nadu, India

Baskaran Ganesh Kumar
Department of Chemistry
PSR Arts and Science College

and

Department of Science and Humanities
PSR Engineering College
Tamil Nadu, India

R. V. Mangalaraja
Department of Materials Engineering
University of Concepción
Concepción, Chile

and

Technological Development Unit (UDT)
University of Concepción
Coronel, Chile

Herman Murillo
Department of Chemical Engineering
University of Santiago de Chile
Santiago, Chile

Florinella Muñoz
Department of Nuclear Sciences
Escuela Politécnica Nacional
Quito, Ecuador

Carlos Navas-Cárdenas
Department of Chemical Engineering, Biotechnology and Materials
University of Chile
Santiago, Chile

and

Biomass to Resources Group
Amazon Regional University Ikiam
Tena, Ecuador

J. Nimita Jebaranjitham
P.G. Department of Chemistry
Women's Christian College
Chennai, India

Rajesh Parsanathan
Department of Pediatrics
Louisiana State University Health Sciences Center
Shreveport, Louisiana

Kriveshini Pillay
Department of Chemical Sciences
University of Johannesburg
Johannesburg, South Africa

Eswaran Prabakaran
Department of Chemical Sciences
University of Johannesburg
Johannesburg, South Africa

Eimmy Ramírez
Department of Analytical and Inorganic Chemistry
University of Concepción
Concepción, Chile

T. A. Revathy
Department of Nuclear Physics
University of Madras
Chennai, India

Cesar Ron
Department of Geology & Geophysics
University of Utah
Salt Lake City, Utah

Maibelin Rosales
Advanced Mining Technology Center (AMTC)
University of Chile
Santiago, Chile

Paola Santander
Center of Biotechnology
University of Concepción
Concepción, Chile

and

Millennium Nucleus on Catalytic Processes towards Sustainable Chemistry (CSC)
Santiago, Chile

M. Sánchez-Cardenas
Instituto Tecnológico El Llano Aguascalientes
Tecnológico Nacional de México
Aguascalientes, México

L. A. Sánchez-Olmos
Universidad Politécnica de Aguascalientes
Aguascalientes, México

Claudio Sandoval
Department of Analytical and Inorganic Chemistry
University of Concepción
Concepción, Chile

K. Sathish-Kumar
Instituto Tecnológico El Llano Aguascalientes (ITEL)
Aguascalientes, Mexico

and

Tecnológico Nacional de México (TecNM)
Mexico City, Mexico

S. V. Sheen Mers
Indian Institute of Technology
Chennai, India

T. Sivaranjani
Department of Nuclear Physics
University of Madras
Chennai, India

Contributors

Moorthy Babu Sridharan
Crystal Growth Centre
Anna University
Chennai, India

Ananthakumar Soosaimanickam
Crystal Growth Centre
Anna University
Chennai, India

and

Institute of Materials (ICMUV)
University of Valencia
Valencia, Spain

R. Suresh
Department of Analytical and Inorganic
   Chemistry
University of Concepción
Concepción, Chile

A. Stephen
Department of Nuclear Physics
University of Madras
Chennai, India

V. Tharmaraj
Department of Analytical Chemistry
National Chung-Hsing University
Taichung, Taiwan

Jorge Yáñez
Department of Analytical and Inorganic
   Chemistry
University of Concepción
Concepción, Chile

1 Electrochemistry: Different Materials and Applications: An Overview

D. Durgalakshmi, Saravanan Rajendran, and Abdullah M. Al-Enizi

CONTENTS

1.1 Introduction to Electrochemistry: History 1
1.2 Oxidation and Reduction 4
1.3 Material Synthesis by Electrode Position Method 5
    1.3.1 Electrodeposition of Metals and Alloys 5
    1.3.2 Electrodeposition of Organic Materials...10
    1.3.3 Electrodeposition of Metals as Sacrificial Coatings..10
    1.3.4 Electrophoretic Deposition 10
1.4 Electrochemistry Application for Health, Energy, and the Environment 11
    1.4.1 Hydrogen Generation...11
    1.4.2 Chemiresistive Gas Sensor 12
    1.4.3 Biosensor..15
        1.4.3.1 Glucose 15
        1.4.3.2 H_2O_2 Sensor 16
    1.4.4 Wastewater Treatment 17
1.5 Summary and Future Directions 17
References..18

1.1 INTRODUCTION TO ELECTROCHEMISTRY: HISTORY

The field of electrochemistry has received intense attention in applied and fundamental research areas. The history of electrochemistry helps us to understand the technological applications in both industrial production and daily life in the 21st century (Table 1.1). The important discoveries in electrochemistry originated with Antoine Lavoisier, who is also called the "father of modern chemistry," from his discovery of oxygen and explanation of the oxidation chemical process during respiration (Ihde 1984). The story of electrochemistry began at the end of the 18th century when Luigi Galvani demonstrated that the transmission of nerve impulses was associated with an electric current (Bensaude-Vincent and Stengers 1996). Later, Alessandro Volta, to challenge Luigi Galvani's claims to have demonstrated that animals produce electricity, announced his invention of the voltaic pile (Mertens 1998). The voltaic pile produced a continuous current and thus opened two new areas of study: the chemical production of electricity and the effects of electricity on chemicals. In May 1800, Nicholson and Carlisle modeled the first ever hydrogen production experiment by the electrolysis of water with the help of battery power. Based on this, the theory of electrolytes and the concept of breaking of water molecules into O and H due to the separation of charges was proposed by Grotthuss in 1807. Just prior to this, in 1806, Sir Humphry Davy presented the electrical

TABLE 1.1
Notable Inventors and Inventions in the History of Electrochemistry

Year/Era	Inventor	Invention
18th century	Luigi Galvani	Transmission of nerve impulses associated with an electric current
18th century	Alessandro Volta	Voltaic pile—Chemical production of electricity and the effects of electricity on chemicals
1800	Nicholson and Carlisle	Electrolysis of water
1806	Humphry Davy	Electrical theory of chemical affinity
1807	Theodor Grotthuss	Theory of electrolytes
1821	Michael Faraday	First electric motor
1831	Michael Faraday	First dynamo
1834	Michael Faraday	Stated first and second laws of electrolysis
1926	Henri Becquerel	Hydrogen evolution electrodes
1936	John Frederic Daniell	Daniell cell
1939	Sir William Grove	First fuel cell
1874	Friedrich Kohlrausch	Redefined the theory on the conductivity of electrolytes
1887	Svante Arrhenius	Arrhenius theory on acid–base reaction
1897	Cart Bottger	Development of H_2 electrodes and measurement of pH
1902	Frederick Gardner Cottrell	Electrode kinetics by diffusion
1905	Julius Tafel	Law of electrode overpotential
1914	Thomas Edison	Nickel/Iron battery
1914	Nikola Tesla	Alternating current
Beginning of the 20th century	John Alfred Valentine Butler, Max Volmer, Heinz Gerisher, Rudolph A. Marcus and others	Studies on the transport of charged species and thermodynamics considerations
Middle of the 20th century	—	Understanding the chemical and electronic structure of solids–solution interfaces
		Studies on *in situ* and *ex situ* spectroscopic techniques with electrochemical experiments
1932–1942	James J. Drumm	Zinc–nickel alkaline battery
1932	Francis Thomas Bacon	First fuel cell device
—	Mond and Langer	Improved hydrogen–oxygen cell
1948	Arne Tiselius	Sophisticated electrophoretic apparatus
1937	Arne Tiselius	Protein electrophoresis
—	Wiktor Kemula	Hanging mercury drop electrode (HMDE)
1950	—	Launch of modern electroanalytical chemistry in Warsaw

theory of chemical affinity (Davy 1840). Davy, of the Royal Institution in London, realized that the production of electricity by the voltaic pile depended on the existence of chemical reactions, not just on the contact of different types of metals. Davy's student and successor, Michael Faraday, followed the relationship between electricity and magnetism and invented the first electric motor (in 1821) and the first dynamo (in 1831). He also published two laws of electrochemistry in 1834 and predicted the quantity of product that results from passing a certain amount of current through a chemical compound or its solution, a process that he named "electrolysis" (Aftalion 2001). Based on this method, a lot of industries produced metals in large amounts. The two-compartment battery was proposed based on the observation of hydrogen evolution in electrodes by Becquerel in 1926 and developed by Daniell in 1936, giving birth to the Daniell cell. The reversibility of the water electrolysis reaction was tried by Grove in 1939, which is the basis of the first fuel cell. At the end of the 19th century, theoretical aspects of electrochemistry were refined due to the development of the theory on conductivity of electrolytes by Kohlrausch in 1874, followed by Arrhenius's theory on acid–base reaction in 1887; the development of the hydrogen electrode and measurement of pH by Bottger in 1897; electrode kinetics with mass transport by diffusion written by Cottrell in 1902, shedding a light onto many electrochemical processes; and the empirical law of electrode overpotential discovered by Tafel in 1905. The first electrodeposition of alloys probably took place at about the same time that cyanides were introduced into electroplating. De Ruolz is generally credited with having been the first to deposit brass and bronze. The bronze bath that he described in 1842 apparently was similar to the modern bath in that it contained a cyanide copper complex and a stannate. The electrodeposition of brass apparently was the only alloy plating process that had any commercial application up to modern times, although some mention is also made of bronze plating in the early books on plating published before 1870 (Brenner 2013). The only other commercial alloy plating, other than brass plating, apparently was the deposition of gold alloys, which was practiced to obtain different shades of gold, as is done today. The use of additions of copper or silver to gold cyanide plating baths to obtain various shades of red, green, and green-white was described by L. Eisner in "Die galvanische Vergoldung und Versilberung," published in Leipzig in 1851 (Blaschko 1889). Before 1900, the knowledge of alloy plating was very limited and entirely empirical, as no systematic or scientific research on the subject had been carried out. Probably the first scientific work on alloy deposition was the classic work of Fritz Spitzer published in 1905, which dealt with the cathode potentials involved in the electrodeposition of brass. The textbooks of the period around 1910 showed very little appreciation of the factors governing alloy deposition. Among the more satisfactory ones were Field's text, *The Principles of Electro-Deposition* (Field 1911), which incorporated some of his work on brass plating, and a book by M. Schlötter, titled *Galvanostegie* (Brenner 1963). In 1914, Tesla was giving a contradictory opinion to Edison on alternating current. In the 20th century, electrochemistry was mostly portrayed for the studies related to charge carrier transport of materials and thermodynamic perspectives. With the pioneering work carried out by Butler, Volmer, Gerischer, Marcus, and others, kinetic aspects of electrochemistry have become more important in electrochemical research. These theories are the strong pillars of electrochemistry, helping researchers understand the electrochemical concepts on new organic, inorganic, and biological systems. In the middle of the 20th century there was much interest in understanding the electrochemical aspects of the solid–solution interface. After 1920, research on the electrodeposition of alloys greatly increased, and while much effort was devoted to devising plating baths for particular alloys and to studying the variations in the composition of the alloys with plating conditions, very little of a general theoretical nature was uncovered. The development of new plating baths has remained empirical up to the present. These studies have been accelerated by the application of numerous *in situ* and *ex situ* spectroscopic techniques, which have been combined with electrochemical experiments. Probably the most important advance of this period was the application of X-rays to the elucidation of the structure of electrodeposited alloys. This was first done by Nakamura in 1925 and was followed by the work of Roux and Cournot in 1928. Their research showed that electrodeposited alloys had structures similar to those of the corresponding thermally prepared alloys. Notable inventions are the Drumm traction battery (zinc–nickel alkaline battery) invented by James J. Drumm, which was successfully employed to power a suburban train in Ireland (1932–1942) (Allmand 1932). In 1932, inventions by engineer Francis Bacon resulted in the first successful fuel cell devices, which were further improved by Mond and Langer with a hydrogen–oxygen cell. The Nobel Prize was awarded to Arne Tiselius during 1948 for his work on the invention of a sophisticated electrophoretic apparatus and for his work in protein electrophoresis 1937 (Bacon 1969). In 1950, the hanging mercury drop electrode (HMDE) was developed by Wiktor Kemula, which plays a central role in the field of modern electroanalytical chemistry (Jindra and Heyrovský 2015). Accordingly, a huge number of research articles have been reported in the field of amalgam electrochemistry, cyclic voltammetry, anodic and cathodic stripping of inorganic and organic substrates, and electron transfer rate. The field of electrochemical research provides various potential application areas, starting from qualitative analysis to quantitative analysis, including batteries, fuel cells, photovoltaic devices, supercapacitors, gas sensors, biosensors, corrosion of materials and purification of metals.

1.2 OXIDATION AND REDUCTION

The following definitions are taken from the *American Standard Definitions of Electrical Terms*, which was revised in 1956 by a committee of about a dozen men representing several technical organizations. "Electrochemistry is that branch of science and technology which deals with interrelated transformations of chemical and electric energy" (Lefrou et al. 2012). Oxidation–reduction is a notion first introduced by Lavoisier (Roco et al. 2011) in which the reaction occurs between the dioxygen and the chemical species and is denoted by the oxidation reaction, e.g., $2Hg + O_2 \rightarrow 2HgO$. The modern notion, related to electron transfer, was entrenched at the early 20th century. The formation of redox reaction (oxidation–reduction) engages altering electron shifts at the core-level regions. When a chemical species loses one or more electrons in a reaction, this species is known as oxidation and the reaction undergoes oxidation reaction. When the species gains electrons, it is known as reduction reaction. The following are the overall reaction for Ox/Red half couple reaction,

$$Red \underset{\text{reduction}}{\overset{\text{oxidation}}{\rightleftharpoons}} Ox + ne^-$$

The term *Ox* refers to oxidant, which means able to gain electrons, and the term *Red* refers to reductant, which means able to donate electrons.

The basic criterion for choosing material for electrochemistry is if it is able to conduct electricity. Various materials are used in the field of electrochemical research including both organic and inorganic materials. The direct current power supply is usually connected to the two terminals of the electrochemical materials, which is controlled by the external devices either in the form of voltage or current. The chemical species or electrons can be transferred from one terminal to another terminal of the electrochemical material via electrolyte solution. The term *electrode* is generally used in the field of electrochemistry to mean a metallic strip used as a conductive terminal of the electrochemical devices. Sometimes, the word *modified electrode* or *positive electrode* is used to refer to a metal electrode covered with a thin film of conducting material over the surfaces. Finally, the term *reference electrode* will be used for an electrochemical half-cell. The *anode* is the electrode material where oxidation reactions take place. The electrode material with the reduction reaction is called the *cathode* (Crow 2017).

In the electrochemical system, the power supply is essential for the operation of the electrochemical cell, which is referred to as the electrolyzer mode. The most frequently used power sources in the operation of electrochemical devices are current and voltage. A simple and changeable carrier resistance can be sufficient to study an electrochemical system in the power supply mode. While observing the current in the electrochemical system, it is termed as the intensiostatic or galvanostatic method. This concerns when the imposed current is time-dependent, and the subsequent measurements are called potentiometry. In this instance, the voltages in the system are dignified, while current is executed. When the voltage is executed at the terminals, this denotes a potentiostatic setup. If the executed voltage is time-dependent, the consequent measurements are called amperometry. In this instance, while current is increased, the voltage in the system is amplifed. The electrochemical experimental details can be classified into four main groups, which are shown in Figure 1.1. A similar table could be drawn up for the usual potentiometry techniques, apart from voltammetry, because its equivalent in potentiometry is not used.

Electrochemistry: Different Materials and Applications

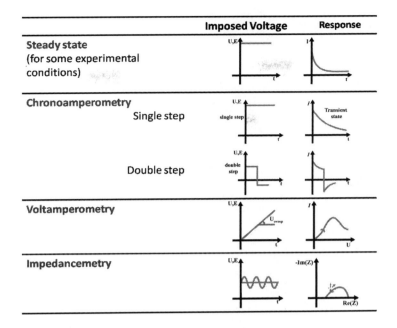

FIGURE 1.1 Brief classification of four main amperometry experiments.

1.3 MATERIAL SYNTHESIS BY ELECTRODE POSITION METHOD

1.3.1 Electrodeposition of Metals and Alloys

The electrodeposition by applying a pulse signal of nanocrystalline metals like silver, gold, nickel, cobalt, iron, platinum, and copper, ranging in size from 5 nm to 100 nm, has been reported (Mohanty 2011). Various morphologies of electrodeposited materials can be modified by choosing the parameters like desired electrolyte, applied potential, current density, pH of the solution, pulsed signal, and duration of the electrodeposition process. For example, gold nanoparticles were uniformly deposited on the carbon electrodes by an electrolytic bath consisting of sulfuric acid and gold auricchloride precursors (Toney et al. 1995; Wang et al. 2013). Surfactants like CTAB, PVP, and TTABr will help to modify the shape of the nanoparticles deposited on the substrates by electrodeposition techniques. Apart from the noble metal nanomaterials, nanocomposites have also been prepared using electroposition with the use of noble metals and some conducting polymers. Porous metal nanostructures were developed by electrodeposition onto anodic alumina disc or polycarbonate membranes with the help of shape modifiers, such as sodium sulfate or potassium hydrogen phosphate (Tiano et al. 2010; Stine 2019). Alloys have received high interest in the technological field due to their ability to engineer the properties beyond the pure metals and to alter them specifically through composition or phase selection. The preparation of alloys using electrodeposition or by thermal treatment is more or less similar in terms of structural or crystallographic phases. In terms of application, magnetic alloys like Ni–Fe or Fe–Co by electrodeposition show better coercivity value than the other preparation techniques due to homogenous distribution, phase purity, and high crystalline nature (Liu et al. 2000; Rahman et al. 2007). Some of the alloys are commercially available by using electrodeposition techniques like Pb–Sn (Liu et al. 2008), Ni–Co, Fe–Ni (Liu et al. 2008), and Sn–Zn (Kazimierczak and Ozga 2013). Some of the earlier reports on electrodeposition of metals and alloys are given in Table 1.2.

TABLE 1.2
Earlier Reports on Electrodeposition of Metals and Alloys

Compound	Alloy/Experiment	Electrolyte	Potential and Time	Temperature	Reference
Aluminum	Al–Mn alloy coating on Mg alloy	AlCl$_3$–NaCl–KCl–MnCl$_2$ molten salts	Cathode current density (D_k) of 40 mA/cm^2 for 30 min	170°C	Zhang et al. 2009
	Al–Cr and Al–Ni alloy	AlCl$_3$–NaCl–KCl molten salt containing CrCl$_2$ and NiCl$_2$	Constant charge density at 40 C cm^{-2} Cr deposit at −0.1 V Ni deposit at −0.4 V	423 K	Ueda et al. 2007
	Al–Li–Mg alloy	LiCl–KCl–MgCl$_2$ molten salt	Current density −1.337 A cm^{-2} Potential −2.38 V	620°C	113Ye et al. 2010
	Al–Pt alloy	AlCl$_3$–NaCl–KCl molten salt containing PtCl$_2$	At 1.2 V pure Pt is deposited At 1.0 V Pt, intermetallic compounds of AlPt$_2$, and/or AlPt$_3$ are deposited Lower than 0.8 V, intermetallic compounds of AlPt$_2$ and/or AlPt$_3$.	448 K	Ueda et al. 2011
	Al–Sn alloy	AlCl$_3$–NaCl–KCl–SnCl$_2$ molten salt	0.5 V	423 K	Ueda et al. 2013
	Deposition of Al films	AlCl$_3$–dimethyl sulfone	40 mAcm^{-2}	1 h at 403 K	Salman et al. 2019
Magnetic materials (Fe, Ni)	Fe–Si alloys	KCl–NaCl–NaF–SiO$_2$ molten salts	Current density 30 mA/cm^2, different deposition times: 2 min, 5 min, 10 min, 15 min, 20 min and 25 min	1073 K under argon atm	Yang et al. 2012
	Fe–Pd alloy	[Pd(NH$_3$)$_4$]$^{2+}$ (0.02 M) and [Fe-Citrate]$^{2+}$ (0.2 M)	−850 to −1050 mV		Rezaei et al. 2018
	Nanocrystalline quasi-bulk Fe–C material via electrodeposition	Citrate bath (sodium sulfate as a conducting salt and the small amount of citric acid)	At current densities ranging from −20 mA/cm^2 up to about −50 mA/cm^2 $E = -1000$ to -1200 mV and 22 h		Dehgahi et al. 2017
	Ni–aluminide layer on Nb–W and Nb–W–Mo alloys and Al that can be electrodeposited	Molten salt as the electrolyte	200 Am^{-2} for 3.6, 7.2, 10.8 and 12.6 ks Electrodeposition of Al at −1.4 V for 3.6 ks	1023 K	Sato et al. 2016

(Continued)

Electrochemistry: Different Materials and Applications

TABLE 1.2 (CONTINUED)
Earlier Reports on Electrodeposition of Metals and Alloys

Compound	Alloy/Experiment	Electrolyte	Potential and Time	Temperature	Reference
	Tantalum on NiTi alloy	1-butyl-1-methyl-pyrrolidinium bis(trifluoromethylsulfonyl)imide ([BMP]Tf$_2$N) containing TaF$_5$ as a source of tantalum	−1.8 V	RT	El Abedin et al. 2005
	Ni–Al$_2$O$_3$–SiC nanocomposite coatings are electrodeposited on a steel	Watt's bath containing Al$_2$O$_3$ and SiC nanoparticles	Current density 1 A/dm^2 $E = -0.22$ V	RT	Dehgahi et al. 2017
Zinc	Al–Zn on an active magnesium alloy	ZnCl$_2$-containing AlCl$_3$–EMIC ionic liquid	−0.2 V	RT	Pan et al. 2010
	Metallic Zn on Mg alloy substrates	N-butyl-N-methyl-pyrrolidinium dicyanamide (BMP-DCA) ionic liquids containing ZnCl$_2$ at 323 K	−2.5 V	RT	Deng et al. 2011
	Zinc–cobalt alloys	Zinc chloride-1-ethyl-3-methylimidazolium chloride molten salt containing cobalt(II)	0.15 V	80°C	Chen and Sun 2001
	Zn–Fe multilayered alloy coatings are deposited on mild steel substrates	Solution 1: electrodeposition of Zn–1 wt% Fe alloy layers; solution 2: Zn–10 wt% Fe alloy layers	Current density 30 mA/cm^2	40°C ± 2°C	Panagopoulos et al. 2019
	To study the electrochemical behavior of copper	Trimethyl-n-hexylammoniumbis ((trifluoromethyl)sulfonyl)amide (TMHA-Tf$_2$N)	5.6 V	50°C	Murase et al. 2001
Copper	Cu layer was successfully electrodeposited on mild steel	Copper chloride (CuCl$_2$·2H$_2$O)–1-ethyl-3-methylimidazolium chloride [EMIM] Cl–ethylene glycol [EG] ionic liquid	−0.70 V	28°C ± 2°C	Saravanan and Mohan 2013
	Deposition of copper nanoparticles	1-butyl-1-methylpyrrolidinium bis(trifluoromethylsulfonyl)amide ([Py$_{1,4}$] Tf$_2$N) and 1-ethyl-3-methylimidazolium bis(trifluoromethylsulfonyl)amide ([EMIm] Tf$_2$N).	Current of 10 mA	RT	Brettholle et al. 2010
	Copper–tin (Cu–Sn) alloy	1-butyl-3-methylimidazolium chloride ([BMIM]Cl)	Current density = 1.540 mA/cm^2 $E = -1.5 – 1.5$ V	RT	Jie et al. 2019

(Continued)

TABLE 1.2 (CONTINUED)
Earlier Reports on Electrodeposition of Metals and Alloys

Compound	Alloy/Experiment	Electrolyte	Potential and Time	Temperature	Reference
Rare-earth metals	Lanthanum electrodeposition	1-octyl-1-methyl-pyrrolidinium bis(trifluoromethylsulfonyl)imide ionic liquid.	−1.5 V	25°C to 80°C	Legeai et al. 2008
	Electrodeposition of dysprosium (Dy)	Phosphoniumcations with bis (trifluoromethylsulfonyl) amide anions	Potential and the cathode current density were −3.8 V and −3.2 mA cm^{-2}	150°C	Lair et al. 2010
	Cerium oxide thin layer was electrodeposited onto stainless steel	Mixed ionic liquid1-methyl-3-butylimidazolium bis(trifluoromethylsulfonyl)imide)/ethanol solutions	−1.4 V and −0.6 V applied for 2 h	RT	Lisenkov et al. 2010
	Al–Ce alloy on surface of Pt and AA2024 aluminium alloy	AlCl$_3$ in 1-butyl-3-methylimidazolium chloride (BMIM-Cl)	−2.5 V to 2.5 V	Below 80°C	Tachikawa et al. 2009
	To study electrochemical and spectroscopic properties of divalent and trivalent europium species	Hydrophobic ionic liquid, 1-butyl-1-methylpyrrolidinium bis(trifluoromethylsulfonyl)amide (BMPTFSA)	−0.44 V	RT	Tsuda et al. 2001
	Electrochemical behavior of lanthanum ion studied in various AlCl$_3$–EMICl molten salts saturated with LaCl$_3$	LaCl$_3$ in acidic AlCl$_3$–1-ethyl-3-methylimidazolium chloride (AlCl$_3$–EMICl)	−1.95 V	298 K	Kamimoto et al. 2018
	Zinc–rare-earth alloy	Eutectic LiCl–KCl mixture	−1.0 V	473 K for 24 h under vacuum	Jou et al. 2010

(Continued)

TABLE 1.2 (CONTINUED)
Earlier Reports on Electrodeposition of Metals and Alloys

Compound	Alloy/Experiment	Electrolyte	Potential and Time	Temperature	Reference
Nobel metals	Palladium–tin alloys	1-ethyl-3-ethylimidazolium chloride–tetrafluoroborate ionic liquid	0.8 V	120°C	Jou et al. 2009
	Palladium–copper alloys	1-ethyl-3-methylimidazolium chloride–tetrafluoroborate ionic liquid	−0.35 to −0.60 V	120°C	Huang and Sun 2004
	Pt–Zn alloys	40–60 mol% zinc chloride–1-ethyl-3-methylimidazolium chloride ionic liquid containing $PtCl_2$	−0.20 to −0.24 V	90°C	Molodkina et al. 2019
	Silver electrodeposition on Pt (111), Au (111), and Au (100) single-crystal surfaces	Dicyanamide anions: 1-butyl-1-methylpyrrolidinium dicyanamide [BMP][DCA] and 1-butyl-3-methylimidazolium dicyanamide [BMIm][DCA]	0.5 to 0.5 V	23°C ± 1°C	Tavakkoli et al. 2019
	Gold–platinum nanoparticles deposited on an indium tin oxide (ITO) surface	Electrolytes with different Au:Pt ratios (best response is 3:1 ratio of Au:Pt)	0 and −1.5 V	RT	Song et al. 2010

1.3.2 ELECTRODEPOSITION OF ORGANIC MATERIALS

The electrodeposition of organic materials onto single or composite materials will help to protect the surface or decorative purposes for potential applications like self-cleaning windows, antifouling glasses, antibacterial coating on biomaterials, and antifriction in mechanical motors. The main advantage of using organic coating by the electrodeposition technique on the surface of the alloys or a single metal is a uniform and homogenous coating, which helps for protecting steels and alloys from corrosion (Twite and Bierwagen 1998; Hu et al. 2007; Wu et al. 2014). The decorative appearance of the alloys or steels will be enhanced by electrodeposited organic materials onto it. Most of the alloys are decoratively polished to a brighter surface than single metals, which is due to the usage of electrodeposited organic agents. Some of the alloys like Co–Ni and Cu–Sn–Zn (Beattie and Dahn 2005) were used to enhance not only the surface appearance but also to increase the mechanical strength by electrodeposition by organic materials on the surfaces. Due to their diverse properties, organic electrodeposited coatings are used as molds for making plastic utensils or as wear-resistant agents for certain machinery alloys or improving the hardness of certain metals and alloys. In electronic industries, low-melting-point alloys are used as soldering materials for making electronic circuits and devices, and also used to replace the toxic metals in computer chassis.

1.3.3 ELECTRODEPOSITION OF METALS AS SACRIFICIAL COATINGS

The most important use of electrodeposits, whether individual metals or alloys, is as coatings to protect other metals, such as steel, brass, and zinc, from corrosion. For this reason, the main interest in the chemical properties of electrodeposited alloys has centered on the corrosion and tarnish resistance of the alloys with particular emphasis on the protection that alloy coatings afford to steel. To show that the chemical nature of the metal alone does not determine its value as a coating, it may be pointed out that, for the protection of steel against corrosion, metals of diametrically opposite degrees of chemical reactivities are utilized; yet each type satisfactorily affords protection against corrosion but in different applications. The two types of coatings that are applied to steel are designated as cathodic and anodic (Brenner 2013). Cathodic coatings, of which nickel and copper are examples, are chemically less active than steel. In a salt solution, these metals are positive (more noble) to steel. Coatings of these metals protect steel by shielding it from corrosive environments and are effective as long as they completely envelop the steel (Gan et al. 1994; Von Baeckmann et al. 1997). Being less reactive than steel, a given thickness of coating lasts much longer than an equivalent thickness of steel. However, if discontinuities extending down to the steel occur in the coating, the corrosion of the steel is accelerated by the galvanic action that is set up between the coating as the cathode and the steel as an anode in the presence of moisture. In contrast to the cathodic coatings, the anodic coatings, of which zinc and cadmium are the most important, are more active than steel. In a salt solution, these metals are negative (less noble) to steel. Anodic coatings protect steel from corrosion initially, just as do the cathodic coatings, by shielding it from corrosive environments (Sørensen et al. 2009; Shreir 2013), but being chemically active they are attacked more rapidly than cathodic coatings. When, or if, the steel is exposed through discontinuities in the coating, the latter still protects the steel by virtue of galvanic action. In the presence of moisture, the coating becomes the anode of a galvanic cell and is slowly consumed, whereas the steel being cathodic does not corrode until the exposed area becomes so large that parts of it are too far removed from the coated area to receive appreciable current.

1.3.4 ELECTROPHORETIC DEPOSITION

Electrophoretic deposition (EPD) has received great interest among materials scientists for the production of novel inorganic nanostructured and nanoscale materials with various morphologies. Recent advances have been made in the field of electrophoretic deposition and its process route for the fabrication of different varieties of ceramic and metallic nanoparticles, carbon nanotubes, and other nanomaterials. In the 1980s, Hamaker (Besra and Liu 2007) first developed the electrodeposition technique for the preparation of ceramic materials. In the process of EPD coating by applying a direct electric field to the conductive substrate, the charged particles in the liquid medium get attracted and coated on the conductive substrate of opposite charge. The fundamental difference between the electrolytic and electrophoretic deposition techniques are that the deposition takes place from the ionic species (salt solution), and the deposition occurs due to presence of charged particle in the electrolytic medium, respectively. The potential applicability of electrodeposited alloys include the fields of corrosion technology, fuel cells, antimicrobial coating, biomedical implants, wear-resistant materials, load-bearing ceramics, and so on. Some areas are highly enriched in the production of multifunctional coatings with nanostructured materials (Besra and Liu 2007). The EPD technique also provides a unique approach for coating complex structures and advanced ceramic coatings (Boccaccini and Zhitomirsky 2002). The deposition rate and uniformity of the coating are monitored by the stoichiometric ratio of the coating materials. With respect to the applied potential, the driving force of the electrophoretic deposition depends on the charge of the coating materials and the mobility of the electrolyte. By carefully monitoring the driving force, the coating kinetics will be drastically improved. The stability of the coating is directed by the interparticle forces presented in the liquid medium. Some of the forces like repulsive force, van der Waals interaction, and electrostatic interaction acting between the charged particles in the liquid medium control the homogeneity and stability of the deposition. The usage of water in the electrolyte liquid medium is still limited in the electrodeposition technique, which affects the coating process due to oxidation and reduction reactions taking place when potential is applied to the electrodes for coating. This is the major limitation of electrodeposition. All the strategies should be reviewed before fabricating the nanostructures using electrophoretic deposition method (Corni et al. 2008; Gurrappa and Binder 2008; Liu et al. 2015).

1.4 ELECTROCHEMISTRY APPLICATION FOR HEALTH, ENERGY, AND THE ENVIRONMENT

1.4.1 Hydrogen Generation

Design and development for sustainable and renewable energy production in an ecofriendly manner are highly needed for the near future. Concerns about environmental problems arise from the depletion of fossil fuels due to the production of energy sources. In order to avoid the environmental issues developed by the creation of energy, an alternative method is badly needed. Sustainable ways of producing alternative energies in the field of electrochemical techniques have received great interest among material scientists. Among them, the photoelectrochemical reduction of CO_2 and the generation of hydrogen fuels provide the highest energy capacity per unit mass as well as they are ecofriendly energy carriers because they do not create toxins during the production of hydrogen fuels. The most abundant source of hydrogen energy is water, but the splitting of water molecules into the oxygen and hydrogen requires high temperatures, nearly 2000°C, thus making experiments problematic. Photocatalytic water splitting by sunlight radiation is attractive and has been extensively examined in the past few decades. When a photon of light energy is equal or greater than the semiconductor bandgap absorbs, the electron in the conduction band leaves a positive hole in the valence band. In a stipulated time period, photoexcited charge carriers hinder the recombination and travel to the surface, where they react with adsorbed molecules causing reduction and oxidation, respectively (Figure 1.2). These oxidation and reduction reactions occur in the water medium and produce hydrogen and oxygen gases. Suitable semiconducting nanomaterials with transition metal cations have d^0 (Ti^{4+}, Ta^{5+}, and W^{6+}) or d^{10} electron configurations (Ga^{3+}, Ge^{4+}, and Sn^{4+}). Other metal oxides with d orbitals, such as Zn^{2+} and Mo^{6+}, are able to alter the band edge position in a suitable manner to efficiently generate hydrogen fuel.

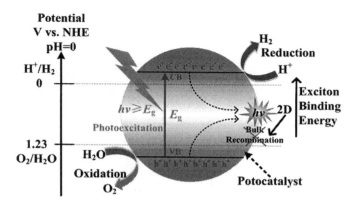

FIGURE 1.2 Schematic illustration of water splitting over semiconductor photocatalysts (Su et al. 2018).

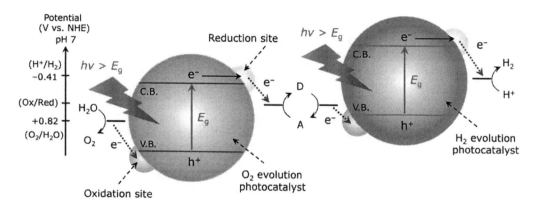

FIGURE 1.3 Schematic representation of artificial direct Z-scheme photocatalyst (Maeda and Domen 2010).

An alternate approach to create hydrogen fuels is called the two-photon system, which resembles Z-scheme photocatalysis. In this system, the flow of electrons shows the signature Z-direction in two types of semiconductor photocatalysts. One semiconducting material acts as an oxidation reactant and the other one acts as a reduction reactant, and the electron can be excited by one of the conduction band and resides at the valence band of another semiconductor. For example, an electron donor/acceptor transport is added to two different photocatalytic sites (Figure 1.3). With the photocatalytic generation of hydrogen, the donor is oxidized to the acceptor by h^+ and this is reduced back to the donor at the O_2 evolution site, while holes oxidize water to O_2. However, at the H_2 evolution photocatalyst, the reduction of H^+ to H_2 and oxidation of the donor to acceptor may stop when the concentration of the acceptor becomes high enough towards reduction of the acceptor competes with the donor. Likewise, the rate of O_2 evolution over the other photocatalyst may decrease when backward oxidation of donor to acceptor becomes competitive.

1.4.2 Chemiresistive Gas Sensor

In monitoring the environment in the aspects of air quality, gas sensing technologies play a crucial role because air quality affects our daily lives. The first commercial metal-based sensor, a hot platinum wire-based gas sensor produced was developed in 1923, and in 1962 Taguchi patented the first oxide-based gas sensor. Various types of gas sensors have been developed, but resistivity-based sensors are widely preferred owing to their inexpensive cost fabrication, smooth operation,

Electrochemistry: Different Materials and Applications

and possible miniaturization. In metal oxide gas sensors, research has been initiated to improve the sensitivity, stability, and selectivity. The first work on metal oxide gas sensors was reported by Seiyama with gas sensors made by ZnO thin films with the basic electrical circuit operating at 485°C (Seiyama et al. 1962). With the pioneering developments in the field of nanotechnology, it has become possible for many researchers to make significant progress with oxide nanostructures, conducting polymers, carbon nanostructures, and 2D materials. Metal oxide semiconductor-based chemiresistive sensors have recently attracted significant attention for a wide variety of applications, including food processing (Neethirajan et al. 2009), environmental monitoring (Fine et al. 2010), the agriculture industry (Lerma-García et al. 2009), and medical diagnosis. Most conventional metal oxide-based gas sensors normally operate at temperatures ranging from 100°C to 400°C. However, owing to an induced growth of metal oxide grains, this will lead to high power consumption, reduced sensor stability, and shorter lifetime. Interestingly, room temperature gas sensors produced with conducting polymers was developed. But they are affected by humidity, are less stable in operating environment, have higher recovery time, and have poor stability. Studies on including surface functionalization with noble metals, using oxide incorporated heterostructures and thermal assistance with ultraviolet (UV) illumination, have been done improve sensitivity and selectivity of these conducting polymers. The gas adoption mechanism and consequent resistance change in a metal oxide semiconductor gas sensor mainly involves three major functions: receptor function, transducer function, and utility factor (Simon et al. 2001; Yamazoe et al. 2003). The function of the receptor is mostly ascribed to the selectivity and sensitivity of the device and response of the device with respect to the surrounding environment (Figure 1.4a). In this stage, the amount of oxygen adsorbed on the surface leading to the depletion of the surface mainly governs the sensing

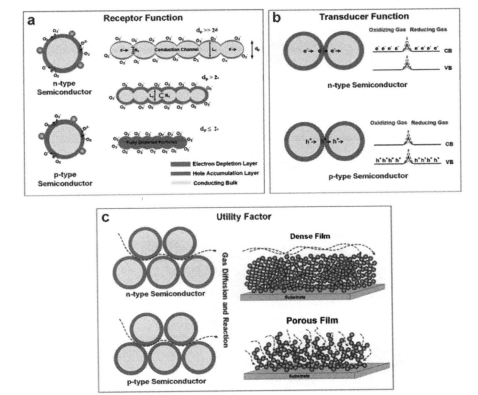

FIGURE 1.4 Three main factors controlling semiconductor gas sensors: (a) receptor function, (b) transducer function, and (c) utility factor (Roco et al. 2011; Nasiri and Clarke 2019).

capabilities of the device, which depends on the particle size (d_p) of the sensing material. If $d_p \gg 2\delta$, the sensing mechanism is determined by transferring electrons at the particle's grain boundary resulting in a low sensitivity. If $d_p > 2\delta$, a large portion of the bulk participates in the sensing mechanism leading to a moderate sensitivity (Tricoli et al. 2010; Nasiri et al. 2015). In contrast, if $d_p \leq 2\delta$ the entire particle is an electron depleted with no mobile charge carrier, leading to significantly high resistance with very low baseline currents.

The transducer function is an interparticle issue related to how the surface phenomenon is transformed into a change in electrical resistance of the sensor (Dey 2018) (Figure 1.4b). The chemical interaction of the semiconductor surface creates an electrical signal in the transducer function, which is mainly controlled by the surface potential and potential barriers formed between grains, trapping states in grain boundaries and defect states in the semiconductor structure. Schottky barriers between two grains impede electrons transferring across the boundary. Therefore, the boundaries between the grains act as transducers when the resistance change by the gas adsorption is amplified. Quite a lot of studies have shown the significance of analyzing the grain boundary for improving the transduction of the surface response (Korotcenkov et al. 2009; Nasiri et al. 2016).

Last, the utility factor (Figure 1.4c) is related to the morphological structure of the metal oxide semiconductor and consequently the diffusion and reaction of target gas through the structure pores (Barsan et al. 2007; Kim et al. 2017). The utility factor determines how the sensing performance is affected by the device structure, with the film porosity as the most important parameter in achieving the highest utility factor. In a very porous structure, the target gas particles penetrate into the lowest layers of the film, resulting in an effective resistance variation of the sensing device (Nasiri et al. 2015). However, the utility factor might be meaningless for monolayer structures such as MoS_2, as the material structure is atomically thin and the gas adsorption is not associated with the diffusion through the material. For such 2D-structured sensors, the utility factor is already maximized, and the highest performance of the device should be achieved by choosing proper receptors and enhancing the interaction between target gas and sensing material (Kim et al. 2017).

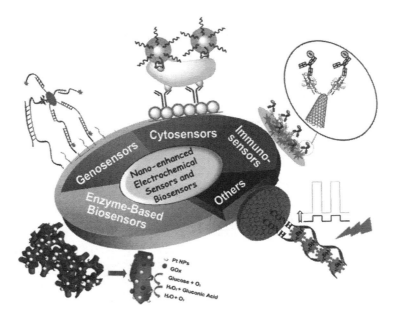

FIGURE 1.5 Schematic illustration of electrochemical-based nanobiosensors (Zhu et al. 2014).

Electrochemistry: Different Materials and Applications

1.4.3 BIOSENSOR

The biosensor is marked as the future of analytical devices one that has a major impact on our everyday lives. It has several advantages including high sensitivity, low cost, rapid response, good selectivity, high-throughput, real-time monitoring, easy to make, and easy to handle (Roco et al. 2011; Teles 2011). Noting these exceptional attributes, considerable attention has been dedicated to the integration of recognition elements with electronic elements, which can be developed as electrochemical-based biosensors. Our day-to-day lives are occupied by using various types of biosensors from medical diagnosis to food analysis, which extend to pathogen determination, glucose monitoring, determination of drug residues in food such as antibiotics and growth promoters, environmental cleaning applications, homeland security, determination of levels of toxic substances in bioremediation, routine analytical measurement of folic acid, biotin, vitamin B_{12} and pantothenic acid, drug discovery, and evaluation of biological activity (Byfield and Abuknesha 1994; Patel 2002; Ahmadalinezhad and Chen 2013). With the help of nanotechnology, nanomaterial-based electrochemical signal amplifications have been achieved with great prospective for improving both selectivity and sensitivity. In electrochemical sensors, it is distinguished that the electrode materials play an important part in detecting target molecules through various analytical principles. By altering the electrode materials with suitable functional nanomaterials, the catalytic activity, conductivity, and biocompatibilitity properties of the electrode can be enhanced and leads to accelerated signal transduction. By choosing suitable modeled signal tags such as biorecognition agents the sensitive of biosensors were notably increased. A detailed review made by Chengzhou et al. (2014) outlines on electrochemical sensors for small molecules, enzyme-based biosensors, immunosensors, genosensors, and cytosensors.

1.4.3.1 Glucose

Over the past 50 years, intense development has been seen in enzymatic glucose sensors. The developments have balanced the benefits versus disadvantages (Khan et al. 1996). Glucose biosensors have developed into a lot of technological advancements such as point-of-care devices, continuous glucose monitoring systems, and noninvasive glucose monitoring systems. The first-generation of glucose sensors developed were based on high oxygen presence, and hence have limitations in use for commercial and practical applications (Wang 2001). The second-generation glucose sensor was not working due to the existence of other electroactive interferences in the sample and also its disadvantages in large-scale production (Scheller et al. 1991). The present

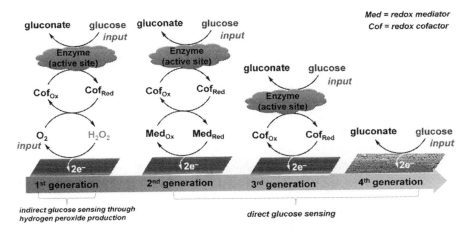

FIGURE 1.6 Summary of glucose electrooxidation mechanisms: presented as enzymatic (first-, second-, and third-generation sensors) and non-enzymatic (so-called herein "fourth-generation" sensors) (Holade et al. 2017).

shift to third-generation sensors is focusing on 3D orientation, low electron tunneling distances, and substrate accessibility of the enzyme by glucose (Khan et al. 1996; Toghill and Compton 2010). Even with all these efforts, ensuring stability is the main issue for the commercialization of analytical devices, and this seriously influences the research on the development of new technology and new materials for electrochemical sensors. The fourth-generation glucose sensors shift toward nonenzymatic electrodes, which are presently identified as a potential candidate for analytical glucose oxidation. In this generation of sensors, the active noble metal nanomaterials experience an oxidation reaction that forms a hydrous oxide layer, OH_{ads}, which arbitrates the oxidation of the adsorbed species. These nonenzymatic materials are abiotic catalysts, which extends their applications to the food industry and to fuel cells and batteries operating in close biological conditions. Quality control in industrial processes is essential, and measuring the number of numerous sugars to validate the quality, authenticity, maturity, and nutritional value of the product is of substantial importance.

1.4.3.2 H_2O_2 Sensor

Hydrogen peroxide (H_2O_2) is an indispensable substance in many intermediate natural reactions, a strong oxidizing agent, and is simply miscible with water. In addition to its many applications in the electronic and food industries, it is an important by-product in the chemical industry, food industry, pharmaceutical industry, clinical medicine, and environmental monitoring (Zhao et al. 2017). Hence, an accurate determination of H_2O_2 is of high importance (Wen et al. 2011). Due to this importance, numerous kinds of H_2O_2 detection and measurement methods have been established, which include fluorometry, spectrophotometry, and electrochemical analysis (Dickinson and Chang 2008). Among these methods, the electrochemical-based analytical technique has been considered in a lot of industries due to its promising approach and excellent advantages which include quick response, ease of operation, and high sensitivity. H_2O_2 biosensors by the electrochemical detection method can be achieved by nonenzymatic and enzymatic methods. In enzyme-based methods, horseradish peroxidase (HRP) and catalase (CAT) are commonly used for H_2O_2 detection (Hernandez-Ruiz et al. 2001). The immobilization and stabilization protocol of the enzyme makes it challenging to attain the direct electron transfer on a bare electrode. This problem was rectified to an extent by the development of excellent support materials, which offer a better environment for enzyme loading. For this, carbon-based nanomaterials, metal nanoparticles, and polymers have been utilized. The introduction of these nanomaterials has shown a notable rise in the enzyme loading area and enhanced the contact area with more detectors, which in turn enhance the electron transfer rate. The working nonenzymatic biosensor is based on the electrocatalytic reduction, and the reaction is accomplished on the surface of the electrode. Conventional electrodes used for electrochemical detection require higher overpotential. In order to enhance work with lower overpotential, efforts were then devoted to using nanomaterial-based modified electrodes. Hence, nonenzymatic biosensors with metal nanoparticle-modified electrodes have been chosen, as they can greatly improve the surface area, mass transport rate, and catalytic efficiency. Many experimental results have shown valid proof that nanoelectronic mediums can advance the electron transfer rate and increase the anti-interference ability of biosensors. Notably, the electrocatalytic performance toward H_2O_2 was efficient when choosing a transition metal oxide, such as CuO (Gu et al. 2010), MnO_2 (Bai et al. 2007) or Co_3O_4 (Wang et al. 2014). Even though these nanomaterials have the advantages of low cost, and being environmentally friendly, and easy to prepare, they have limitations in stability and sensitivity. Hence, the noble metal-based nonenzymatic H_2O_2 biosensor needs more research (Zhao et al. 2017). Recently, Shu-Xia Xu et al. reported on third-generation H_2O_2 sensors using tetraethoxy-silicone (TEOS) sol–gel film for the immobilization of horseradish peroxidase on a multiwalled carbon nanotube (MWNT)-modified glass carbon electrode (GCE) (Xu et al. 2014). Third-generation biosensors can detect H_2O_2 at relatively low applied potentials, and feature the advantages of operation simplicity, no mediator, easy fabrication, and high sensitivity.

Electrochemistry: Different Materials and Applications

1.4.4 Wastewater Treatment

The environment is polluted day-by-day due to either natural or man-made activities (Alqadami et al. 2017; Naushad 2014; Naushad et al. 2015). One of the important natural resources, water, is polluted due to natural disaster, industrial use, and various human activities (Naushad et al. 2016; Sharma et al. 2014). By measuring certain parameters, it can be known up to what extent water is polluted. These parameters include COD (chemical oxygen demand), BOD (biochemical oxygen demand), pH, TOC (total organic carbon), TDS (total dissolved solids), volatile matter, nonvolatile matter, total hardness, color, and odor. Electrolysis is one way of treating wastewater containing a high concentration of metals. Industrial electrochemistry has participated in creating ideal processes and more ecofriendly yield. One important breakthrough in environmental protection applications includes two oxidation processes that help to remove the toxins in our environment (Martínez-Huitle and Panizza 2018):

1. *Direct electrolysis*—In this process, the reaction between the species at the sensing electrode is direct. The electron transfer to or from the undesired diluted pollutant occurs on the electrode surface, which results in reduction or oxidation processes directly on the electrode's surface without the involvement of any intermediate substances (e.g., electron mediators, biocidal species). This direct electrolysis process can be further subdivided into the oxidation and reduction of organic and inorganic pollutants.
2. *Indirect electrolysis*—This electrolysis process involves the invention of active species at the electrode surface and further reaction with the targeted pollutants. The resultant redox reagent plays an intermediary role for shuttling electrons between the electrode and the pollutant substrate. Electrolysis can be done both with reversible and irreversible electrogenerated reagents. Indirect processes include electrocoagulation, electro notation, electro occultation, and advanced oxidation. Among these electro notation is widely used because this process involves the electrolytic production of gases (e.g., H_2) that can be utilized to attach pollutants (e.g., fats and oils) to the gas bubbles and transfer them up to the top of the solution, which can be more effortlessly collected and separated. Indirect electrolysis can be electrogenerated by either the cathodic or anodic process. The most popular cathodic processes method is the electro-Fenton (EF), where H_2O_2 is produced at the cathode with O_2 or air feeding while an iron catalyst is also regenerated on the surface of cathode (Panizza and Cerisola 2009) (Figure 1.7b). Recently, other emerging technologies based on the EF method (Brillas and Martínez-Huitle 2015; Garcia-Segura et al. 2017; Ganiyu et al. 2018), such as solar-PEF (SPEF), photo-EF (PEF), ultrasound irradiation, coagulation based on dissolution of iron anodes (peroxi-coagulation [PC]), dissolution of heterogeneous catalysts that supply Fe^{2+} (heterogeneous-EF), and bioremediation, have also received a great deal of consideration. This process selection depends on the experimental conditions, nature and structure of the electrode material, and electrolyte composition.

1.5 SUMMARY AND FUTURE DIRECTIONS

This chapter is an overview of electrochemistry from its history to various applications and from synthesis to analytical techniques. Even though most of the electrochemical concepts deal with the electron movement in a conductive medium and the oxidation–reduction concept, the concepts of electrochemistry is expanding day by day. During the initial days of the electrochemical concept it was widely used by metal manufacturing industries. However, with the increase in the understanding of analytical chemistry and development of related devices this field has grown from coatings to device designing. Other than industrial applications, due to the advancement in nanotechnology, the interdisciplinary research on biomedical and electrochemistry is also growing with application in various biosensors at low cost. In the future, a lot of low cost, miniature portable analytical devices can be developed utilizing the concepts of electrochemistry.

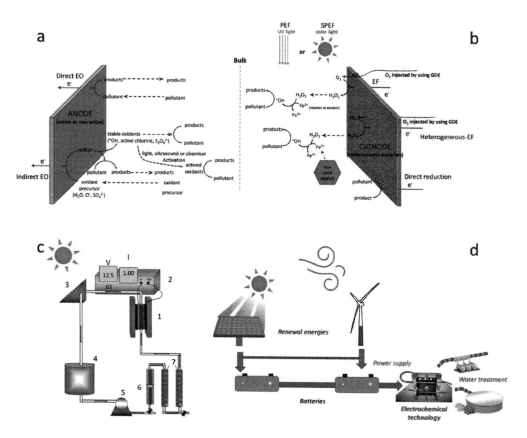

FIGURE 1.7 A conceptual approach to (a) direct EO and mediated EO; (b) EF-based technologies as well as electrochemical emerging wastewater treatments; (c) the 2.5 L pre-pilot plant; and (d) the use of renewable energy sources together with electrochemical treatments (Martínez-Huitle and Panizza 2018).

REFERENCES

Aftalion, F. (2001). *A history of the international chemical industry*, Chemical Heritage Foundation.
Ahmadalinezhad, A. and A. Chen (2013). "Nanomaterial-based electrochemical biosensors." *Nanomedical Device and Systems Design: Challenges, Possibilities, Visions*, CRC Press: 339.
Allmand, A. (1932). *The Drumm Traction Battery*, Nature Publishing Group.
Alqadami, A. A., M. Naushad, M. A. Abdalla, T. Ahamad, Z. Abdullah ALOthman, S. M. Alshehri and A. A. Ghfar (2017). "Efficient removal of toxic metal ions from wastewater using a recyclable nanocomposite: a study of adsorption parameters and interaction mechanism." *Journal of Cleaner Production* **156**: 426–436. doi:10.1016/j.jclepro.2017.04.085
Bacon, F. T. (1969). "Fuel cells, past, present and future." *Electrochimica Acta* **14**(7): 569–585.
Bai, Y.-H., Y. Du, J.-J. Xu and H.-Y. Chen (2007). "Choline biosensors based on a bi-electrocatalytic property of MnO2 nanoparticles modified electrodes to H2O2." *Electrochemistry Communications* **9**(10): 2611–2616.
Barsan, N., D. Koziej and U. Weimar (2007). "Metal oxide-based gas sensor research: how to?" *Sensors and Actuators B: Chemical* **121**(1): 18–35.
Beattie, S. and J. Dahn (2005). "Combinatorial electrodeposition of ternary Cu–Sn–Zn alloys." *Journal of the Electrochemical Society* **152**(8): C542–C548.
Bensaude-Vincent, B. and I. Stengers (1996). *A history of chemistry*, Harvard University Press.
Besra, L. and M. Liu (2007). "A review on fundamentals and applications of electrophoretic deposition (EPD)." *Progress in Materials Science* **52**(1): 1–61.
Blaschko, A. (1889). "Die Berufsdermatosen der Arbeiter." *DMW-Deutsche Medizinische Wochenschrift* **15**(45): 925–927.

Boccaccini, A. R. and I. Zhitomirsky (2002). "Application of electrophoretic and electrolytic deposition techniques in ceramics processing." *Current Opinion in Solid State and Materials Science* **6**(3): 251–260.

Brenner, A. (1963). *Electrodeposition of alloys: principles and practice: general survey, principles, and alloys of copper and silver*, Academic Press.

Brenner, A. (2013). *Electrodeposition of alloys: principles and practice*, Elsevier.

Brillas, E. and C. A. Martínez-Huitle (2015). "Decontamination of wastewaters containing synthetic organic dyes by electrochemical methods. An updated review." *Applied Catalysis B: Environmental* **166**: 603–643.

Byfield, M. and R. Abuknesha (1994). "Biochemical aspects of biosensors." *Biosensors and Bioelectronics* **9**(4–5): 373–399.

Chen, P.-Y. and I.-W. Sun (2001). "Electrodeposition of cobalt and zinc cobalt alloys from a lewis acidic zinc chloride-1-ethyl-3-methylimidazolium chloride molten salt." *Electrochimica Acta* **46**(8): 1169–1177.

Corni, I., M. P. Ryan and A. R. Boccaccini (2008). "Electrophoretic deposition: from traditional ceramics to nanotechnology." *Journal of the European Ceramic Society* **28**(7): 1353–1367.

Crow, D. R. (2017). *Principles and applications of electrochemistry*, Routledge.

Davy, H. (1840). *The collected works of Sir Humphry Davy...: Bakerian lectures and miscellaneous papers from 1806 to 1815*, Smith, Elder and Company.

Deng, M.-J., P.-C. Lin, J.-K. Chang, J.-M. Chen and K.-T. Lu (2011). "Electrochemistry of Zn (II)/Zn on Mg alloy from the N-butyl-N-methylpyrrolidinium dicyanamide ionic liquid." *Electrochimica Acta* **56**(17): 6071–6077.

Dey, A. (2018). "Semiconductor metal oxide gas sensors: a review." *Materials Science and Engineering: B* **229**: 206–217.

Dickinson, B. C. and C. J. Chang (2008). "A targetable fluorescent probe for imaging hydrogen peroxide in the mitochondria of living cells." *Journal of the American Chemical Society* **130**(30): 9638–9639.

El Abedin, S. Z., U. Welz-Biermann and F. Endres (2005). "A study on the electrodeposition of tantalum on NiTi alloy in an ionic liquid and corrosion behaviour of the coated alloy." *Electrochemistry Communications* **7**(9): 941–946.

Field, S. (1911). *The principles of electro-deposition: a laboratory guide to electro-plating*, Longmans, Green.

Fine, G. F., L. M. Cavanagh, A. Afonja and R. Binions (2010). "Metal oxide semi-conductor gas sensors in environmental monitoring." *Sensors* **10**(6): 5469–5502.

Gan, F., Z.-W. Sun, G. Sabde and D.-T. Chin (1994). "Cathodic protection to mitigate external corrosion of underground steel pipe beneath disbonded coating." *Corrosion* **50**(10): 804–816.

Ganiyu, S. O., M. Zhou and C. A. Martínez-Huitle (2018). "Heterogeneous electro-Fenton and photoelectro-Fenton processes: a critical review of fundamental principles and application for water/wastewater treatment." *Applied Catalysis B: Environmental* **235**: 103–129.

Garcia-Segura, S., M. M. S. Eiband, J. V. de Melo and C. A. Martínez-Huitle (2017). "Electrocoagulation and advanced electrocoagulation processes: a general review about the fundamentals, emerging applications and its association with other technologies." *Journal of Electroanalytical Chemistry* **801**: 267–299.

Gu, A., G. Wang, X. Zhang and B. Fang (2010). "Synthesis of CuO nanoflower and its application as a H2O2 sensor." *Bulletin of Materials Science* **33**(1): 17–20.

Gurrappa, I. and L. Binder (2008). "Electrodeposition of nanostructured coatings and their characterization—a review." *Science and Technology of Advanced Materials* **9**(4): 043001.

Hernandez-Ruiz, J., M. B. Arnao, A. N. Hiner, F. García-Cánovas and M. Acosta (2001). "Catalase-like activity of horseradish peroxidase: relationship to enzyme inactivation by H2O2." *Biochemical Journal* **354**(1): 107–114.

Holade, Y., S. Tingry, K. Servat, T. W. Napporn, D. Cornu and K. B. Kokoh (2017). "Nanostructured inorganic materials at work in electrochemical sensing and biofuel cells." *Catalysts* **7**(1): 31.

Hu, J.-M., L. Liu, J.-Q. Zhang and C.-N. Cao (2007). "Electrodeposition of silane films on aluminum alloys for corrosion protection." *Progress in Organic Coatings* **58**(4): 265–271.

Huang, J.-F. and I.-W. Sun (2004). "Electrodeposition of PtZn in a lewis acidic ZnCl2–1-ethyl-3-methylimidazolium chloride ionic liquid." *Electrochimica Acta* **49**(19): 3251–3258.

Ihde, A. J. (1984). *The development of modern chemistry*, Courier Corporation.

Jie, S., M. Ting-yun, Q. Hui-xuan and L. Qi-song (2019). "Electrochemical behaviors and electrodeposition of single-phase Cu-Sn alloy coating in [BMIM] Cl." *Electrochimica Acta* **297**: 87–93.

Jindra, J., and Michael Heyrovský. (2015). *In electrochemistry in a divided world*, Springer, Cham.

Jou, L.-H., J.-K. Chang, T.-J. Whang and I.-W. Sun (2010). "Electrodeposition of palladium–tin alloys from 1-ethyl-3-methylimidazolium chloride–tetrafluoroborate ionic liquid for ethanol electro-oxidation." *Journal of The Electrochemical Society* **157**(8): D443–D449.

Jou, L.-S., J.-K. Chang, T.-J. Twhang and I.-W. Sun (2009). "Electrodeposition of palladium–copper films from 1-ethyl-3-methylimidazolium chloride–tetrafluoroborate ionic liquid on indium tin oxide electrodes." *Journal of The Electrochemical Society* **156**(6): D193–D197.

Kamimoto, Y., T. Itoh, G. Yoshimura, K. Kuroda, T. Hagio and R. Ichino (2018). "Electrodeposition of rare-earth elements from neodymium magnets using molten salt electrolysis." *Journal of Material Cycles and Waste Management* **20**(4): 1918–1922.

Kazimierczak, H. and P. Ozga (2013). "Electrodeposition of Sn–Zn and Sn–Zn–Mo layers from citrate solutions." *Surface Science* **607**: 33–38.

Khan, G. F., M. Ohwa and W. Wernet (1996). "Design of a stable charge transfer complex electrode for a third-generation amperometric glucose sensor." *Analytical Chemistry* **68**(17): 2939–2945.

Kim, T., Y. Kim, S. Park, S. Kim and H. Jang (2017). "Two-dimensional transition metal disulfides for chemoresistive gas sensing: perspective and challenges." *Chemosensors* **5**(2): 15.

Korotcenkov, G., S.-D. Han, B. Cho and V. Brinzari (2009). "Grain size effects in sensor response of nanostructured SnO2-and In2O3-based conductometric thin film gas sensor." *Critical Reviews in Solid State and Materials Sciences* **34**(1–2): 1–17.

Lair, V., J. Sirieix-Plenet, L. Gaillon, C. Rizzi and A. Ringuedé (2010). "Mixtures of room temperature ionic liquid/ethanol solutions as electrolytic media for cerium oxide thin layer electrodeposition." *Electrochimica Acta* **56**(2): 784–789.

Lefrou, C., Pierre Fabry, and Jean-Claude Poignet. (2012). *Electrochemistry: the basics, with examples*, Springer Science & Business Media.

Legeai, S., S. Diliberto, N. Stein, C. Boulanger, J. Estager, N. Papaiconomou and M. Draye (2008). "Room-temperature ionic liquid for lanthanum electrodeposition." *Electrochemistry Communications* **10**(11): 1661–1664.

Lerma-García, M., E. Simó-Alfonso, A. Bendini and L. Cerretani (2009). "Metal oxide semiconductor sensors for monitoring of oxidative status evolution and sensory analysis of virgin olive oils with different phenolic content." *Food Chemistry* **117**(4): 608–614.

Lisenkov, A., M. Zheludkevich and M. Ferreira (2010). "Active protective Al–Ce alloy coating electrodeposited from ionic liquid." *Electrochemistry Communications* **12**(6): 729–732.

Liu, J., G. Mao, X. Ma, L. Li, Y. Guo and G. Liu (2008). "Discovery of Cu-Ni-Zn-Sn-Fe intermetallic compounds and S-bearing alloys in the Zhaishang gold deposit, southern Gansu Province and its geological significance." *Science in China Series D: Earth Sciences* **51**(6): 769–777.

Liu, Q., D. Chen and Z. Kang (2015). "One-step electrodeposition process to fabricate corrosion-resistant superhydrophobic surface on magnesium alloy." *ACS Applied Materials & Interfaces* **7**(3): 1859–1867.

Liu, X., P. Evans and G. Zangari (2000). "Electrodeposited Co-Fe and Co-Fe-Ni alloy films for magnetic recording write heads." *IEEE Transactions on Magnetics* **36**(5): 3479–3481.

Maeda, K. and K. Domen (2010). "Photocatalytic water splitting: recent progress and future challenges." *The Journal of Physical Chemistry Letters* **1**(18): 2655–2661.

Martínez-Huitle, C. A. and M. Panizza (2018). "Electrochemical oxidation of organic pollutants for wastewater treatment." *Current Opinion in Electrochemistry* **11**: 62–71.

Mertens, J. (1998). "Shocks and sparks: the voltaic pile as a demonstration device." *Isis* **89**(2): 300–311.

Mohanty, U. (2011). "Electrodeposition: a versatile and inexpensive tool for the synthesis of nanoparticles, nanorods, nanowires, and nanoclusters of metals." *Journal of Applied Electrochemistry* **41**(3): 257–270.

Molodkina, E. B., M. R. Ehrenburg, P. Broekmann and A. V. Rudnev (2019). "Initial stages of silver electrodeposition on single crystal electrodes from ionic liquids." *Electrochimica Acta* **299**: 320–329.

Murase, K., K. Nitta, T. Hirato and Y. Awakura (2001). "Electrochemical behaviour of copper in trimethyl-n-hexylammonium bis ((trifluoromethyl) sulfonyl) amide, an ammonium imide-type room temperature molten salt." *Journal of Applied Electrochemistry* **31**(10): 1089–1094.

Nakamura, H. (1925). "X-ray analysis of electrolytic brass." *Scientific Papers of the Institute of Physical and Chemical Research (Tokyo)* **2**: 287–292.

Nasiri, N. and C. Clarke (2019). "Nanostructured chemiresistive gas sensors for medical applications." *Sensors* **19**(3): 462.

Nasiri, N., R. Bo, F. Wang, L. Fu and A. Tricoli (2015). "Ultraporous electron-depleted ZnO nanoparticle networks for highly sensitive portable visible-blind UV photodetectors." *Advanced Materials* **27**(29): 4336–4343.

Nasiri, N., R. Bo, H. Chen, T. P. White, L. Fu and A. Tricoli (2016). "Structural engineering of nano-grain boundaries for low-voltage UV-photodetectors with gigantic photo-to dark-current ratios." *Advanced Optical Materials* **4**(11): 1787–1795.

Naushad, M. (2014). "Surfactant assisted nano-composite cation exchanger: development, characterization and applications for the removal of toxic Pb2+ from aqueous medium." *The Chemical Engineering Journal* **235**: 100–108. doi:10.1016/j.cej.2013.09.013

Naushad, M., A. Mittal, M. Rathore and V. Gupta (2015). "Ion-exchange kinetic studies for Cd(II), Co(II), Cu(II), and Pb(II) metal ions over a composite cation exchanger." *Desalination and Water Treatment* **54**: 2883–2890. doi:10.1080/19443994.2014.904823

Naushad, M., S. Vasudevan, G. Sharma, A. Kumar and Z. A. Alothman (2016). "Adsorption kinetics, isotherms, and thermodynamic studies for Hg2+ adsorption from aqueous medium using alizarin red-S-loaded amberlite IRA-400 resin." *Desalination and Water Treatment* **57**: 18551–18559. doi:10.1080/19443994.2015.1090914

Neethirajan, S., D. Jayas and S. Sadistap (2009). "Carbon dioxide (CO 2) sensors for the agri-food industry—a review." *Food and Bioprocess Technology* **2**(2): 115–121.

Pan, S.-J., W.-T. Tsai, J.-K. Chang and I.-W. Sun (2010). "Co-deposition of Al–Zn on AZ91D magnesium alloy in AlCl3–1-ethyl-3-methylimidazolium chloride ionic liquid." *Electrochimica Acta* **55**(6): 2158–2162.

Panagopoulos, C. N., K. G. Georgarakis and K. I. Giannakopoulos (2019). "Zn–Fe multilayered alloy coatings produced by electrodeposition." *SN Applied Sciences* **1**(1): 114.

Panizza, M. and G. Cerisola (2009). "Direct and mediated anodic oxidation of organic pollutants." *Chemical Reviews* **109**(12): 6541–6569.

Patel, P. (2002). "(Bio) sensors for measurement of analytes implicated in food safety: a review." *TrAC Trends in Analytical Chemistry* **21**(2): 96–115.

Rahman, I., M. Khaddem-Mousavi, A. Gandhi, T. Lynch and M. Rahman (2007). "Growth of electrodeposited Ni-Co and Fe-Co magnetic films on Cu substrates." *Journal of Physics: Conference Series* **61**: 523–528.

Rezaei, M., D. F. Haghshenas, M. Ghorbani and A. Dolati (2018). "Electrochemical behavior of nanostructured Fe-Pd alloy during electrodeposition on different substrates." *Journal of Electrochemical Science and Technology* **9**(3): 202–211.

Roco, M. C., C. A. Mirkin and M. C. Hersam (2011). *Nanotechnology research directions for societal needs in 2020: retrospective and outlook*, Springer Science & Business Media.

Roux, A. and J. Cournot (1928). "Etude cristallographique par rayons X de la structure de dépôts électrolytiques simultanés de deux métaux." *Comptes rendus de l'Académie des Sciences Paris* **186**: 1733–1736.

Salman, S., S. Kim, K. Kuroda and M. Okido (2019). "Influence of amine additives on the electrodeposition of aluminum from AlCl 3-dimethyl sulfone electrolytes." In C. Chesonis, ed., *Light Metals 2019*, Springer: 115–119.

Saravanan, G. and S. Mohan (2013). "Nucleation of copper on mild steel in copper chloride (CuCl 2· 2H 2 O)–1-ethyl-3-methylimidazolium chloride [EMIM] Cl–ethylene glycol (EG) ionic liquid." *New Journal of Chemistry* **37**(8): 2564–2567.

Sato, N., M. Fuji, N. Kohiruimaki, M. Fukumoto and M. Hara (2016). "Preparation of Ni aluminide/Ni bilayer coating on Nb–W alloys by molten salt electrodeposition and oxidation resistance." *Oxidation of Metals* **85**(1–2): 29–38.

Scheller, F. W., F. Schubert, B. Neumann, D. Pfeiffer, R. Hintsche, I. Dransfeld, U. Wollenberger, R. Renneberg, A. Warsinke and G. Johansson (1991). "Second generation biosensors." *Biosensors and Bioelectronics* **6**(3): 245–253.

Seiyama, T., A. Kato, K. Fujiishi and M. Nagatani (1962). "A new detector for gaseous components using semiconductive thin films." *Analytical Chemistry* **34**(11): 1502–1503.

Sharma, G., D. Pathania, M. Naushad and N. C. Kothiyal (2014). "Fabrication, characterization and antimicrobial activity of polyaniline Th(IV) tungstomolybdophosphate nanocomposite material: efficient removal of toxic metal ions from water." *The Chemical Engineering Journal* **251**: 413–421. doi:10.1016/j.cej.2014.04.074

Shreir, L. L. (2013). *Corrosion: corrosion control*, Newnes.

Simon, I., N. Bârsan, M. Bauer and U. Weimar (2001). "Micromachined metal oxide gas sensors: opportunities to improve sensor performance." *Sensors and Actuators B: Chemical* **73**(1): 1–26.

Song, Y., Y. Ma, Y. Wang, J. Di and Y. Tu (2010). "Electrochemical deposition of gold–platinum alloy nanoparticles on an indium tin oxide electrode and their electrocatalytic applications." *Electrochimica Acta* **55**(17): 4909–4914.

Sørensen, P. A., S. Kiil, K. Dam-Johansen and C. E. Weinell (2009). "Anticorrosive coatings: a review." *Journal of Coatings Technology and Research* **6**(2): 135–176.

Spitzer, F. (1905). "Über das elektromotorische Verhalten von Kupfer und Zink gegenüber ihren cyankalischen Lösungen." *Zeitschrift für Elektrochemie und angewandte physikalische Chemie* **11**(23): 345–368.

Stine, K. J. (2019). "Biosensor applications of electrodeposited nanostructures." *Applied Sciences* **9**(4): 797.

Su, T., Q. Shao, Z. Qin, Z. Guo and Z. Wu (2018). "Role of interfaces in two-dimensional photocatalyst for water splitting." *Acs Catalysis* **8**(3): 2253–2276.

Tachikawa, N., Y. Katayama and T. Miura (2009). "Electrochemical and spectroscopic studies of europium (II/III) species in a hydrophobic room-temperature ionic liquid." *Electrochemistry* **77**(8): 642–644.

Tavakkoli, N., N. Soltani, Z. K. Tabar and M. R. Jalali (2019). "Determination of dopamine using the indium tin oxide electrode modified with direct electrodeposition of gold–platinum nanoparticles." *Chemical Papers* **73**(6): 1–12.

Teles, F. S. R. R. (2011). "Biosensors and rapid diagnostic tests on the frontier between analytical and clinical chemistry for biomolecular diagnosis of dengue disease: a review." *Analytica chimica acta* **687**(1): 28–42.

Tiano, A. L., C. Koenigsmann, A. C. Santulli and S. S. Wong (2010). "Solution-based synthetic strategies for one-dimensional metal-containing nanostructures." *Chemical Communications* **46**(43): 8093–8130.

Toghill, K. E. and R. G. Compton (2010). "Electrochemical non-enzymatic glucose sensors: a perspective and an evaluation." *International Journal of Electrochemical Science* **5**(9): 1246–1301.

Toney, M. F., J. N. Howard, J. Richer, G. L. Borges, J. G. Gordon, O. R. Melroy, D. Yee and L. B. Sorensen (1995). "Electrochemical deposition of copper on a gold electrode in sulfuric acid: resolution of the interfacial structure." *Physical Review Letters* **75**(24): 4472–4475.

Tricoli, A., M. Righettoni and A. Teleki (2010). "Semiconductor gas sensors: dry synthesis and application." *Angewandte Chemie International Edition* **49**(42): 7632–7659.

Tsuda, T., T. Nohira and Y. Ito (2001). "Electrodeposition of lanthanum in lanthanum chloride saturated AlCl3–1-ethyl-3-methylimidazolium chloride molten salts." *Electrochimica Acta* **46**(12): 1891–1897.

Twite, R. and G. P. Bierwagen (1998). "Review of alternatives to chromate for corrosion protection of aluminum aerospace alloys." *Progress in Organic Coatings* **33**(2): 91–100.

Ueda, M., H. Hayashi and T. Ohtsuka (2011). "Electrodeposition of Al–Pt alloys using constant potential electrolysis in AlCl3–NaCl–KCl molten salt containing PtCl2." *Surface and Coatings Technology* **205**(19): 4401–4403.

Ueda, M., R. Inaba and T. Ohtsuka (2013). "Composition and structure of Al–Sn alloys formed by constant potential electrolysis in an AlCl3–NaCl–KCl–SnCl2 molten salt." *Electrochimica Acta* **100**: 281–284.

Ueda, M., H. Kigawa and T. Ohtsuka (2007). "Co-deposition of Al–Cr–Ni alloys using constant potential and potential pulse techniques in AlCl3–NaCl–KCl molten salt." *Electrochimica Acta* **52**(7): 2515–2519.

Von Baeckmann, W., W. Schwenk and W. Prinz (1997). *Handbook of cathodic corrosion protection*, Elsevier.

Wang, J. (2001). "Glucose biosensors: 40 years of advances and challenges." *Electroanalysis: An International Journal Devoted to Fundamental and Practical Aspects of Electroanalysis* **13**(12): 983–988.

Wang, Y., E. Laborda, A. Crossley and R. G. Compton (2013). "Surface oxidation of gold nanoparticles supported on a glassy carbon electrode in sulphuric acid medium: contrasts with the behaviour of 'macro' gold." *Physical Chemistry Chemical Physics* **15**(9): 3133–3136.

Wang, Y., T. Zhou, K. Jiang, P. Da, Z. Peng, J. Tang, B. Kong, W. B. Cai, Z. Yang and G. Zheng (2014). "Reduced mesoporous Co3O4 nanowires as efficient water oxidation electrocatalysts and supercapacitor electrodes." *Advanced Energy Materials* **4**(16): 1400696.

Wen, F., Y. Dong, L. Feng, S. Wang, S. Zhang and X. Zhang (2011). "Horseradish peroxidase functionalized fluorescent gold nanoclusters for hydrogen peroxide sensing." *Analytical Chemistry* **83**(4): 1193–1196.

Wu, L.-K., X.-F. Zhang and J.-M. Hu (2014). "Corrosion protection of mild steel by one-step electrodeposition of superhydrophobic silica film." *Corrosion Science* **85**: 482–487.

Xu, S.-X., J.-L. Li, Z.-L. Zhou and C.-X. Zhang (2014). "A third-generation hydrogen peroxide biosensor based on horseradish peroxidase immobilized by sol–gel thin film on a multi-wall carbon nanotube modified electrode." *Analytical Methods* **6**(16): 6310–6315.

Yamazoe, N., G. Sakai and K. Shimanoe (2003). "Oxide semiconductor gas sensors." *Catalysis Surveys from Asia* **7**(1): 63–75.

Yang, H. L., L. Wu, G. Z. Tang, Y. G. Li and Y. Z. Zhang (2012). *Growth characteristics of pulse electrodeposition Fe-Si layer*. Advanced Materials Research, Trans Tech Publ.

Ye, K., M. L. Zhang, Y. Chen, W. Han, Y. De Yan and P. Cao (2010). "Electrochemical codeposition of Al-Li-Mg alloys at solid aluminum electrode from LiCl-KCl-MgCl 2 molten salt system." *Metallurgical and Materials Transactions B* **41**(3): 691–698.

Zhang, J., C. Yan and F. Wang (2009). "Electrodeposition of Al–Mn alloy on AZ31B magnesium alloy in molten salts." *Applied Surface Science* **255**(9): 4926–4932.

Zhao, Z., Q. Ou, X. Yin and J. Liu (2017). "Nanomaterial-based electrochemical hydrogen peroxide biosensor." *International Journal of Biosensors & Bioelectronics* **2**: 25–28.

Zhu, C., G. Yang, H. Li, D. Du and Y. Lin (2014). "Electrochemical sensors and biosensors based on nanomaterials and nanostructures." *Analytical Chemistry* **87**(1): 230–249.

2 Controlled Electrochemical Deposition for Materials Synthesis

T. Sivaranjani, T. A. Revathy, and A. Stephen

CONTENTS

2.1	Introduction	26
2.2	Fundamentals of Electrodeposition	27
	2.2.1 Thickness of the Deposition	28
	2.2.2 Electrodes	28
	2.2.3 Standard Electrode Potential	29
	2.2.4 Electrolyte–Electrode Surface	29
	2.2.5 Mechanism of Electrodeposition	29
	2.2.6 Electrical Double Layer	31
	2.2.6.1 Helmholtz Double Layer	31
	2.2.6.2 Gouy-Chapman Double Layer	31
	2.2.6.3 Stern Layer	32
	2.2.6.4 Grahame Layer	32
	2.2.6.5 Bockris-Devanathan-Müller (BDM) Model	32
	2.2.6.6 Trasatti-Buzzanca Model	32
	2.2.6.7 Conway Model	32
	2.2.6.8 Marcus Model	32
2.3	Parameters Affecting Electrodeposition	32
	2.3.1 Effect of Current Density and Distribution	32
	2.3.2 Effect of Bath Temperature	33
	2.3.3 Effect of Bath Concentration	33
	2.3.4 Complexing Agent	33
	2.3.5 Effect of pH	33
	2.3.6 Addition of Agents	33
	2.3.7 Effect of Plating Time	33
	2.3.8 Effect of Agitation	34
	2.3.9 Hydrogen Embrittlement	34
2.4	Tunable Properties of the Material by Electrodeposition Parameters	34
	2.4.1 Composition	35
	2.4.2 Structure	37
	2.4.3 Morphology	40
2.5	Multilayer Coatings	41
2.6	Composite Coatings	43
2.7	Conclusion	43
Acknowledgments		44
References		44

2.1 INTRODUCTION

Electrodeposition (ED), or electrochemical deposition as the name suggests, is a chemical reduction process induced by electric current. In the electrodeposition technique, electrical energy is passed through an electrolytic cell setup, thereby transferring the energy to ions in the electrolyte solution so as to prepare the coating of the material present in the electrolyte over the surface of another material or substrate (cathode). In the ED method, a two-electrode or three-electrode setup is conventionally used. In the two-electrode setup, there is a cathode and anode (anode or counter electrode), whereas in the three-electrode system there is an additional reference electrode. The electrodes are dipped inside the conducting electrolyte. The electrical energy is applied through a power supply connected parallel to the electrodes. The electrolyte used in the deposition is an important component since the conductivity of the whole circuit is based on the ions present in the electrolytic solution. There are different types of electrolytes that are used in the electrodeposition process such as liquid electrolytes, molten/ionic melt electrolytes, polymer electrolytes, and solid-state electrolytes (Plieth 2008). The type of electrolyte decides the mobility and conducting mechanism of ions in the circuit. As stated by Arrhenius, the dissolution of the salt in water forms the basis of electrodeposition (Arrhenius 1889). According to Arrhenius, adding a corresponding salt of a metal to any solvent like water makes it an appropriate electrolyte for electrodeposition. The electrolytes that are made of water or other organic solvents are branched under liquid electrolytes. $NiSO_4$ dissolved in water is one of the best examples for aqueous electrolytes. For example, consider $NiSO_4$ when dissolved in water, the ions dissociate into the ions Ni^{2+} and SO_4^{2-}. Then, on applying the electric current, the positive ions move toward the cathode and catch up with two electrons to form a nickel atom on the substrate. This is a simple example to show that electrodeposition is an atom-by-atom aligning process (Kim, Dick, and Bard 2016; Al-Osta et al. 2017). In the case of ionic melts, ions are separated and driven toward the electrodes at suitable elevated temperatures and depositions are made (Malyshev 2012; Endres, Abbott, and MacFarlane 2017). The polymer electrolytes are an important class of electrolytes. On mixing the polymer with salt, the polymers help the ionic conductance of the metal concerned. Based on the defects and disorder in the crystallographic structure, the conductance of ions through solids is also made possible.

The choice of suitable electrode is indeed an essential aspect of electrochemistry. The electrode material is chosen based on the deposition potential range. A suitable electrode should not lead to the dissolution or formation of an unwanted layer or formation of any oxide over the surface of the electrode (Brett and Brett 1993). The cathode is the one on which the deposition will occur. The anode is to connect the positive terminal and close the circuit. There are two types of conventional anodes used in electrodeposition, namely, dissolvable (or sacrificial) anodes and nondissolvable anodes. The dissolvable anodes continuously replenish metal in the bath (Erb, Palumbo, and McCrea 2011). Nickel and silver are the best examples of sacrificial anodes. Nondissolvable anodes do not replenish the metal in the bath, but the terminal closes the circuit, e.g., carbon-based electrodes like graphite. At the anode, oxidation takes place and the valence state of the metal increases.

$$\text{At anode}: M \rightarrow M^+ + e^- \text{(oxidation)} \tag{2.1}$$

The reduction process takes place at the cathode. Electrons produced near the anode are carried to the cathode and are consumed at the cathode surface. During the reduction process, unlike the anodic process, the valence of the material decreases and it reaches zero state during deposition.

$$\text{At cathode}: M^+ + e^- \rightarrow M \left(\text{reduction}\right) \tag{2.2}$$

The reference electrode is the one that has a standard electrode potential.

The power supplied to the electrodeposition arrangement is in the form of current/potential. There are two different types of deposition based on the applied current/potential and they are direct deposition (DC) and pulsed electrodeposition (PED). In direct current/potential plating, electrodeposition is performed on a plating unit that is connected to a DC source. Therefore, the average current density equals the peak current density. In PED, the potential is alternated swiftly between two different values. This results in a series of pulses of equal amplitude, duration, and polarity, separated by zero current. Each pulse consists of an on-time (T_{ON}) during which the potential/current is applied, and an off-time (T_{OFF}) during which zero current is applied. In certain cases, bipolar power sources are also employed. Various types of wave and pulse forms are given by Morton Schwartz (Bunshah 2001). It is possible to control the deposited film's composition and thickness by regulating the pulse amplitude, resulting in unique composition and microstructure. This helps in producing various nanocrystalline metals, alloys, and composites successfully. It controls the microstructure and composition of deposited alloys by varying the pulse potential/current on-time (T_{ON}), the pulse current/potential off-time (T_{OFF}), peak current density (I_P), average current density (I_A), the duty cycle (θ), and pulse frequency (f):

$$I_A = \frac{I_P \, T_{ON}}{T_{ON} + T_{OFF}} \tag{2.3}$$

$$\theta = \frac{T_{ON}}{T_{ON} + T_{OFF}} \tag{2.4}$$

$$\text{Pulse frequency} = \frac{T_{ON} + T_{OFF}}{T_{ON}} \tag{2.5}$$

The electrodeposition process starts with nucleation on a kink site and develops into a metal coating/alloy/semiconductor film (Plieth and Georgiev 2006). Hence, to synthesize a material by adopting electrodeposition, the amount of electric current needs to be applied and the chemical properties of the particular material should first be well understood. The electrodeposition mechanism follows reduction processes by supplying electrons at the cathode terminal. Electrodeposition is a strong method for coating on the surface. The main properties of the electrodeposited materials are coating adhesion, hardness, abrasion resistance, and corrosion resistance. All these properties and characteristics can be changed by selecting one or more variables such as the bath's concentration, current density, temperature of the bath, pH, using the additives in electrolytic bath, time of plating (Jokar and Aliofkhazraei 2017).

Erb and his group are pioneers in the field of electrodeposition and studied the various effects on electrodeposits such as size, ductility, corrosion resistance, and other functional properties (Brooks et al. 2011; Facchini et al. 2009). Their group has worked to prepare alloys and composites both at bulk and nanolevel (Erb, Palumbo, and McCrea 2011). They have also successfully deposited hydrophobic materials (Tam, Palumbo, and Erb 2016; Victor, Facchini, and 24Erb 2012).

2.2 FUNDAMENTALS OF ELECTRODEPOSITION

A rule that governs the amount of the substance deposited and gives a key idea on the electrodeposition process is Faraday's first and second laws of electrolysis. Faraday's law relates the current flow, time, and the equivalent weight of the metal with the weight of deposit and may be stated as follows:

(i) The amount of chemical change at an electrode is directly proportional to the quantity of electric current passing through the solution.
(ii) The amount of different substances liberated at an electrode by a given quantity of electric current is proportional to their chemical equivalent weights.

Faraday's laws may be expressed quantitatively:

$$W = \frac{I.t.E_q}{F} \tag{2.6}$$

where W is weight of deposit in grams, I is current flow in amperes, t is time in seconds, E_q is equivalent weight of deposited element, and F is Faraday's constant (96,500 coulombs).

$$E_q = \frac{\text{Atomic weight}}{\text{Valence change}} \tag{2.7}$$

where valence change is the number of electrons involved in the reaction.

2.2.1 Thickness of the Deposition

The average deposit thickness (T) is calculated from Faraday's law (Palli and Dey 2016)

$$T = \frac{M\,j\,t}{n\,F\,\rho} \tag{2.8}$$

where M is the molar mass, j is applied current density, t is total deposition time, n is the number of electrons transferred in the electrode reaction, F is the Faraday constant, and ρ is density. The synthesis mechanism of metal/alloy by electrodeposition is unique when compared to the other electrochemical processes wherein a new solid phase is formed from liquid electrolyte.

2.2.2 Electrodes

In the electrodeposition setup, the electrode should be a conducting material allowing the electric field or electric current to flow in a closed circuit. In electrochemical processes the electrodes are linked via the electrolyte. The electrode can be either a metal or a semiconductor material.

There are different kinds of electrodes available according to the necessity of the experiments in which they are used. The major types of electrodes are

- Nonpolarizable electrodes
- Polarizable electrodes

The electrodes at which the transfer of electrons and ions take place are called nonpolarizable electrodes where there is free movement of ions at the electrode–electrolyte interface. The electric current that flows due to the charge transfer across the electrode–electrolyte boundary is called the faradaic current. The electrodes at which no electron transfer takes place are called polarizable electrodes. It means that the electrons form a layer at the cathode surface. The positive ions form a layer near the electrode surface. There is no transfer between the electrode–electrolyte surfaces. The current that flows to maintain the electrostatic equilibrium at the electrical double layer (EDL) or interfacial double layer on both polarizable and nonpolarizable electrodes is called the non-faradaic current or otherwise known as the transient current.

The term *polarizable* means that the potential of the electrode is fixed and will not change. These kinds of electrodes act as mere capacitors. These electrodes are not soluble in nature, for example, platinum electrodes. The charge–transfer equilibrium is not achieved in polarizable electrodes since there is no movement of ionic or electron species; but electrostatic equilibrium is maintained, producing charge of equal amount with opposite signs on both sides termed as the electrical double layer. Every electrode material has a specific potential when there is no electric flow. The deviation of potential from this specific potential is known as polarization in electrochemistry. Anodic polarization is termed as the shift in electrode potential from the specific potential of the electrode in the anodic direction, and cathodic polarization means the shift of the electrode potential toward the cathodic direction. The nonpolarizable electrodes cannot be polarized since the charge transfer occurs in them with a large value of electric current. In such a reaction there is no change in electrode potential. At the nonpolarizable electrode, the electron or ion transfer reaction will be in equilibrium. For the electrode reaction to be in equilibrium at the interface of nonpolarizable electrodes, an appreciable concentration of redox particles or potential determining ions must exist in the electrolyte. In polarizable electrodes, no charge transfer (no electron or ion transfer) across the EDL occurs, and the electrode can be readily polarized.

2.2.3 Standard Electrode Potential

The electrode potential and standard electrode potential of the concerned metal(s) play a major role in the process of electrodeposition. The metal and the standard hydrogen electrode are immersed in the conducting electrolyte. Then the potential drop across the concerned metal and the standard hydrogen electrode gives the standard electrode potential of that metal. The electrode potential represents not only the electrostatic potential but also the energy of free electrons or ions in the electrode–electrolyte system. The electron energy level in the electrode depends on the chemical potential of electrons in the electrode, the potential between the electrode and electrolytic surface, and the surface potential of the electrolyte solution. For a given system, i.e., the combination of electrode–electrolyte, the chemical potential of electrons and the surface potential remain constant. Hence, the electrode potential or real potential of the electrode depends on the interfacial potential between the electrode and electrolytic surface. Since the electrode–electrolyte interface plays a major role in measuring the electrode potential, study of the electrical double layer has become prevalent.

2.2.4 Electrolyte–Electrode Surface

The electrolytic cell reaction is studied in two separate sets of reactions. The cathode is merely a half-cell in the whole setup where the reduction takes place and vice versa, i.e., oxidation takes place at the anode. The deposition potential and deposition rate of the same metal differs when a different type of electrolytic bath is employed. In a galvanostatic setup, where the cell is controlled by current, the electrode potential is the potential drop observed in the electrode–electrolyte interface. This potential drop is defined as galvanostatic potential difference, which cannot be measured experimentally since this is not constant and changes from time to time. When a suitable electrolyte and appropriate electrode is chosen, the electrode–electrolyte interface controls the whole electrodeposition process. When a solute (e.g., salt "MA") is added to the electrolyte, the ions will dissociate into M^+ and A^-. The electrolyte is in liquid phase and the substrate (electrode) is in solid phase. When exchange of M^+ ions from solution to the substrate happens, the reduction phenomenon will occur at the cathode by the addition of electrons produced by the negative charge of the cathode. This step in electrodeposition is not as simple as we think. At the electrode surface, which is in a solid and liquid heterogeneous phase, the behavior of the double layer has local variations, which possibly affects the deposition kinetics and hence the overall reaction might be altered.

2.2.5 Mechanism of Electrodeposition

On applying electric field, the positive ions in the electrolytic bath move toward the cathode. Chemical steps occurring near the electrode involve ion dehydration or ligand dissociation. The movement of ions in the bulk electrolyte takes different forms as described next (Zangari 2015).

Mass transfer is defined as the movement of material in a solution from one location to another. The different modes of mass transfer include:

- *Migration*—The motion of a charged particle due to an applied electric field is called migration. Only those ions that are near the interfacial region are likely to undergo migration.
- *Diffusion*—In response to the concentration gradient, ionic motion throughout the bulk of the solution occurs mostly by diffusion.
- *Convection*—The electrodeposition process is essentially due to the movement of ions that constitute current through electrolyte and is called convection. Charge transfer causes the formation of a charge cloud near electrodes.

At the atomic scale, three different steps occur in ED, namely, ionic migration, electron transfer, and incorporation. These processes have been described in schematic diagrams (Figure 2.1 and Figure 2.2). The deposition starts with the nucleation of the first atom on the kink site on the cathode surface used. The nucleation of the metal/alloy is based on the nature and energy of the ions present in the electrolyte (Sobha Jayakrishnan 2012).

The behavior of solvated ions in electrolyte changes especially when complexing agents are added and they transform the ions into complex ions. The metal ions are associated with the solvent molecules and complexed with additives either by electrostatic attraction or by covalent bond. Electrical energy is applied in order to transfer the metal ion out of solution to the growing crystal lattice. In the nucleation and growth of metal/alloys at the substrate surface a great technology is actually involved. The ions will cross the double layer and get adsorbed on the surface of the substrate. On the substrate surface these adsorbed ions or adatoms or adions wander before they position themselves in a kink site. More adsorbed ions combine to form clusters of the adatoms. The clusters grow in size depending on the transfer energy and surface energy of these adatoms and reach a critical size (N_c) (Paunovic and Schlesinger 2006):

$$N_c = \frac{bs\varepsilon^2}{ze\eta} \tag{2.9}$$

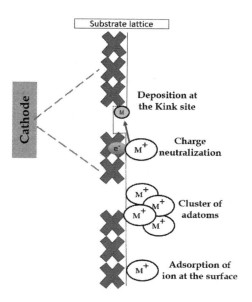

FIGURE 2.1 Deposition mechanism at electrolyte–electrode surface.

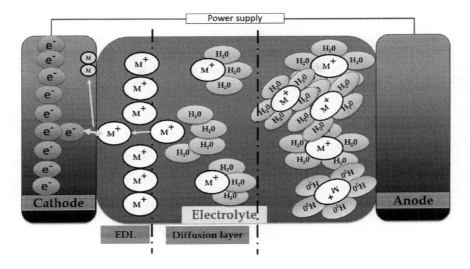

FIGURE 2.2 Steps involved in electrodeposition from electrolyte.

where b is the factor relating the surface area of the nucleus to its perimeter, s is the area occupied by one atom on the surface of the nucleus, η is the overpotential, and ε is the edge energy. The critical size forms a nuclei for a new phase. This nuclei has equal probability for growth and dissolution.

2.2.6 Electrical Double Layer

A double layer (DL) or electrical double layer (EDL) is a structure that appears on the surface of an object when it is exposed to a fluid. The object might be a solid particle, a gas bubble, a liquid droplet, or a porous body. The DL refers to two parallel layers of opposite charges surrounding the medium/object. The explanation about the electrical double was given by various scientists and they were named after them.

2.2.6.1 Helmholtz Double Layer

Helmholtz was the first to give a theoretical explanation about this double layer charging effect. He gave the simplest approximation to this double layer effect stating that the electrons form a rigid layer in the cathode, and positive ions form another layer in the electrolyte near the electrode. When the potential applied is not enough to decompose the layer of ions, the stored charge in EDL is linearly dependent on the voltage applied. Once the critical potential for deposition is reached, the surface charge on the surface of the cathode will be neutralized by the counter ions placed at a distance from the substrate surface, thereby the surface potential is distributed to the counter ions linearly. This idea of rigid double layer interprets the differential capacitance and thickness of EDL but does not necessarily explain all the features of the double layer phenomenon.

2.2.6.2 Gouy-Chapman Double Layer

Gouy and Chapman independently put forth the idea of the diffuse double layer. The ideas of these two scientists contradict Helmholtz in that the capacitance value is not constant and changes with the applied field and composition of the electrolyte containing metallic ion species. The electrochemical behavior of electrodes plays an important role in electrical double layer. This model suggested that the Maxwell-Boltzmann statistic has to be followed to calculate the charge distribution of ions as a function of distance from the conducting surface. The approximation of Gouy-Chapman gave a better value of interfacial potential but not accurate.

2.2.6.3 Stern Layer

Stern gave an improved and combined hypothesis of Helmholtz and Gouy-Chapman double layers. The Stern layer assumed the ions are of finite size and a closed ion approximates the ionic radius. The limitations of the Stern layer include (i) this theory considered the ions as point charges and only columbic interactions are studied, and (ii) dielectric permittivity is taken as a constant.

2.2.6.4 Grahame Layer

The Grahame layer is a modified form of the Stern layer. Grahame proposed that some ions near the electrode will lose their solvation shell and diffuse through the Stern layer. These ions, which crossed the Stern layer, get adsorbed to the electrode surface. Most important is that the Grahame theory gave the idea of presence of three regions, namely, inner Helmholtz plane, outer Helmholtz plane, and diffuse layer.

2.2.6.5 Bockris-Devanathan-Müller (BDM) Model

Bockris, Devanathan, and Müller suggested that the attached molecules of the solvent to the ions, say, water, is influenced and has a particular alignment toward the electrode surface. The orientation of these solvated molecules is dependent on the ionic charge and applied electric field. This orientation would have a great influence on the permittivity of the solvent which varies with field strength.

2.2.6.6 Trasatti-Buzzanca Model

This model gave the basic idea to understand the concept of pseudocapacitance. Trasatti and Buzzanca demonstrated that at lower applied voltage, specific adions exhibit behavior similar to capacitors expressing a partial charge–transfer mechanism.

2.2.6.7 Conway Model

Conway coined the term *supercapacitor*, which explained the increased capacitance value possessed by the ions and electrode system. His supercapacitors stored electrical charge partially in the Helmholtz double layer and partially as the result of faradaic reactions with pseudocapacitance charge transfer of electrons and protons between the electrode and electrolyte.

2.2.6.8 Marcus Model

The Marcus theory explains the rates of electron transfer reactions—the rate at which an electron can move from one chemical species to another. Marcus received the Nobel Prize in Chemistry in 1992 for this theory.

2.3 PARAMETERS AFFECTING ELECTRODEPOSITION

During the electrodeposition process, the preparative parameters significantly affect the structural, morphological, and optical properties of the deposits. Some of the preparative parameters are discussed here. These parameters can be controlled to obtain uniform, smooth, and stoichiometric electrodeposits.

2.3.1 Effect of Current Density and Distribution

In the electrodeposition process, the uniform coating of the specimen depends upon the current distribution parameter. The metal ions are attached to the cathode at certain favored sites. This condition results due to the presence of discontinuities in the form of pores, cracks, or other irregularities. The current density over a cathode will vary from point to point. Current density will be more at the edges leading to changes in the average current density. Current density must be held within the proper interval with respect to bath composition and temperature. Insufficient current for the given specimen will result in poor coating. Increasing current density results in a decrease in the crystallite size of the deposit. On the other hand, applying higher current density will lead to hydrogen evolution.

Controlled Electrochemical Deposition for Materials Synthesis

2.3.2 Effect of Bath Temperature

In general, an increase in bath temperature causes an increase in the crystallite size. Also, an increase in bath temperature increases solubility and thereby increases the transport number, which in turn leads to increased conductivity of the solution. It also decreases the viscosity of the solution, thereby replenishing the double layer relatively faster. High bath temperature usually leads to less adsorption of hydrogen on the deposits and it reduces stress and the tendency toward cracking. The effect of temperature on electrodeposits has been studied by T. Chan et al. (2011).

2.3.3 Effect of Bath Concentration

Bath concentration plays an important role in the plating performance. In normal plating conditions, the increase of bath concentration will increase the concentration of metal ions in the solution. Therefore it will increase the plating rate of the deposition process.

2.3.4 Complexing Agent

The complex ions react with the metal ions and form metal complexes, thereby altering the potential of the metal. The electrode potential of the more noble metal gets shifted toward the negative potential due to the formation of complex ions. In the ED technique a minor drawback in making the film is that the coating thickness is not uniform all over the surface of deposition area. The deposition rate is higher at the edges and corners of the substrate, whereas lower at other parts of the substrate. To some extent, this can be avoided by adding complexing agents to the bath (Sahoo, Das, and Davim 2017). While depositing alloys, complexing agents favor bringing in the standard reduction potential of constituent metals of the alloy fairly close to each other.

2.3.5 Effect of pH

Low pH causes hydrogen evolution, which penetrates the layer, and the deposited material becomes harder, and it introduces a lot of stress to the layer. The electrodeposition process should be conducted at a stable pH. During electrolysis of an aqueous solution, hydrogen ions are discharged together with the ions of the metal being deposited. The hydrogen evolved has often unfavorable effects on the structure and properties of the electrodeposits. Change in pH value affects the current efficiency of the electrodeposition process.

2.3.6 Addition of Agents

Additives are commonly used in plating baths to improve the adhesion and fine grain growth of the deposits. The additives do not react with any ions in the electrolyte. The addition of agents will induce more sites for nucleation rather than the existing growth sites. They adsorb onto the surface and accelerate the rate of the deposition within trenches, thus enabling depositions of smooth and bright deposits.

2.3.7 Effect of Plating Time

According to Faraday's law, quantity of charge flow (Q) in the solution is proportional to the current flow (I), and also the flow time (t) as shown in the following equation (Dahotre and Sudarshan 1999):

$$Q = I * t \tag{2.10}$$

In general, the plating thickness increases and is directly proportionate to the plating time and current.

2.3.8 Effect of Agitation

In general, agitation provides sufficient mixing of the individual metal salts so that the chemicals become intimate and react with each other. Agitation of the solution replenishes metal salts at the diffusion layer. It reduces the gas bubbles, which may otherwise cause pits. It helps in increasing the operating current density and thereby permits a higher operating current density. These factors influence the structure of the deposited metal ion and also increase in the concentration of the metal, since it more rapidly compensates for the loss of the metal ions through discharge at the cathode. Agitation promotes the deposition of the noble metal on the cathode process. In brief, an agitation system can greatly improve the deposition performance since it provides sufficient mixing of the metal salt for the plating solution.

2.3.9 Hydrogen Embrittlement

The preceding elementary discussion of deposition reaction was aimed at giving the essentials of how metal is deposited. But there is also simultaneous reduction reaction of two or more electro active species in the electrolyte during electroplating. This arises from the fact that there is always less metal deposited than it should be, based on Faraday's laws. In some cases, there is almost no deposit at all, which is mainly due to the hydrogen evolution. From a technological point of view, one seeks to deposit a metal without wasting electrical energy on hydrogen evolution. But practically, more current density is needed than the limiting current density for the deposition of the metal to occur due to the hydrogen evolution.

In aqueous solutions, hydrogen evolution is unpreventable. If the total deposition current is made sufficiently high, the hydrogen may come not only from H^+ (of limited concentration) but also from the water itself. The effects of hydrogen codeposition on the crystallography of metal deposits are widespread and rather complicated. The entry of hydrogen into the structure of the metal can lead to hydrogen embrittlement. As the hydrogen rearranges itself within the structure, the internal stress set up are often enough to cause a part fracture with no external force being applied. Even if the part does not fracture on its own, it may easily do so with the application of minimal external stress. But even removal of H^+ ions from the diffusion layer near the electrode changes its properties, too; it makes the solution at the interface become more alkaline. If the hydrogen reaction leads to production of a sufficiently alkaline solution at the interface, this would cause precipitation of hydroxide and may contaminate the deposition. Thus, hydrogen codeposition is usually destructive. The hydrogen evolution reaction from aqueous solution can be presented as, in acid solutions,

$$2\,H^+ + 2e^- \rightarrow H_2 + 2\,H_2O \tag{2.11}$$

$$2\,H_2O + 2e^- \rightarrow H_2 + 2\,OH^- \tag{2.12}$$

And in alkaline solutions, it is only

$$2\,H_2O + 2e^- \rightarrow H_2 + 2\,OH^- \tag{2.13}$$

2.4 TUNABLE PROPERTIES OF THE MATERIAL BY ELECTRODEPOSITION PARAMETERS

The challenge in the electrochemical reduction of alloys is in choosing the optimum value of current density and appropriate bath with its additional agents. Choosing an irrelevant value of current without considering the basic properties of materials, results in deposition of only one metal retarding the reduction of other metal(s) at the cathode. In the case of binary alloys, for both the metals to

Controlled Electrochemical Deposition for Materials Synthesis

be deposited simultaneously with atomic level mixing, the potential range at which each metal gets deposited must be close to each other. If in the case when the potential is not in the closer window, then complexing agents must be added to reduce the potential of one metal or increase the potential range for deposition of the other. In the deposited materials, structure and morphology depend on the thermodynamic and kinetic condition of the reaction. The availability of phase diagrams for binary alloys help in studying the alloy systems. One can predict the resultant phase from phase diagrams, but the electrodeposited alloys do not merely follow it in certain cases (Beattie and Dahn 2003; Bockris, Conway, and White 2013).

In almost all materials synthesis methods, the subjected material can be controlled in terms of composition, structure, size, and morphology. There are many properties that get influenced by the electrochemical phenomenon and parameters applied. When it comes to electrodeposition, all the aforementioned characteristics can be tuned by simple use of variable parameters including current density/potential, composition of the electrolytic bath, adding agents, and temperature by using a conventional electrolytic cell setup.

2.4.1 Composition

The composition of the material is readily tunable in electrochemical deposition. Almost all the parameters play a significant role in the composition and phase of the sample. Most important, the influence of current density changes the composition and hence in some cases alters the phase of the sample.

The current density decides which material from the electrolyte has to be deposited and how much of that same material is to be deposited. In the case of alloys and intermetallic compounds, the electrolyte contains more than one kind of ionic substance. Hence depending upon the standard electrode potential of the substance under study, one material might get deposited earlier and the other metal does not have many favorable conditions to get deposited.

For example, Co–Fe alloy electrodeposited by Heydarzadeh Sohi (2012) exists at the body-centered cubic (bcc) phase at limited current density. But as the current density is increased, the amount of deposition of iron is increased, which induces the face-centered cubic (fcc) phase structure. An increase in the current density increases the iron content of the substance and follows a phase transition from bcc to fcc dominance. Hence, one can state that the change in phase is coherently linked with the composition of the sample. With an increase in the current density, the content of the iron increases and gets saturated at one point and decreases drastically at further higher current density. This value tells us that it is important to note that there is a working range limit for all the elements. The metals will get deposited prominently within that range.

The composition of the precursor in the electrolyte has a direct influence on the composition of the final product. Effect of bath composition on electrodeposited Ag–Co was reported by Santhi et al. (2017). Ag–Co was deposited on stainless steel substrate from two bath compositions at three different current densities. The ratio of cobalt to silver increased a lot for the alloy deposited at same current density from a different bath composition. The reported results clearly show that the electrolyte composition governs the atomic compositions of Ag–Co alloy. In both compositions, the cobalt content increases with an increase in current density, reaching a maximum at 30 mA/cm^2. The bath concentrations and the current density play a major role in determining the magnetic properties more than the actual Co content (Santhi et al. 2017). It is clear from the results obtained that the variations in structure, morphology, and magnetic properties of the deposits are attributable to the electrolyte composition and the plating conditions, and especially the current density. The changes in magnetic properties are discussed in Section 2.4.2.

The nickel–zinc alloy deposits are well known for their corrosion-resistant properties. The phase diagram of Ni–Zn alloy evidences for phase transitions and has high stability over a wide temperature range. In the electrodeposition of Ni–Zn, the deposition weight percentage of nickel increases with increasing current density and duty cycle. The deposition of nickel percentage has increased by 25% on increasing the applied current from 0.1 A to 0.3 A. In the combination of nickel and zinc, the nonnoble zinc deposition is favored when compared to nickel. At lower current density, anomalous behavior is observed in that the less noble zinc deposition is favored with lower cathodic current efficiency. The effect of electrolytic pH plays an important role in determining the basic characteristics of the deposited material. This is due to the increase in the local pH near the electrode area, and formations of zinc hydroxides inhibit nickel deposition. Also the length of the flakes of Ni–Zn varies with applied current density and duty cycle. Consider the formation of Ni–Sn alloy through electrochemical synthesis. The standard reduction potential of nickel is −0.25 V and for tin it is −0.14 V. Though the reduction potential of nickel and tin differs only by 0.11 V, predominant deposition of Sn is found even when varying the current density and duty cycle (Dhanapal 2016).

In cobalt electrodeposition, the hexagonal close-packed (hcp) phase is predominant and the fcc phase vanishes with increasing pH. The impedance studies reveal that the hcp phase of cobalt has better electrochemical properties (Ali and Salim 2013). In copper selenide deposition, with an increase in the value of pH, the cathodic current efficiency of deposition decreased and deposition does not occur at higher pH value (Ganjkhanlou et al. 2015). Proper pH for a metal can be anywhere from acidic to neutral to alkaline, depending on the solubility of that metal in the given solvent. Nickel and copper plating in acid baths are optimized around 3.8 to 4.2 and complications at higher pH values. For electrodeposition of metals like gold, zinc, and chromium alkaline pH is preferred. The suitable pH was decided by electrochemical measurements. For example, if it is below, the Cu/In ratio declines. If it is more, the distribution is lost. So, it will suit deposition only at the optimum pH (Yang et al. 2008). Hence, the current density, composition, and pH play an important role during the synthesis of materials. Additionally, some electroplating processes may benefit from the use of a buffer, such as boric acid. The buffer ensures that pH near the cathode remains stable at the optimum level for metal plating. The optimum pH for an electroplating process depends on the metal to be plated as well as the solvent and any other additives.

Temperature is one of the important parameters in materials synthesis. During the deposition of nickel from type watt bath it is concluded that the cathodic efficiency increases with temperature and decreases with current density (Poroch-Seri et al. 2009). The increase of current density has certain disadvantages in deposition technique. An increase in current density above a particular value leads to the formation of hydrogen bubbles and decreases the current efficiency.

The electrodeposition of ternary alloys is rather an interesting phenomenon to discuss. In the case of the Ni–Zn–Fe system, at moderate and high current density, nickel deposition is favored (Abou-Krisha 2012). In the ternary system Ni–Zn–P, the effect of pH is studied between 8.5 and 10.5. In this range, upon increasing the pH, the amount of Ni content increases and at the same time, the content of zinc and phosphorous reduces. Hence, in the given pH range, Ni deposition is favored (Srivastava et al. 2017). In case of the ternary system of Zn–Fe–Mo it is very clear that an increase in pH shows the hydrogen evolution peaks. This indicates that the deposition is controlled by the electrodeposition parameters. Hence, only the lower pH favors this ternary alloy deposition (Winiarski et al. 2016). The deposition of the copper tin selenide films was done at ambient conditions and the pH variation was in the small range from 1 to 1.5. At 1.5 the material has minimum thickness with improved properties suitable for many applications (Kassim et al. 2009). The formation of the compound Cu_2ZnSnS_4 and its phase is consistent with the amount of complexing agent used (Pawar et al. 2010).

Oxide electrodeposition is an important phenomena in this era since the semiconducting oxides find potential applications in various fields of development and electronic devices today. In depositing copper oxide films, the interfacial charging effects play an important role and pH plays an important role in deciding that property (Kafi et al. 2018). Reports clearly show that the composition and phase of the sample change with current density.

2.4.2 Structure

The structure of each material is unique, and the important phenomenon that decides the type and name of the material is its phase. The materials that are used in different fields of technology are structure dependent (Erb 2012; C. Chan et al. 2011). Hence, fabricating a material with notable structure is an important aspect in materials science.

Stephen et al. (2008) exclusively worked on the structure of various alloys by altering the electrodeposition parameters. The Ni–Mn alloys were electrodeposited on a stainless steel substrate from simple acidic sulfate baths at room temperature with pH of 3.0 with deposition current density of 30 mA cm^{-2}. The deposited Ni–Mn alloys were formed within a narrow compositional range, independent of the bath composition with constant current density of 30 mA cm^{-2}. The films of Ni$_3$Mn had a nanocrystalline fcc structure and 76.2 at. % Ni deposition indicating a higher degree of ordering in the as-deposited condition. The ordering was visible for electrodeposited Ni–Mn with Ni ranging from 70% to 85%. The alloy with 76.2 at. % Ni was the most stable with pure ferromagnetic phase. Superlattice reflection confirmed from selected area electron diffraction (SAED) patterns supported the L1$_2$ type of atomic ordering in Ni–Mn films (Arumainathan et al. 2008). The scanning electron microscope (SEM) image of electrodeposited Ni$_{78}$Mn$_{22}$ alloy is shown in Figure 2.3a. The transmission electron microscope (TEM) and SAED images of electrocrystals of Ni–Mn alloys are depicted in Figure 2.3b and c, respectively. It is evident from the TEM picture that electrocrystals can be grown from the deposition technique. The pictured electrocrystal is of 800 × 600 µm in size.

In the case of electrodeposition of Ni–Mn alloys, the change in magnetic property of the alloy with respect to nickel composition has been studied (Stephen 1999). Figure 2.4a represents the change in saturation magnetization value of the alloys before and after annealing. The maximum saturation value is reached during the Ni$_3$Mn phase evolution leading to maximum nickel ratio. Figure 2.4b depicts the change in Curie temperature (T_C) with nickel composition. There is an

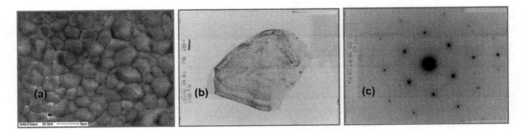

FIGURE 2.3 (a) SEM image of Ni–Mn alloy. (b) HRTEM image of Ni–Mn electrocrystal. (c) SAED image of Ni–Mn alloy.

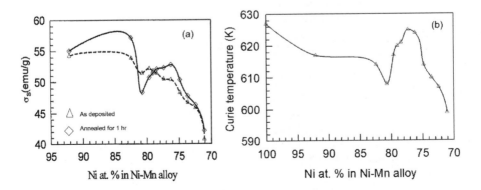

FIGURE 2.4 Magnetic studies on Ni–Mn alloy.

exponential decrease in the Curie temperature explaining the disordering in Ni–Mn alloy. The Curie temperature reaches a maximum of 625 K when the nickel concentration is 76.2 at. % and below this percentage the T_C reduces gradually representing the ordering of Ni$_3$Mn grains. This behavior is due to the manganese (Mn) composition and alloying behavior of Ni–Mn (Stephen et al. 2000; Stephen, Nagarajan, and Ananth 1998). The magnetic property is an important phenomenon in the case of iron group metals and alloys. For the synthesis of iron group materials, electrodeposition is a unique method. The magnetic properties are influenced by the electrochemical parameters applied while synthesizing the material. Thus, it can be concluded that the change in microstructure decides the magnetic property of the material (Encinas et al. 2002).

Santhi et al. (2012) studied the effect of bath composition and current density on an Ag–Ni system from nontoxic gluconate bath using pulsed electrodeposition. Ag–Ni alloys were deposited at different current densities from two bath compositions. Though these parameters do not have adverse effects on the atomic composition of prepared alloy, there is a notable change over crystal structure and morphology. Alloys deposited at lower current densities from electrolytes having lower concentration of nickel exhibit both solid solution of Ag and Ni along with Ag–Ni alloy phase. Alloys deposited at high current densities from electrolytes having higher concentration of nickel form solid solution of Ag and Ni alone. Lowering of the saturation magnetization value evidences the occurrence of atomic level mixing of silver and nickel. From magnetic thermogravimetric analysis, alloys deposited at different current densities depicted weight loss at different temperatures due to magnetic transitions. It is concluded that the changes in weight was due to structural and compositional variations (Santhi et al. 2012). The structural change in electrodeposited Ag–Fe alloy on altering the current density and bath composition has been studied elaborately. The samples deposited from lower Fe salt concentration in the bath contain more of an Ag-rich fcc phase along with the Fe rich bcc phase, while the samples obtained from higher Fe salt concentration exhibit only the bcc Fe structure. Also it has been found that current density has a considerable effect on Fe content in the sample. The Fe content in the samples is primarily bath-concentration dependent for the lower bath concentration. But at the higher bath concentration, variation due to current density is less pronounced (Santhi, Narayanan, and Arumainathan 2014). The magnetization properties of Ag–Co samples prepared from two different electrolytic baths containing precursors of Ag and Co in the ratio of 1:7 and 1:14 at the same current density. The magnetization of the sample deposited from 1:7 ratio bath is very poor (1.9 A-m^2/kg) when compared with the sample (40.2 A-m^2/kg) deposited from 1:14 bath. This property is due to the orientation of the grains in that particular sample. Also, for the same system it is observed that the magnetic properties vary, though composition of cobalt is almost the same. The magnetic properties not only depend on the composition of the sample but also on the structural and domain orientation (Santhi et al. 2017). This similar structure–property related to grain boundary effects has been discussed by Gino Palumbo et al. (Palumbo et al. 2011).

Palladium alloys are well-known due to their paramount activity in catalysis and other properties. Electrolyte composition, current density, and complexing agent of the synthesis technique has been suitably optimized to deposit phase pure Pd–Ni alloy. At the same time, in this system the structure remains the same even when the atomic ratio of Pd and Ni are changed following the phase diagram (Revathy et al. 2018). In the case of the copper–indium alloy, the bath composition is varied while keeping all other parameters constant. The change in bath composition has a direct influence of the preferred orientation of crystallites and hence the phase of the prepared material. At 60:40 ratio of precursors (Cu:In), dual phases of alloy exist, whereas at 40:60 ratio, one phase diminishes and the other phase predominates in the sample (Sivaranjani et al. 2018). The copper indium selenide (CIS) films made of electrodeposition has a tunable bandgap due to the phase formation (Karthikeyan et al. 2017). The nickel-phosphorous has an important influence in the field of electrochemistry. The effect of duty cycle in pulsed electrodeposition plays a major role in determining the amount of phosphorous content in the system and in the formation of Ni$_3$P phase. The phase and composition of the phosphorous in turn control the saturation magnetization value. The duty cycle makes an impression on the coercivity and saturation magnetization value. The duty cycle makes an impression on the coercivity and retentivity values for the same applied current (Dhanapal, Narayanan and Arumainathan 2015).

The formation process of CoNiFe ternary alloy is depicted in Figure 2.5. The diagram explains that higher content of Ni is deposited with fcc structure. The higher content of Fe deposited corresponds to mixed bcc/fcc structure and the higher ratio of Co/Fe deposits are to be bcc structured. The same has been depicted in Figure 2.6. The same effect has been studied with influence of current density on composition by Yufang Yang (2015). The films deposited at higher current densities are less porous and smooth. The Ni content of these coatings is increased from 10 to 17 wt%. The Fe-rich alloy deposits obtained at low current density surface is rough, highly porous, and exhibited a columnar appearance ($Co_{64}Ni_{11}Fe_{25}$). With an increase in the current density, the Fe content in film decreases with a corresponding increase in the Ni content. At current densities <5 mA/cm², bcc is the main phase; at 5 mA/cm², fcc is the main phase with a small amount of bcc phase; whereas at current densities >10 mA/cm², fcc is the main phase. Following the transition from bcc to fcc phase

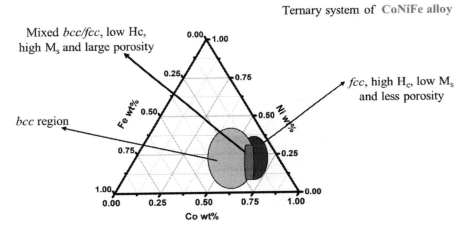

FIGURE 2.5 Structural variation in ternary alloy system with composition.

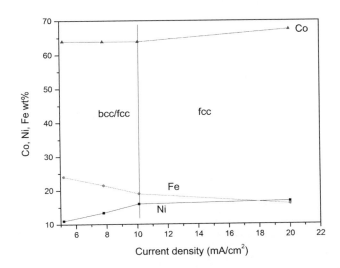

FIGURE 2.6 The change in composition of Co, Ni, and Fe elements in the CoNiFe alloy deposits with corresponding change in current density.

with increase in current density, the grain size is decreased from 16 to 3 nm (Baskaran 2006). The morphology of the same system is discussed in Section 2.4.3.

Studies have been made on the electrodeposited iron, and its mechanical properties are evaluated based on the pH of the electrolyte solution. It is clear that with the increase in pH above 3.8, there is a sudden increase in grain size of the electrodeposits (Levy and Macinnis 1968).

In case of the nickel deposition on gold film substrate, the lattice strain gradually decreased on changing the pH from 1.4 to 2. The lattice constants also tend to decrease as the pH level is increased (Boubatra et al. 2012). At the same time, pH of the sample prominently determines the phase of the Co–Fe alloy. The lower pH favors the formation of fcc (gamma) phase, whereas the higher pH leads to the formation of beta phase of Fe–Co (Lu et al. 2013). This purely indicates that pH of the electrolytic bath plays a major role in the formation of the alloy substance. The surface forces are the key factors in the influence of material deposition on the substrate. The effect of additives greatly influences the pH and hence also influences the surface force of the cations. The zeta potential of the charged particles helps predict colloidal stability and electrostatic interaction of the particles. The zeta potential represents these repulsive forces between the particles. With an increase in pH, the zeta potential increases negatively and the particle size decreases as in the case of nickel-layered silicate nanocomposites. The composite deposited at pH 3 shows improved corrosion performance (Tientong et al. 2013).

2.4.3 Morphology

Materials with interesting morphologies have been chosen for energy and catalytic applications. Pd–Ni alloy electrodeposited using galvanostatic pulsed power supply using stainless steel substrate was reported by Revathy et al. (2018). It has been reported that the effects of bath composition and current density play a major role on composition and morphology. The alloys form dendritic structures at lower current density, whereas when increasing the current density, the structure gets distorted. On correlating the composition results and morphological analysis, it is understood that palladium enhances dendritic growth while nickel suppresses it (Revathy et al. 2018). Silver–nickel (Ag–Ni) alloy prepared by PED method using anodic alumina membrane (AAM) template was reported by Dhanapal et al. (2013). Ag–Ni was deposited from gluconate bath at pH 4 with the current density of 40 mA/cm^2. The commercially available Whatman anodisc of diameter 2 cm with a porous diameter of 20 nm and depth 100 μm was used as the template. The morphology of the Ag–Ni alloy was found to be double-dumbbell shaped. The average length of the dumbbell was found to be 390 nm.

The morphology of CoNiFe ternary alloys is depicted in Figure 2.7. The Fe-rich alloy deposits obtained at low current density surface are rough, highly porous, and exhibit a columnar appearance. The morphological change has been coherent with the structural change in the alloys, where Figure 2.7a depicts the sample with fcc structure, and Figure 2.7b and c represent the bcc and bcc/fcc mixed phases, respectively. NiCoFe ternary alloy rods were deposited from simple sulfate baths at room temperature with pH of 3.0. The anodic aluminium oxide (AAO) membrane was employed

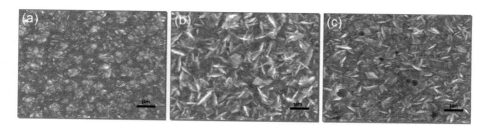

FIGURE 2.7 Morphology of CoNiFe alloys: (a) $Co_{67}Ni_{17}Fe_{16}$. (b) $Co_{54}Ni_{10}Fe_{36}$. (c) $Co_{64}Ni_{11}Fe_{25}$.

Controlled Electrochemical Deposition for Materials Synthesis 41

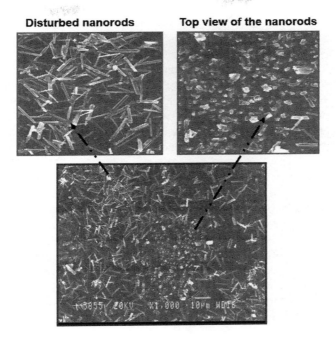

FIGURE 2.8 Morphology of NiCoFe magnetic nanorods.

for deposition, whereas the other side of the membrane was masked using a silver-coated conducting resin. PED was employed with a current density of 5 mA/cm². The plating times were $T_{ON} = 1$ ms and $T_{OFF} = 9$ ms. The duty cycle is 10% with a frequency of 100 Hz. After the deposition the AAO membrane with the NiCoFe, nanorods were rinsed with deionized water, then dried. Subsequently, the AAO membrane is chemically etched using a mixture of 1.5 ml of sulfuric acid, 12.5 ml monophosphoric acids, and 10 g of chromium trioxide. The typical SEM images of NiCoFe alloy rods are shown in Figure 2.8 with dissipation of rods and top view of the rods.

In the presence of the ethyl vanillin, the morphology has been completely changed (Nayana and Venkatesha 2014). The change in applied potential typically changes the morphology of Cu_2O films prepared by electrodeposition (Bijani et al. 2011). With controlled measure of pH and other additives, a single layer of atoms can also be deposited by the electrodeposition method (Switzer 2012). Gold was electrodeposited at a potential of −0.5 V such that the critical nuclei size is calculated to be less than one atomic layer (Hussein et al. 2018). With increased application in various fields, it has become very important to produce nanoparticles of regular arrays with shapes having more surface area (Prida et al. 2015). Palladium arrays have been deposited on Al film (Bera et al. 2003).

2.5 MULTILAYER COATINGS

Multiple layers of coating are of great interest today because of their excellent properties in optics and electronics owing to their two-dimensional structures. Multilayers are typically made by sputtering, evaporation, chemical vapor deposition, molecular beam epitaxy, and other vacuum deposition processes. Among the other techniques, ED can be employed to get phase pure alloy (Csik et al. 2009; Schwarzacher and Lashmore 1996). Multilayer films are typically coatings of two different alternating materials (Ross 1994). The combination of magnetic and nonmagnetic layers lead to the overwhelming growth in spintronics. In the case of magnetic and nonmagnetic combination, the magnetic property will be dependent on the applied magnetic field direction, and on the formation of films the magnetic property is greatly tuned. Hence the anisotropic property will become important. The thickness of the layers also determine the formation of perpendicular anisotropy.

The structure, growth condition, and layer composition determines the magnetic property of the layers. The overall anisotropy in the layers is influenced by the crystallographic texture within the layers and the degree of mixing between the layers (Masrour et al. 2010). The advancements in giant magneto resistance (GMR) effects produced by materials prepared from electrodeposition was enlisted by Schwarzacher and Lashmore (1996). Nickel bilayers and trilayers have been fabricated and their magnetic properties explored by our group. The Ni/Pd/Ni trilayer is depicted in Figure 2.9. The magnetic anisotropy of the Ni/Cu/Ni layer (Figure 2.10a) and Ni/Ag/Ni have been studied by Dhanapal et al. (2014, 2017). The Ni/Ag bilayer has been depicted in Figure 2.10b. Nickel–chromium layers were electrodeposited and tested for their hardness behavior (Etminanfar and Heydarzadeh Sohi 2012).

The importance of the Cu–Co multilayers and other similar growth of films is because of the giant magnetoresistance (GMR) effect shown by these structured layers on combining a nonmagnetic layer with a ferromagnetic layer. Bakonyi and Péter (2010) wrote a review about the GMR effect produced by the electrodeposited films made of various materials and has produced a detailed report on various parameters including homogeneity of the film. From the Central Electrochemical Research Institute (CECRI) of India, works have been carried out to study the effect of Cu thickness affecting the magnetic property in Co/Cu multilayers (Rajasekaran et al. 2014; Rajasekaran and Mohan 2012). Like every other parameter, the addition of additives has a significant influence on the film formation. The addition of NaCl greatly influences the formation of the Co–Cu/Cu layer (Péter et al. 2001). Deposition of multilayers upon one another leads to the formation of superlattice

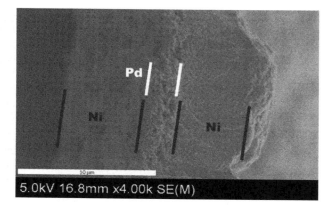

FIGURE 2.9 Cross-section SEM image of Ni/Pd/Ni layers.

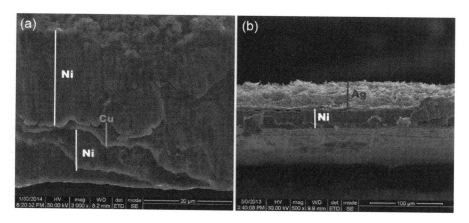

FIGURE 2.10 Cross-section image of (a) Ni/Cu/Ni layer and (b) Ni/Ag layer.

in some cases. The electrodeposited film on simple annealing changes its behavior from GMR to AMR (anisotropic magnetoresistance) and then back to GMR (Schwarzacher et al. 1997).

2.6 COMPOSITE COATINGS

Surface coatings have a wide application because of their improved mechanical, corrosion, and wear-resistance properties. Corrosion and wear behavior of pulsed electrodeposited Ni–Al$_2$O$_3$ nanocomposite coating assisted with ultrasound shows improved wear-resistance behavior (Majidi et al. 2016; Allahyarzadeh et al. 2017). In the formation of Ni–B alloy film from optimized electrolytic bath, the preferred orientation of nickel changes from fcc (200) to fcc (111) due to the incorporation of boron elements. The boron elements are incorporated into the nickel matrix by adsorption followed by catalytic reduction of dimethylamine borane (Krishnaveni, Tsn, and Seshadri 2003; Krishnaveni, Sankara Narayanan, and Seshadri 2006). The zinc–zinc phosphate coatings are done by galvanostatic electrodeposition, which is an energy efficient method for preparing coatings with improved corrosion resistance (Kavitha et al. 2014). The nitride compounds are chemically stable compounds, but they can still be electrodeposited by dispersing them in the electrolytic bath. Ni–B–Si$_3$N$_4$ coatings are prepared by electrodeposition method at proper electrochemical conditions (Krishnaveni, Sankara Narayanan, and Seshadri 2008; Krishnaveni, Narayanan, and Seshadri 2009). Electrodeposition has its own scope in the medical field also. Hydroxyapatite is one of the important materials present in bones, and teeth have been electrodeposited on magnesium by Jamesh et al. (2012).

The deposition of collagen is a difficult thing since it has less conductivity toward the cathode. Though various pH solutions have been used, it is possible only at pH 5 toward the titanium electrode (Kamata et al. 2011). Recent advances in the electrodeposition include the deposition of rare earth oxides and composites (Tam, Erb, and Azimi 2018; Tam et al. 2017). Diamond compounds and other nonconducting composites have been prepared to address the corrosion prevention at high temperature. These composites improve the thermal and wear properties of the metal matrix materials (Jun Cho et al. 2019; Jun Cho et al. 2016).

2.7 CONCLUSION

Electrodeposition is a conventional and an effective route to synthesize materials with challenging and tunable physicochemical properties. It is clear that via the electrodeposition technique one can synthesize single-oriented films, bulk and nanocrystalline metals, alloys, semiconductors, and multilayers composites. The mechanism of electrodeposition is a quite easy procedure to follow, but surface migration and lattice development of a particular material changes with the material and the material's composition. The parameters, such as current density, pH, and temperature, play a major role in forming the final product of the reaction. And it is important to note that the pulsed electrodeposition enhances nucleation rather than growth resulting in lower grain size. Other than the experimental and environmental conditions the important factor that decides the deposition of a material is its standard electrode potential. The deposition behavior is greatly influenced by addition of complexing agents, which will shift the deposition potential for a particular material and can be notified with the help of cyclic voltammetry. The deposited material's composition, structure, morphology, and other properties are significantly varied on varying a particular parameter. It is possible to produce different intermetallic phases at room temperature through electrodeposition at room temperature. Besides all the advantages of the electrodeposition, hydrogen evolution during the electrodeposition at the cathode paves way for hydrogen embrittlement, which is an unwanted thing in preparing a material. All the parameters have to be optimized by a plausible number of trial-and-error experiments so as to reduce hydrogen formation and avoid the internal stress in the deposited material. When the pulsed electrodeposition technique is followed, there is an appropriate value of pulse frequency and the duty cycle for each material at which the composition of that material is maximum in the deposit and the anticipated ordered phase is formed leading to tailored-made properties for special applications. The change of pulse frequency may lead to the deterioration of materials' properties. The purity of the samples prepared by electrodeposition is very high and does not need any further purification. The advancements in electrodeposition led to fabrication of materials with functional properties. Novel materials including composites and multilayers synthesized from electrodeposition has led to potential applications in the field of science and technology.

ACKNOWLEDGMENTS

The authors extend their gratitude to the University of Madras and acknowledge Dr. K. Dhanapal, Dr. Kalavathy Santhi, and Dr. I. Baskaran for their scientific contributions.

REFERENCES

Abou-Krisha, M. M. 2012. "Effect of pH and current density on the electrodeposition of Zn–Ni–Fe alloys from a sulfate bath." *Journal of Coatings Technology and Research* 9 (6):775–783. doi: 10.1007/s11998-012-9402-1.

Ali, S., and M. Salim. 2013. "Effect of pH at early formed structures in cobalt electrodeposition." *Asian Journal of Chemistry* 25 (8). doi: 10.14233/ajchem.2013.11031.

Allahyarzadeh, M., M. Aliofkhazraei, A. S. Rouhaghdam, and V. Torabinejad. 2017. "Electrodeposition of multilayer Ni–W and Ni–W–alumina nanocomposite coatings." *Surface Engineering* 33 (5):327–336. doi: 10.1080/02670844.2016.1277640.

Al-Osta, A., B. S. Samer, V. V. Jadhav, U. T. Nakate, R. S. Mane, and M. Naushad. 2017. "NiO@CuO@Cu bilayered electrode: two-step electrochemical synthesis supercapacitor properties." *Journal of Solid State Electrochemistry* 21 (9):2609–2614. doi: 10.1007/s10008-016-3489-8.

Arrhenius, S. 1889. "Über die Dissociationswärme und den Einfluss der Temperatur auf den Dissociationsgrad der Elektrolyte." *Zeitschrift für Physikalische Chemie* 4U (1):96–116.

Arumainathan, S., F. Rossi, L. Nasi, C. Ferrari, P. Nagamony, M. V. Ananth, and V. Ravichandran. 2008. "Induced ordering in electrodeposited nanocrystalline Ni–Mn alloys." *Journal of Applied Physics* 103:053511.

Bakonyi, I., and L. Péter. 2010. "Electrodeposited multilayer films with giant magnetoresistance (GMR): progress and problems." *Progress in Materials Science* 55 (3):107–245. doi: 10.1016/j.pmatsci.2009.07.001.

Baskaran, I. 2006. *Preparation and characterization of nanocrystalline nickel based alloy deposits*, Department of Nuclear physics, University of Madras, Chennai.

Beattie, S. D., and J. R. Dahn. 2003. "Single-bath electrodeposition of a combinatorial library of binary Cu1 − x Sn x Alloys." *Journal of The Electrochemical Society* 150 (7):C457–C460.

Bera, D., S. C. Kuiry, S. Patil, and S. Seal. 2003. "Palladium nanoparticle arrays using template-assisted electrodeposition." *Applied Physics Letters* 82 (18):3089–3091. doi: 10.1063/1.1572465.

Bijani, S., R. Schrebler, E. A. Dalchiele, M. Gabás, L. Martínez, and J. R. Ramos-Barrado. 2011. "Study of the nucleation and growth mechanisms in the electrodeposition of micro- and nanostructured Cu2O thin films." *The Journal of Physical Chemistry C* 115 (43):21373–21382. doi: 10.1021/jp208535e.

Bockris, J. O. M., B. E. Conway, and R. E. White. 2013. *Modern aspects of electrochemistry*. Springer US.

Boubatra, M., A. Azizi, G. Schmerber, and A. Dinia. 2012. "The influence of pH electrolyte on the electrochemical deposition and properties of nickel thin films." *Ionics* 18 (4):425–432. doi: 10.1007/s11581-011-0642-3.

Brett, C. M. A., and A. M. O. Brett. 1993. *Electrochemistry: principles, methods, and applications*: Oxford University Press.

Brooks, I., G. Palumbo, G. D. Hibbard, Z. Wang, and U. Erb. 2011. "On the intrinsic ductility of electrodeposited nanocrystalline metals." *Journal of Materials Science* 46 (24):7713–7724.

Bunshah, R. F. 2001. *Handbook of hard coatings: deposition technologies, properties and applications*: Elsevier.

Chan, C., J. L. McCrea, G. Palumbo, and U. Erb. 2011. "Microstructural and mechanical characterization of multilayered iron electrodeposits." *Advanced Materials Research* 409:474–479.

Chan, T., D. Backman, R. Bos, T. Sears, I. Brooks, and U. Erb. 2011. "Temperature changes during deformation of polycrystalline and nanocrystalline nickel." *Advanced Materials Research* 409:480–485.

Csik, A., K. Vad, G. Langer, G. Katona, E. Toth-Kadar, and L. Peter. 2009. "Reverse depth profiling of electrodeposited Co/Cu multilayers by SNMS." *arXiv*:0902.1672.

Dahotre, N. B., and T. S. Sudarshan. 1999. *Intermetallic and ceramic coatings*: Taylor & Francis.
Dhanapal, K. 2016. *Magnetic & catalytic properties of Ni-X (X = Sn, Zn, P) binary alloys and magnetic anisotropy behaviour of electrodeposited Ni based films*, Department of Nuclear Physics, Univeristy of Madras, Chennai.
Encinas, A., M. Demand, J. George, and L. Piraux. 2002. "Effect of the pH on the microstructure and magnetic properties of electrodeposited cobalt nanowires." *IEEE Transactions on Magnetics* 38 (5):2574–2576. doi: 10.1109/tmag.2002.801956.
Endres, F., A. Abbott, and D. R. MacFarlane. 2017. *Electrodeposition from ionic liquids*: John Wiley & Sons.
Erb, U. 2012. "Size effects and structure-sensitivity of properties in electrodeposited nanomaterials." *Materials Science Forum* 706–709:1601–1606.
Erb, U., G. Palumbo, and J. L. McCrea. 2011. "The processing of bulk nanocrystalline metals and alloys by electrodeposition." In *Nanostructured metals and alloys*, edited by Sung H. Whang, 118–151. Woodhead Publishing.
Etminanfar, M., and M. H. Sohi. 2012. "Hardness study of the pulse electrodeposited nanoscale multilayers of CR-NI." *International Journal of Modern Physics: Conference Series* 5:679–686.
Facchini, D., J. L. McCrea, F. González, G. Palumbo, K. Tomantschger, and U. Erb. 2009. "Nanostructured metals and alloys: Use of nanostructured cobalt phosphorus as a hard chrome alternative for functional applications." *Galvanotechnik* 3:523–534.
Ganjkhanlou, Y., T. Ebadzadeh, M. Kazemzad, A. Maghsoudipour, and M. Kianpour-Rad. 2015. "Effect of pH on the electrodeposition of Cu(In, Al)Se2 from aqueous solution in presence OF citric acid as complexing agent." *Surface Review and Letters* 22 (05):1550057. doi: 10.1142/s0218625x15500572.
Heydarzadeh Sohi, M. 2012. "Effects of sodium citrate and current density on electroplated nanocrystalline cobalt-iron thin films." *International Journal of Modern Physics: Conference Series* 5:696–703.
Hussein, H. E. M., R. J. Maurer, H. Amari, J. J. P. Peters, L. Meng, R. Beanland, M. E. Newton, and J. V. Macpherson. 2018. "Tracking metal electrodeposition dynamics from nucleation and growth of a single atom to a crystalline nanoparticle." *ACS Nano* 12 (7):7388–7396. doi: 10.1021/acsnano.8b04089.
Jamesh, M., S. Kumar, and T. S. N. S. Narayanan. 2012. "Electrodeposition of hydroxyapatite coating on magnesium for biomedical applications." *Journal of Coatings Technology and Research* 9 (4):495–502. doi: 10.1007/s11998-011-9382-6.
Jokar, M., and M. Aliofkhazraei. 2017. "3.20 surface preparation and adhesion tests of coatings." In *Comprehensive materials finishing*, edited by M. S. J. Hashmi, 306–335. Oxford: Elsevier.
Jun Cho, H., J. Tam, M. Kovylina, Y.-J. Kim, and U. Erb. 2016. "Thermal conductivity of bulk nanocrystalline nickel-diamond composites produced by electrodeposition." *Journal of Alloys and Compounds* 687:570–578.
Jun Cho, H., D. Yan, J. Tam, and U. Erb. 2019. "Effects of diamond particle size on the formation of copper matrix and the thermal transport properties in electrodeposited copper-diamond composite materials." *Journal of Alloys and Compounds* 791:1128–1137.
Dhanapal, K., M. Vasumathi, K. Santhi, V. Narayanan, and S. Arumainathan. 2013. "Double dumbbell shaped AgNi alloy by pulsed electrodeposition." *AIP Conference Proceedings* 1576 (1):95–97.
Dhanapal, K., T. A. Revathy, M. Anand Raj, V. Narayanan, and S. Arumainathan. 2014. "Magnetic anisotropy studies on pulsed electrodeposited Ni/Ag/Ni trilayer." *Applied Surface Science* 313:698–703.
Dhanapal, K., V. Narayanan, and S. Arumainathan. 2015. "Effect of phosphorus on magnetic property of Ni-P alloy synthesized using pulsed electrodeposition." *Materials Chemistry and Physics* 166:153–159.
Dhanapal, K., D. Prabhu, R. Gopalan, V. Narayanan, and S. Arumainathan. 2017. "Role of Cu layer thickness on the magnetic anisotropy of pulsed electrodeposited Ni/Cu/Ni tri-layer." *Materials Research Express* 4 (7):075040.
Kafi, F. S. B., K. M. D. C. Jayathileka, R. P. Wijesundera, and W. Siripala. 2018. "Effect of bath pH on interfacial properties of electrodeposited n-Cu2O films." *Physica Status Solidi (b)* 255 (6):1700541. doi: 10.1002/pssb.201700541.
Kamata, H., S. Suzuki, Y. Tanaka, Y. Tsutsumi, H. Doi, N. Nomura, T. Hanawa, and K. Moriyama. 2011. "Effects of pH, potential, and deposition time on the durability of collagen electrodeposited to titanium." *Materials Transactions* 52 (1):81–89. doi: 10.2320/matertrans.M2010311.
Karthikeyan, N., T. Sivaranjani, S. Dhanavel, V. K. Gupta, V. Narayanan, and A. Stephen. 2017. "Visible light degradation of textile effluent by electrodeposited multiphase CuInSe2 semiconductor photocatalysts." *Journal of Molecular Liquids* 227:194–201. doi: 10.1016/j.molliq.2016.12.019.
Kassim, A., S. Nagalingam, W. Tan, A. M. Shariff, D. Kuang, M. J. Haron, and H. Min. 2009. "Effects of pH value on the electrodeposition of Cu_4SnS_4 thin films." *Analele Universităţii din Bucureşti – Chimie, Anul XVIII (serie nouă)* 1:59–64.

Kavitha, C., T. S. N. Sankara Narayanan, K. Ravichandran, and M. H. Lee. 2014. "Deposition of zinc–zinc phosphate composite coatings on aluminium by cathodic electrochemical treatment." *Surface and Coatings Technology* 258:539–548. doi: 10.1016/j.surfcoat.2014.08.040.

Kim, J., J. E. Dick, and A. J. Bard. 2016. "Advanced electrochemistry of individual metal clusters electrodeposited atom by atom to nanometer by nanometer." *Accounts of Chemical Research* 49 (11):2587–2595. doi: 10.1021/acs.accounts.6b00340.

Krishnaveni, K., T. S. N. S. Narayanan, and S. K. Seshadri. 2003. "Electrodeposited Ni-B alloy films: Preparation and structural aspects." *Transactions: Indian Institute of Metals* 56:341–346.

Krishnaveni, K., T. S. N. S. Narayanan, and S. K. Seshadri. 2006. "Electrodeposited Ni–B coatings: formation and evaluation of hardness and wear resistance." *Materials Chemistry and Physics* 99 (2):300–308. doi: 10.1016/j.matchemphys.2005.10.028.

Krishnaveni, K., T. S. N. S. Narayanan, and S. K. Seshadri. 2008. "Electrodeposited Ni–B–Si3N4 composite coating: preparation and evaluation of its characteristic properties." *Journal of Alloys and Compounds* 466 (1):412–420. doi: 10.1016/j.jallcom.2007.11.104.

Krishnaveni, K., T. S. N. S. Narayanan, and S. K. Seshadri. 2009. "Corrosion resistance of electrodeposited Ni–B and Ni–B–Si3N4 composite coatings." *Journal of Alloys and Compounds* 480 (2):765–770. doi: 10.1016/j.jallcom.2009.02.053.

Levy, E. M., and R. D. Macinnis. 1968. "Effect of pH on structure and mechanical properties of electrodeposited iron." *Journal of Applied Chemistry* 18 (9):281–284. doi: 10.1002/jctb.5010180906.

Lu, W., C. Ou, P. Huang, P. Yan, and B. Yan. 2013. "Effect of pH on the structural properties of electrodeposited nanocrystalline FeCo films." *International Journal of Electrochemical Science* 8:8218–8226.

Majidi, H., M. Aliofkhazraei, A. Karimzadeh, and A. S. R. Aghdam. 2016. "Corrosion and wear behaviour of multilayer pulse electrodeposited Ni–Al2O3 nanocomposite coatings assisted with ultrasound." *Bulletin of Material Science* 39 (7):1691–1699.

Malyshev, V. 2012. "Electrodeposition of chromium from ionic melts." *Russian Journal of Non-Ferrous Metals* 52 (6):473–480.

Masrour, R., M. Hamedoun, A. Benyoussef, and H. Lassri. 2010. "Study of the magnetic properties of Ni/Ag superlattice." *Journal of Superconductivity and Novel Magnetism* 23 (4):433–436. doi: 10.1007/s10948-009-0604-4.

Nayana, K. O., and T. Venkatesha. 2014. "Effect of ethyl vanillin on ZnNi alloy electrodeposition and its properties." *Bulletin of Material Science* 37 (5):1137–1146.

Palli, S., and S. R. Dey. 2016. "Theoretical and experimental study of copper electrodeposition in a modified hull cell." *International Journal of Electrochemistry* 2016:13. doi: 10.1155/2016/3482406.

Palumbo, G., E. M. Lehockey, P. Lin, U. Erb, and K. T. Aust. 2011. "A grain boundary engineering approach to materials reliability." *Materials Research Society Symposium Proceedings* 458:273–282.

Paunovic, M., and M. Schlesinger. 2006. *Fundamentals of electrochemical deposition*. Wiley.

Pawar, B. S., S. M. Pawar, S. W. Shin, D. S. Choi, C. J. Park, S. S. Kolekar, and J. H. Kim. 2010. "Effect of complexing agent on the properties of electrochemically deposited Cu2ZnSnS4 (CZTS) thin films." *Applied Surface Science* 257 (5):1786–1791. doi: 10.1016/j.apsusc.2010.09.016.

Péter, L., Z. Kupay, Á Cziráki, J. Pádár, J. Tóth, and I. Bakonyi. 2001. "Additive effects in multilayer electrodeposition: properties of Co–Cu/Cu multilayers deposited with NaCl additive." *The Journal of Physical Chemistry B* 105 (44):10867–10873. doi: 10.1021/jp011380t.

Plieth, W. 2008. "1 - Electrolytes." In *Electrochemistry for materials science*, edited by Waldfried Plieth, 1–26. Amsterdam: Elsevier.

Plieth, W., and G. Georgiev. 2006. "Residence times in kink sites and Markov chain model of alloy and intermetallic compound deposition." *Russian Journal of Electrochemistry* 42 (10):1093–1100.

Poroch-Seri, M., An, G. Gutt, and T. Severin. 2009. "Study on the influence of current density and temperature about electrodepositions of nickel by electrolytes of type watts." *Annals of Suceava University: Food Engineering* 2: 16–23.

Prida, V. M., V. Vega, J. García, L. Iglesias, B. Hernando, and I. Minguez-Bacho. 2015. "1 - Electrochemical methods for template-assisted synthesis of nanostructured materials." In *Magnetic nano- and microwires*, edited by Manuel Vázquez, 3–39. Woodhead Publishing.

Rajasekaran, N., L. Pogány, Á. Révész, B. G. Tóth, S. Mohan, L. Péter, and I. Bakonyi. 2014. "Structure and giant magnetoresistance of electrodeposited Co/Cu multilayers prepared by two-pulse (G/P) and three-pulse (G/P/G) plating." *Journal of the Electrochemical Society* 161 (6):D339–D348.

Rajasekaran, N., and S. Mohan. 2012. "Giant magnetoresistance in electrodeposited films: current status and the influence of parameters." *Critical Reviews in Solid State and Materials Sciences* 37 (3):158–180. doi: 10.1080/10408436.2011.613490.

Revathy, T. A., K. Dhanapal, S. Dhanavel, V. Narayanan, and A. Stephen. 2018. "Pulsed electrodeposited dendritic Pd-Ni alloy as a magnetically recoverable nanocatalyst for the hydrogenation of 4-nitrophenol." *Journal of Alloys and Compounds* 735:1703–1711. doi: 10.1016/j.jallcom.2017.11.264.

Ross, C. A. 1994. "Electrodeposited multilayer thin films." *Annual Review of Materials Science* 24 (1):159–188. doi: 10.1146/annurev.ms.24.080194.001111.

Sahoo, P., S. K. Das, and J. Paulo Davim. 2017. "3.3 surface finish coatings." In *Comprehensive materials finishing*, edited by M. S. J. Hashmi, 38–55. Oxford: Elsevier.

Santhi, K., K. Dhanapal, V. Narayanan, and S. Arumainathan. 2017. "Electrochemical alloying of immiscible Ag and Co for their structural and magnetic analyses." *Journal of Magnetism and Magnetic Materials* 433:202–208.

Santhi, K., S. N. Karthick, H.-J. Kim, M. Nidhin, V. Narayanan, and S. Arumainathan. 2012. "Microstructure analysis of the ferromagnetic Ag-Ni system synthesized by pulsed electrodeposition". *Applied Surface Science* 258 (7):3126–3132.

Santhi, K., V. Narayanan, and S. Arumainathan. 2014. "Dendritic Ag-Fe nanocrystalline alloy synthesized by pulsed electrodeposition and its characterization." *Applied Surface Science* 316:491–496.

Schwarzacher, W., and D. S. Lashmore. 1996. "Giant magnetoresistance in electrodeposited films." *IEEE Transactions on Magnetics* 32 (4):3133–3153. doi: 10.1109/20.508379.

Schwarzacher, W., K. Attenborough, A. Michel, G. Nabiyouni, and J. P. Meier. 1997. "Electrodeposited nanostructures." *Journal of Magnetism and Magnetic Materials* 165 (1):23–29. doi: 10.1016/S0304-8853(96)00465-9.

Sivaranjani, T., T. A. Revathy, S. Dhanavel, K. Dhanapal, V. Narayanan, and A. Stephen. 2018. "Effect of dendritic Cu–In alloy on Cr(VI) reduction synthesized via pulsed electrodeposition." *Chemistry Select* 3 (44):12613–12619. doi: 10.1002/slct.201802651.

Sobha Jayakrishnan, D. 2012. "5 - Electrodeposition: the versatile technique for nanomaterials." In *Corrosion protection and control using nanomaterials*, edited by Viswanathan S. Saji and Ronald Cook, 86–125. Woodhead Publishing.

Srivastava, C., S. K. Ghosh, S. Rajak, A. K. Sahu, R. Tewari, V. Kain, and G. K. Dey. 2017. "Effect of pH on anomalous co-deposition and current efficiency during electrodeposition of Ni-Zn-P alloys." *Surface and Coatings Technology* 313:8–16. doi: 10.1016/j.surfcoat.2017.01.043.

Stephen, A. 1999. *Electrochemical and physics properties of electrodeposited nanocrystalline nickel-manganese*, Department of Nuclear Physics, University of Madras, Chennai.

Stephen, A., M. V. Ananth, V. Ravichandran, and B. R. V. Narashiman. 2000. "Magnetic properties of electrodeposited nickel–manganese alloys: Effect of Ni/Mn bath ratio." *Journal of Applied Electrochemistry* 30 (11):1313–1316. doi: 10.1023/a:1026524907978.

Stephen, A., F. Rossi, L. Nasi, C. Ferrari, N. Ponpandian, M. V. Ananth, and V. Ravichandran. 2008. "Induced ordering in electrodeposited nanocrystalline Ni–Mn alloys." *Journal of Applied Physics* 103 (5):053511. doi: 10.1063/1.2844211.

Stephen, A., T. Nagarajan, and M. V. Ananth. 1998. "Magnetization behaviour of electrodeposited Ni–Mn alloys." *Materials Science and Engineering: B* 55 (3):184–186. doi: 10.1016/S0921-5107(98)00201-3.

Switzer, J. A. 2012. "Atomic layer electrodeposition." *Science* 338 (6112):1300. doi: 10.1126/science.1231853.

Tam, J., U. Erb, and G. Azimi. 2018. "Non-wetting nickel-cerium oxide composite coatings with remarkable wear stability." *MRS Advances* 3 (29):1647–1651.

Tam, J., G. Palumbo, and U. Erb. 2016. "Recent advances in superhydrophobic electrodeposits." *Materials* 9 (3):151.

Tam, J., G. Palumbo, U. Erb, and G. Azimi. 2017. "Robust hydrophobic rare earth oxide composite electrodeposits." *Advanced Materials Interfaces* 4 (24):1700850.

Tientong, J., C. R. Thurber, N. Souza, A. Mohamed, and T. D. Golden. 2013. "Influence of bath composition at acidic pH on electrodeposition of nickel-layered silicate nanocomposites for corrosion protection." *International Journal of Electrochemistry* 2013:8. doi: 10.1155/2013/853869.

Victor, J., D. Facchini, and U. Erb. 2012. "A low-cost method to produce superhydrophobic polymer surfaces." *Journal of Materials Science* 47 (8):3690–3697.

Winiarski, J., A. Leśniewicz, P. Pohl, and B. Szczygieł. 2016. "The effect of pH of plating bath on electrodeposition and properties of protective ternary Zn–Fe–Mo alloy coatings." *Surface and Coatings Technology* 299:81–89. doi: 10.1016/j.surfcoat.2016.04.073.

Yang, H. X., Z. L. Song, Y. W. Song, and H. Zhang. 2008. "Effect of pH value on the electrodeposition potential of Cu-In alloy film." *Key Engineering Materials* 373–374:216–219. doi: 10.4028/www.scientific.net/KEM.373-374.216.

Yang, Y. 2015. "Preparation of Fe-Co-Ni ternary alloys with electrodeposition." *International Journal of Electrochemical Science* 10:5164–5175.

Zangari, G. 2015. "Electrodeposition of alloys and compounds in the era of microelectronics and energy conversion technology." *Coatings* 5 (2):195–218.

3 Photovoltaic Energy Generation System: Material, Device, and Fabrication

Ananthakumar Soosaimanickam and Moorthy Babu Sridharan

CONTENTS

3.1	Introduction	49
3.2	Applications of Electrochemical Synthesis for Inorganic Heterojunction and Organic Solar Cells	50
3.3	Electrochemical Synthesis of Materials for Hybrid Solar Cells	52
3.4	Electrochemical Synthesis of Materials for Dye-Sensitized Solar Cells	54
3.5	Electrochemical Synthesis of Materials for Quantum Dot-Sensitized Solar Cells	56
3.6	Use of Electrochemical Synthesis in the Construction of Perovskite Solar Cells	57
3.7	Conclusion and Summary	58
References		59

3.1 INTRODUCTION

Evolution in the development of synthesis of inorganic semiconductor nanomaterials, organic conducting polymers, organic/organometallic dye complexes, and other organic and inorganic components have delivered amazing results in achieving alternative energy resources for future energy-related applications (Rui Yu et al. 2012; Nozik 2010). The contribution of organic and inorganic materials in the construction of future energy devices has always been attributed with novel approaches and innovations that could be made through these materials. Use of vacuum-based deposition approaches such as physical vapor deposition (PLD), sputtering, atomic layer deposition (ALD), and pulsed-laser deposition (PLD) have delivered higher efficiency in terms of producing large-scale applications (Green 2007; Kaelin et al. 2004). However, when we look from a cost point of view, non-vacuum approaches such as spin coating, drop casting, dip coating, and the electrochemical deposition method are widely used for the fabrication of the modern generation of solar cells. In order to make effective harvesting of photons, the nanomaterials surface structure has to be modified or functionalized using several approaches. Such tailored nanosurfaces act as potential candidates for the charge transfer where the large improvement in efficiency is possible. For the formation of nanostructures and nanohybrids on the conducting substrate, several approaches are followed. Solution-based approaches such as spin coating, dip coating, and spray deposition are used to deposit the nanoparticles and semiconducting polymers either separately or as a hybrid on a conducting substrate (Habass et al. 2010). Among the other approaches, electrochemical deposition of nanomaterials and polymers on the conducting substrate for solar cell application are quite important techniques to realize high efficiency. Electrochemical deposition is an easy experimental process and has more advantages over other methods. Also, through electrochemical deposition, we could easily control the quality of layers through optimizing parameters such as applied voltage, applied current density, concentration of the precursors in the medium, and nature of electrolyte used for the deposition (Lincot 2005). Also, the quality of the electrochemically deposited layers is equal with the vacuum-deposited layers, which requires consumption of less energy. Since the surface plays a major role in achieving higher efficiency of fabricating solar cells, through electrochemical deposition, it is possible to achieve highly smoothed, pinhole-free layers of active materials (D. Wei et al. 2010). Moreover, formation of highly ordered nanostructures is essential in order to facilitate the charge transport efficiently in the structure. Ordered growth of nanomaterials is not only used to improve the charge transport, but also to ease the difficult process associated with the assembly of the nanomaterials for the large-scale production. The chemical kinetics and thermodynamics of the formation of the nanoparticles layer on the substrate is essential since active surface sites present in the nanoparticles play a key role in the charge transfer. The other advantage of electrochemical deposition of materials is that it can be used for large-scale applications. This large scale of fabricating solar cells using the electrochemical method is already successfully commissioned and several other attempts are trying to enlarge it in a different way. Adopting such a large-scale technology for other emerging approaches such as perovskite solar cells is indeed useful for commercializing it for future purposes. The following sections discuss the contributions of electrochemical deposition to each category of modern generation solar cells.

3.2 APPLICATIONS OF ELECTROCHEMICAL SYNTHESIS FOR INORGANIC HETEROJUNCTION AND ORGANIC SOLAR CELLS

Heterojunction solar cells are made through the deposition of two organic or inorganic nanomaterials that have different energy levels and semiconducting properties. Such solar cells are fabricated mostly using the evaporation process through vacuum-based approaches. For the non-vacuum-based approaches, the electrochemical method has been used in most cases because of its reliability for large-scale applications. In the case of inorganic heterojunction, usually an n-type and p-type semiconductor layer is fabricated using the electrochemical deposition process through their precursors. Nanostructures of these materials not only improve the charge generation but also lead to fabrication on the flexible substrates. Formation of a transparent, conductive, and pinhole-free layer is the essential condition for highly efficient photoconduction. In the heterojunction architecture, most of the studies were oriented with the Cu_2O/ZnO heterojunction. In this, Cu_2O is the p-type semiconductor with a bandgap of 2.0 eV and ZnO has a bandgap of ~3.4 eV. In the heterojunction, Cu_2O serves as the absorber layer and ZnO serves as the window layer. Fabrication of p-type Cu_2O layer through electrochemical deposition was analyzed by several researchers, and conditions for the highly conductive Cu_2O layer for the fabrication of solar cells was optimized through their investigations (Pagare, Kanade, and Torane 2017; Brandt et al. 2017; H. M. Wei et al. 2012; Yoon et al. 2016). However, it is important to fabricate a heterojunction and analyze the effect of it for solar cells. When the growth of a heterojunction takes place in solution, the formation of an active interface is quite important since the defects present in the first layer will affect the transformation of carriers into the second layer, which is deposited to form the heterojunction. By coupling ZnO with Cu_2O, it is found that the absorption edge of the heterojunction could be extended to the visible region (Lahmar et al. 2017). Kang et al. (2016) fabricated a heterojunction solar cell based on the electrochemically deposited aluminum doped-ZnO (n-type) and copper oxide (p-type) as the heterojunction. The authors used simple precursors such as $Al(NO_3)_2$ and $Zn(NO_3)_2$ in the medium of dimethyl sulfoxide (DMSO) and the reaction was carried out at 85°C and by passing 0.3 C/cm² at −1.0 V versus Ag/AgCl. The authors observed a 20% increase of J_{sc} values through doping aluminum with the zinc oxide. Though the observed efficiency is quite lower (~1%), these kinds of efforts will be a platform for other low-cost heterojunction solar cells that could deliver higher efficiency. Musselman et al. (2010b) fabricated this heterojunction by electrodepositing Cu_2O arrays. The Cu_2O arrays were formed using a $CuSO_4$–lactic acid mixture in the electrochemical bath at 40°C with pH 12.5. The arrays were deposited on free-standing ZnO nanowires (1 μm length) that were grown by Zn seeds on the ITO–glass substrate. These arrays were found to be quite compatible with

Photovoltaic Energy Generation System

the already grown ZnO nanowires, and through this structure a continuous junction was produced. The authors were able to achieve an internal photo conversion efficiency (IPCE) of about 85% with the power conversion efficiency of 0.5%. Also, this research group has done extensive analysis of Cu_2O-based nanostructured materials along with ZnO nano heterostructures (Musselman et al. 2010a, 2012). Cui and Gibson (2010) fabricated a nanopillar of ZnO on the ZnO thin layer through electrodeposition, which was then used to couple with the Cu_2O to form a heterojunction (Figure 3.1). This nanopillar junction is typically used to improve the contact area and charge collection. These nanopillars were grown using an equimolar solution of $ZnNO_3$ and hexamine solution at 9°C with −2.5 V of applied potential. The observed efficiency in this case was 0.88%, which was higher than the planar structured solar cell (0.55%). This study clearly emphasized growing such one-dimensional structures for the large-scale processing of solar cells. A hybrid approach of growing Cu_2O/ZnO nanostructures is also beneficial for the construction of solar cells. Makhloufeet al. (2017) used a hybrid approach of fabricating Cu_2O/ZnO nanostructures in which growth of the ZnO seed layer was carried out through the ALD method, and growth of ZnO nanowires with different lengths and thicknesses, and deposition of Cu_2O nanostructure was carried out through the electrochemical deposition method at solution pH 12.5. It was found that the increase of length of the nanowires is attributed to the thickness of the seed layer of ZnO nanostructures.

Other than the heterojunction, electrochemical deposition has been successfully applied to deposit compound semiconductors such as copper zinc tin sulfide (Cu_2ZnSnS_4) on the conductive substrate. Ahmed et al. (2012) fabricated a solar cell structured Mo/MoS_2/CZTS/CdS/i-ZnO/ITO. Here the authors deposited a Cu_2ZnSnS_4 layer through the metal stacks of copper, tin, and zinc on the molybdenum layer on soda lime glass. Under highly acidic conditions, these metals are electroplated using direct current in the presence of an organic additive. As a final result, the authors were able to achieve an efficiency of 7.3% with the V_{oc} of 0.56 V.

Unlike inorganic materials, deposition of organic semiconductors using the electrochemical method needs a lot of attention for efficient layer formation. Electrochemical deposition of polymers had received lots of attention in the past and several findings are applying this method of deposition under different potential currents. As usual, a three-electrode system is used to deposit a conducting polymer in the electrochemical polymerization method. Usually, a working electrode serves as the electrode for the deposition; the counter electrode is a metallic based one and it is typically based on Pt or Au. In most cases, a saturated calomel electrode Ag/AgCl is generally used as the reference electrode for the polymer deposition process. The supporting electrolyte is generally a quaternary ammonium salt having the general formula R_4N_X (X=methyl, ethyl, etc. or phenyl radicals, X=Cl, Br, I, PF_6^-, ClO_4^-, BF_4^-, etc.). In addition to the conducting purpose, this supporting electrolyte also serves as the dopant for the polymers. Polymers such as polyaniline, polyhexyl thiophene heterocyclic polymers polyfuran, polypyrrole, and poly ethylenedioxythiophene (PEDOT) have been successfully deposited and analyzed in the past years (Gurunathan et al. 1999). The schematic

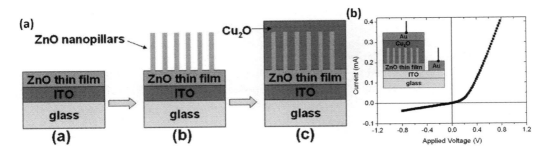

FIGURE 3.1 Growth processing of ZnO nanopillars on the ZnO thin film and the current–voltage (I–V) curve of the fabricated solar cell. (Reprinted with permission from Cui et al. Copyright 2010 American Chemical Society.)

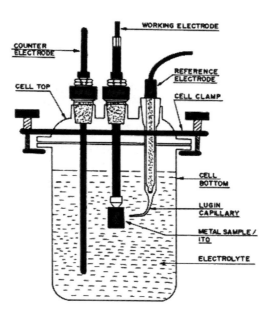

FIGURE 3.2 Schematic diagram of the experimental arrangement used for the electrochemical deposition of polymers. (Reprinted from Gurunathan et al. Copyright 1999, with permission from Elsevier.)

diagram of the experimental arrangement used for the electrochemical process of polymers is given in Figure 3.2. Zaban and Diamant (2000) first deposited crystalline organic polymer perylene bis(phenethylimide) (PPEI) on the mesoporous titanium dioxide (TiO$_2$) layer. A 0.4 M solution of tetrabutylammonium chlorate (TBAClO$_4$) was used as the electrolyte and the reaction was carried out in the presence of four electrodes. It was found that the deposited TiO$_2$/PPEI layer had a very good optical density, which could be useful for the potential applications. In another report, the same authors found that this reaction took place only if the TBAClO$_4$ is used as an electrolyte, and formation of PPEI ions took place at the source electrode (Diamant and Zaban 2001). The authors further revealed that the deposition starts forming with a thin layer coated on the mesoporous substrate and further deposition and growth of the layer on the weak points of this thin layer.

3.3 ELECTROCHEMICAL SYNTHESIS OF MATERIALS FOR HYBRID SOLAR CELLS

The device structure of a hybrid solar cell is adopted from the structure of an organic solar cell in which the donor component is replaced with the semiconducting organic polymers like poly 3-hexyl thiophene (P3HT) and poly[2-methoxy-5-(2′-ethylhexyloxy)-p-phenylene vinylene] (MEH-PPV). The energy level of these polymers is varied through substituting various donor/acceptor groups in their structural moiety. When light strikes the hybrid of the polymer/nanoparticle composite, there is an exciton generation that takes place in the nanocomposite blend. This exciton has to separate out, and the respective charges are collected by the opposite electrodes. Preparation of nanoparticle/polymer hybrids for energy-related applications is one of the challenging and interesting approaches to achieving different morphologies. A reference and counter electrode is required in presence of a suitable electrolyte to carry out such processes. Xi et al. (2008) demonstrated the growth of vertically grown CdS nanorod arrays (20 to 40 nm diameter, height 100 to 300 nm) using a template-free electrochemical approach. On these nanorods, polybithiophene polymer was deposited using *in situ* electrochemical polymerization. This polymerization was carried out using an Ag/AgCl reference electrode and platinum sheet counter electrode. Here the bithiopene molecules diffuse through electrolyte solution and get oxidized. Using this method, the authors constructed a solar cell with an Au/CdS-polybithiophene/Ca/Al hybrid solar cell that delivered about 0.84 as V_{oc}. Formation of uniform films is essential to achieve successful deposition using these approaches. The other advantage of electrochemical deposition is slow incorporation of polymers on the developed inorganic nanostructure is feasible. Shankar et al. (2008) developed a device structure consisting of TiO_2 nanotubes in which the polymers phenyl-C61-butyric-acid-methyl-ester (PCBM) and P3HT are incorporated on it to make a hybrid junction (Figure 3.3). The TiO_2 nanotubes are grown on the barrier layer of the TiO_2 layer, which is formed on a conducting substrate using a formamide-based electrolyte consisting of 0.27 M of tetrabutylammonium fluoride and 5% water. Here, the authors constructed two types of device configurations: front side and back side illumination structures. Out of these two configurations, the device with front-side illumination resulted in high efficiency of 4.1% under air mass (AM) 1.5 configuration. Synthesis and growth of such highly ordered TiO_2 nanotubes by electrochemical methods is discussed in the next section. Formation of these kinds of double heterostructures through the infiltration of polymer compounds is beneficial for extending the light absorption to fabricate future generation solar cell architectures.

Yakup Hames et al. (2010) grew ZnO nanorods on the ZnO film to fabricate a hybrid solar cell. The authors first carried out ZnO thin film through the electrochemical deposition process in the presence of ZnCl2 (5 mM) and KCl (0.1 M) with a three-electrode system. Then by modifying the concentration of $ZnCl_2$ (6 mM) and KCl (7 mM) at the optimized temperature of $80°C \pm 1°C$ and cathodic voltage of 0.9 V, the authors were able to grow ZnO nanorods on the ITO substrate with the radius of 250–300 nm. A solar cell was fabricated further with the configuration of ITO/ZnO (buffer layer)/ZnO(NRs)/P3HT:PCBM/Ag and achieved a 2.44% efficiency. Therefore, it is very clear that for effective layer formation and synthesizing different nanoarchitectures, optimizing physical

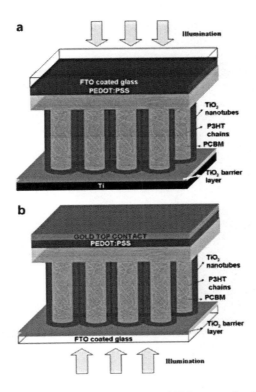

FIGURE 3.3 Schematic diagram of device configuration of TiO_2 nanotubes/P3HT/PCBM double heterojunction with (a) front-side geometry and (b) back-side geometry. (Reprinted from Shankar et al. Copyright 2008, with permission from Elsevier.)

parameters are essential in the electrochemical deposition process. Therefore, these findings are clearly illustrating that the electrochemical method is not only versatile for the development of inorganic nanostructures, but also for the development of the hybrids of them, which is quite useful for other potential applications.

3.4 ELECTROCHEMICAL SYNTHESIS OF MATERIALS FOR DYE-SENSITIZED SOLAR CELLS

Dye-sensitized solar cells (DSSCs) belong to one of the important configurations of the modern generation solar cell architecture. The basic structure composes a photoanode that has a thin layer of nanocrystalline TiO_2 nanoparticles deposited on a conducting substrate, and after a sensitization process in an efficient adsorbing dye, the device structure is completed with a photocathode and the electrolyte. The electrolyte is typically a redox system and is generally an iodide/triiodide couple. Surface states, pore size, nature of ligands on the surface, shape, and size of the TiO_2 severely affect the performance of DSSCs, and several research articles elegantly deal with the kinetics of the charge transfer process of DSSCs. For DSSCs, the electrochemical method is mainly used to prepare the counter electrode to replace the costly platinum electrode and to prepare metal oxide nanostructures. It is possible to prepare the highly aligned, vertically oriented TiO_2 nanotubes using the electrochemical anodization method. Many researchers have worked on this and succeeded in fabricating highly efficient DSSCs. A simple experimental arrangement helps to achieve this goal. Here usually a well-polished titanium (Ti) foil is taken as the anode, and a platinum or a graphite electrode is fixed as the cathode. Different electrolytes and different physical parameters such as anodization voltage, duration of anodization, and distance between the electrodes strongly influence the growth of TiO_2 nanotubes in solution. A schematic diagram of the experimental arrangement used for the synthesis of TiO_2 nanotubes is given in Figure 3.4.

The anodization process was carried out on aluminum (Al) before anodizing titanium (Ti) metal. In the presence of an acidic etchant, the process is controlled by various physical parameters. There are three classifications of the generation of synthesis of TiO_2. The first generation growth of TiO_2 nanotubes is based on hydrofluoric acid (HF) through which about 500–600 nm of thickness was achieved (Macak et al. 2007). The second generation of TiO_2 nanotubes is achieved in the presence of NaF or NH_4F, and the growth of more than 2 µm was achieved. Using water-free electrolytes, the third generation of TiO_2 nanotube growth achieved over 7 µm. Thus, it is clear that optimizing reaction conditions and reactivity of solvents are playing a superior role in the determination of

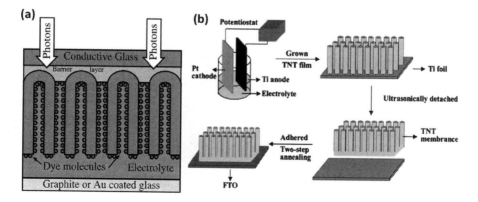

FIGURE 3.4 Schematic diagram of the (a) solar cell structure based on the anodized TiO_2 nanotubes. (Reprinted with permission from Mor et al. Copyright 2006 American Chemical Society.) (b) Experimental steps used for the electrochemical anodization process of TiO_2. (Reprinted with permission from Lei et al. Copyright 2010 American Chemical Society.)

well-aligned growth of the TiO$_2$ nanotubes. The as-synthesized TiO$_2$ nanotubes are in the amorphous form, and to make it crystalline in nature, the nanotubes are annealed at higher temperature. Once annealed, the nanotubes layer is then used for the sensitization of dye or quantum dots for the fabrication of solar cells. The as-synthesized nanotubes are collectively called a membrane. Scanning electron microscope (SEM) images of the typical anodized TiO$_2$ nanotube membrane at various resolutions is given in Figure 3.5.

Other than fabricating TiO$_2$ nanotube arrays, other metal oxides such as ZnO nanostructures are also fabricated using the electrochemical deposition method. In one case, it was observed by Abd-Ellah et al. (2013) that the use of anions in the electrolyte strongly affects the morphology of the ZnO nanostructure and different morphologies were obtained by carefully incorporating the supporting electrolyte with different anions in the reaction medium. Here, the zinc source ZnCl$_2$ is dissolved in water and then forms a hydroxide (Zn(OH)$_2$), which ultimately forms ZnO after the dehydration process. Interestingly, it was found that when the monovalent ions (Cl$^-$, NO$_3^-$, ClO$_4^-$) were used in the electrolyte, ZnO nanotubes (100–200 nm diameter) were produced, whereas for the divalent anions (SO$_4^{2-}$, C$_2$O$_4^{2-}$) ZnO nanorod formation was observed. It was also noticed that through KCl, it was able to prepare both ZnO nanorods and ZnO nanotubes under the optimized conditions. The structural directing ability of these kinds of ions during the growth of the nanomaterials in solution is further motivating synthesis of different morphologies for fabricating high-efficiency solar cells. The authors were also able to fabricate a DSSC using the as-synthesized ZnO nanotubes and achieved an efficiency of 1.6%, which was found to be comparable to similar nanotubes produced through the atomic layer deposition (ALD) method. Because of the high cost,

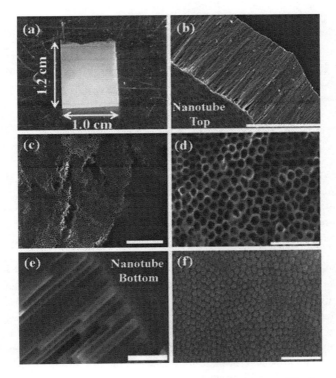

FIGURE 3.5 (a) Digital image of a freestanding TiO$_2$ nanotube membrane. The lateral dimension is 1.2 × 1.0 cm^2. (b) SEM image of as-prepared freestanding TiO$_2$ nanotube arrays (cross-sectional view, scale bar = 50 µm). (c–f) SEM images of the freestanding TiO$_2$ nanotube arrays after mild ultrasonication. (c) low magnification (scale bar = 2 µm) and (d) high magnification SEM images (scale bar = 1 µm) (top view); (e) cross-sectional view (scale bar = 1 µm); and (f) bottom view (scale bar = 1 µm). (Reproduced from Wang et al. with permission from The Royal Society of Chemistry.)

the Pt-based counter electrode in the dye-sensitized solar cells is replaced using electrochemically deposited materials. Because of the high surface area, graphene layers are chosen over other compounds on their surface and thereby formation of nanocomposites is successfully achieved. It is possible to exfoliate high-quality graphene from the graphite flakes using the electrochemical method, and in such processes, graphite foil serves as the working electrode (Belekoukia et al. 2016). Xu et al. (2013) fabricated a counter electrode based on the negatively charged graphene oxide and positively charged poly diallyldimethylammonium chloride through a layer-by-layer approach in the fabrication of the dye-sensitized solar cell. Through a cyclic voltammetry oriented electrochemical reduction procedure, the graphene oxide was converted to graphene. Through this process the electrocatalytic sites present in the graphene was enhanced and the authors finally achieved 9.5% efficiency. Liu et al. (2014) fabricated a similar kind of reduced graphene oxide (RGO) by the cyclic voltammetry method and polypyrrole-based nanocomposites as the counter electrode and achieved 6.45% efficiency. Importantly, in this case the authors fabricated a plastic dye-sensitized solar cell using an indium tin oxide (ITO)/poly ethylene naphthalene (PEN) substrate and achieved 4.25% efficiency. Thus, the electrochemical deposition of materials is not only useful for nonflexible substrates, they perform quite well in the case of flexible substrates too and this has proven beneficial for large-scale applications. Interestingly, these electrochemically deposited graphene/polypyrrole nanocomposites on carbon steel through a potentiostatic technique are also shown useful for anticorrosion applications (Li et al. 2017). This implies that through electrochemical deposition of these kinds of layers, it is possible to extend the applications to various sectors. Along with two layers of the graphene/polypyrrole, it is possible to deposit another polymer layer, like PEDOT, through the electrodeposition method. Here, first the formed graphene/polypyrrole layers are annealed at higher temperature, and finally deposition of PEDOT is carried out on these layers. Ramasamy et al. (282015) used this approach through a one-step electrochemical deposition method of graphene/polypyrrole layers and fabricated a counter electrode by depositing PEDOT as an additional layer and achieved about 7.1% efficiency using these trilayer films as the counter electrode. These results ensure that electrochemical deposition has certainly emerged as a promising technique for low-cost, simple precursors based on deposition of layers.

3.5 ELECTROCHEMICAL SYNTHESIS OF MATERIALS FOR QUANTUM DOT-SENSITIZED SOLAR CELLS

Quantum dots (QDs) are extremely small size nanoparticles having a size smaller than their exciton Bohr radius value. Because of their small size, QDs possess exciting optical properties such as impact ionization, multiple exciton generation, high extinction coefficient values, efficient carrier transfer in solution, and are capable of forming long-range assembly through ligand interactions. The essential requirement of fabricating active layers using QDs by any kind of method is high coverage of QDs on the surface. When nanoparticles are not much closer to each other or when there is a poor deposition of nanoparticles, quantum dot-sensitized solar cells (QDSSCs) suffer by the recombination process. This recombination process also depends on other parameters such as size, shape, and ligand molecules attached on the QDs surface. Use of fabrication of the QDs layer through the electrochemical method is investigated in constructing metal oxide/chalcogenide nanoparticle hybrid junctions. These nanostructures are mostly fabricated through either successive ionic layer adsorption and reaction (SILAR) or chemical bath deposition (CBD) methods. Compared with single junction, heterojunction of semiconductor nanomaterials acts as the driving force for the efficient charge transfer process. For example, through the electrochemical deposition method, the heterojunction of TiO_2/CdS and TiO_2/CdSe nanoparticles can be fabricated a through simple approach. Compared with biheterojunction, triheterojunction makes the electron transport at a much faster rate. Xiao-Yun Yu et al. (2011) fabricated heterojunction of TiO_2 hierarchical nanostructures/CdS/CdSe nanoparticles assembly through an *in situ* electrochemical deposition method. Here, the hierarchical TiO_2 nanostructure is screen printed on the conducting substrate, then the electrochemical deposition process is carried out in the presence of electrolytes such as 0.02 M of Cd(CH$_3$COO)$_2$ and 0.2 M of thiourea in a 1/1 v/v of DMSO/water and maintained at 9°C in a water bath for the deposition of CdS nanoparticles and 0.02 M of Cd(CH$_3$COO)$_2$ and 0.04 M of ethylene diamine tetraacetic acid (EDTA) together with 0.02 M of sodium selenosulfate (Na$_2$SeSO$_3$) for the CdSe nanoparticles. While passing current through the solution, the dissociation of sulfide ions (S^{2-}) from thiourea and selenide (Se^{2-}) ions from Na$_2$SeSO$_3$ takes place in solution, which finally convert into nanoparticles on the substrate by combining with Cd^{2+} ions from the respective source. With J_{sc} = 18.23 mAcm^{-2} and 489 mV, the authors achieved 4.81% efficiency through fabricating TiO$_2$/CdS/CdSe heterojunction. Though the electrochemical approach is used to synthesize QDs, a hybrid approach of synthesizing semiconductor nanoparticles has also been demonstrated. Penner (2000) successfully demonstrated the hybrid method, which describes use of a wet chemical/electrochemical approach to synthesize semiconductor nanomaterials. Here, the formation of the nanoparticle process consisted of three steps. First, the metal nanoparticles are deposited on the graphite substrate. After the oxidation process of these metal nanoparticles, the respective negative ions are added to form metal sulfide, metal iodide, and metal oxide nanoparticles. Using hexagonal symmetry oriented (SKU-1) mesoporous silica as the template, Kim et al. (2010) synthesized cadmium selenide (CdSe) QDs with the potential of −0.7 eV at 5°C on the graphite basal plane. The authors used a three-electrode cell for this process, which was SKU-1–coated graphene as the working electrode, Ag/AgCl as the reference electrode, and Pt as the counter electrode. Using atomic force microscopy (AFM) measurements, the authors found that the height of the CdSe deposits varied from 7 to 9 nm. It is claimed that this method has enabled the direct contact of CdSe nanoparticles with the graphite in order to enhance their properties. These kinds of efforts will eliminate the troubles associated with the fabrication of a hybrid structure for the solar cells.

3.6 USE OF ELECTROCHEMICAL SYNTHESIS IN THE CONSTRUCTION OF PEROVSKITE SOLAR CELLS

Perovskite solar cells are solar cells that have recently been demonstrated with an efficiency of over 20% and commercialization of perovskite solar cells have just begun with seminal efforts. The material properties, surface modifications, and fabrication using non-vacuum strategies have been analyzed by several research groups and significant achievements have been attained. In this scenario, according to its properties, hybrid perovskites are ambipolar and because of this intrinsic ion-conducting feature, deposition of perovskite layers through the electrochemical method has been highly concentrated on in recent years. Using the electrochemical method, Hong et al. (2018) fabricated a large-area perovskite module through the electropatterning process through which series connection is achieved by creating metal filamentary nanoelectrodes. By this method, the authors achieved about 14% efficiency for a total area of 9.06 cm^2. Here during the forward bias, the ionized perovskite charges are redistributed on the surface of the ITO and Ag electrodes. This patterning was also confirmed through using techniques such as x-ray photoelectron spectroscopy (XPS), conductive atomic force microscopy (C-AFM), and scanning transmission electron microscopy (STEM). Metal oxide nanostructures such as ZnO and TiO$_2$ also could be used to couple with perovskite solar cells in order to enhance the charge transport. Using transparent carbon nanotubes as the electrode, Xiaoyan Wang et al. (2015) fabricated a flexible solar cell based on the TiO$_2$ nanotubes as the scaffold and methyl ammonium iodide (CH$_3$NH$_3$PbI$_3$) as the electron conductor. The TiO$_2$ nanotubes were grown using a one-step anodization method and the length of the obtained TiO$_2$ nanotubes was ~300 nm and the diameter was ~60 nm. The carbon nanotubes served as the hole collector together with the traditionally used spiro-OMeTAD polymer. The authors achieved 8.31% efficiency for the 25 μm TiCl$_4$-treated Ti foil, which is the highest for flexible TiO$_2$ nanotube-based solar cells. Inorganic hole-transporting layers also could be fabricated through the electrochemical deposition method since inorganic materials are more adoptable for the ionization process. Park et al. (2017) fabricated an NiO$_x$ hole transport layer for constructing large-area (1 cm^2) p–i–n planar type perovskite solar cells through a three-electrode-based electrochemical deposition process. Here the authors first deposited the nickel hydroxide layer (Ni(OH)$_2$) through its precursors by varying the deposition potential under different current densities in solution and while annealing the deposited Ni(OH)$_2$ film, the authors were able to achieve fine morphology of the NiO$_x$ layer for constructing perovskite solar cells. Other than growing TiO$_2$ nanostructures for perovskite solar cells, growth of ZnO nanomaterials by electrochemical deposition also has been reported. Zhang et al. (2014) prepared a nanostructured ZnO layer with a thickness of around 800–850 nm using zinc chloride medium, and formation of nanowires morphology was observed with increasing deposition temperature and decreasing Zn (II) precursor concentration. Another thin layer of n-type doped ZnO (described as i-ZnO) was also fabricated from the solution using zinc nitrate as the reaction medium to cover the as-grown nanowires/nanorods and the authors observed that this i-ZnO reduced the recombination between ZnO nanostructures and perovskite material in the device, leading to the efficiency over 10% with the J_{sc} ~22 mA cm^{-2} for a solid-state perovskite solar cell. This way of growing nanostructures through the electrochemical method is always advantageous for effective carrier transport. The main advantage is the thickness and diameter of the nanostructures can be manipulated through optimizing voltage, current, pH, and precursor and electrolyte concentration in the reaction medium. Other than inorganic nanostructures, growth of organic polymers on the influence of electrical potential was also carried out for perovskite solar cell applications. Yan et al. (2014) carried out formation of poly-3-hexyl thiophene films on the glass/FTO substrate through the electrochemical deposition method using a three-electrode system at the voltage of +1.30 V. Using distilled boron trifluoride-diethyl etherate (BFEE) with 30 mM thiophene as the electrolyte, the authors fabricated different thicknesses of P3HT films, which functioned as the hole-transport layer, and assembled a device with the configuration of ITO/polythiophene/CH$_3$NH$_3$PbI$_3$/C60/BCP/Ag. For the optimized thickness of 18 nm, the best efficiency of 11.8% was achieved with $J_{sc} = 16.2$ mAcm^{-2} and $V_{oc} = 1.03$ V.

3.7 CONCLUSION AND SUMMARY

From the discussion, it is clear that the electrochemical method is an important technique for the construction of highly efficient solar cells with the use of hybrid nanostructures. Use of electrochemical methods to synthesize TiO$_2$ and ZnO nanostructures have indeed opened a new promising direction to construct them for large-scale applications. However, new kinds of electrolytes and solvents are necessary to achieve a fine layer of deposition for the fabrication of high-efficiency solar cells. Fabrication of these nanostructures by the electrochemical method will be quite useful not only for solar cell applications, but also for other parallel promising technologies such as photocatalysis and supercapacitors. Though there are different structural and morphological characterizations available to execute the perfect nature of these materials for solar cell applications, still much improvement is needed for large-scale applications. The well-organized, well-oriented layer of inorganic nanostructures on a conducting substrate is quite useful in absorbing a large number of photons and converting them into useful ways of producing electricity. However, the limited number of solvents and electrolytes available are creating setbacks in the development of this field. Extending the fabrication of tandem nanoarchitectures through the electrochemical method will be more helpful for constructing high-efficiency future generation solar cells. Also, much work has to be devoted to developing highly efficient perovskite solar cells through the electrochemical deposition of precursors. Optimization of such structures should be expanded with respect to solvent, thickness, and other physical parameters. It is essential to form a common platform to demonstrate the overall possible ways to produce high-quality thin films through electrochemical methods. Deposition of pure inorganic perovskites such as cesium lead halides (CsPbX$_3$, X = Cl, Br, I) and cesium lead-free halides (CsMX$_3$, where M = Sn, Ge) are to be carried out by means of electrochemical methods and efficiency of the fabricated solar cells under various circumstances to be evaluated. Fabrication of efficient layers of new inorganic semiconductor nanomaterials with simple precursors under the optimized electrochemical conditions needs to be developed in order to improve them with different configurations. Though the efforts in these directions are at the initial level, it is expected in the future that achieving these goals will create new pathways in the field.

REFERENCES

Abd-Ellah, Marwa, Nafiseh Moghimi, Lei Zhang, Nina F. Heinig, Liyan Zhao, Joseph P. Thomas, and K. T. Leung. 2013. "Effect of electrolyte conductivity on controlled electrochemical synthesis of zinc oxide nanotubes and nanorods". *J. Phys. Chem. C* no:117(13):6794–6799.

Ahmed, Shafaat, Kathleen B. Reuter, Oki Gunawan, Lian Guo, Lubomyr T. Romankiw, and Harikila Deligianni. 2012. "A high efficiency electrodeposited Cu_2ZnSnS_4 solar cell". *Adv. Ener. Mater.* no:2(2):253–259.

Belekoukia, Meltiani, Madeshwaran Sekkarapatti Ramasamy, Sheng Yang, Xinliang Feng, Georgios Peterakis, Vassilios Dracopoulos, Costas Galiotis, and Panagiotis Lianos. 2016. "Electrochemically exfoliated graphene/PEDOT composite films as efficient Pt-free counter electrode for dye-sensitized solar cells". *Electrochimica Acta* no:194:110–115.

Brandt, Riley, Rachel C. Kurchin, Vera Steinmann, Daniil Kitchaev, Chris Roat, Sergiu Levcenco, Gerbrand Ceder, Thomas Unold, and Tonio Buonassisi. 2017. "Rapid photovoltaic device characterization through Bayesian parameter estimation". *Joule* no:1:843–856.

Cui, Jingbiao and Ursula J. Gibson. 2010. "A simple two-step electrodeposition of Cu_2O/ZnO nanopillar solar cells". *J. Phys. Chem. C* no:114(14):6408–6412.

Diamant, Yishay and Arie Zaban. 2001. "Electrochemical deposition of organic semiconductors. The method, mechanism, and critical deposition parameters". *J. Electrochem. Soc.* no:148(11):C709–C714.

Green, M. A. 2007. "Thin-film solar cells: review of materials, technologies and commercial status". *J. Mater. Sci: Materials in Electronics* no:18:15–19.

Gurunathan, K., Murugan, A. V., Marimuthu, R., Mulik, U. P., and Amalnerkar, D. P. 1999. "Electrochemically synthesized conducting polymeric materials for applications towards technology in electronics, optoelectronics and energy storage devices". *Mater. Chem. Phys.* no:61(3): 173–191.

Habas, S. E., Heather A. S. Platt, Maikel F. A. M. van Hest, and David S. Ginley. 2010. "Low-cost inorganic solar cells: from ink to printed device". *Chem. Rev.* no:110(11):6571–6594.

Hames, Yakup, Zuhal Alpaslan, Arif Koseman, Sait Eren San, Yusuf Yerli. 2010. "Electrochemically grown ZnO nanorods for hybrid solar cell applications". *Solar Energy* no:84(3):426–431.

Hong, Soonil, Jinho Lee, Hongkyu Kang, Geunjin Kim, Seyoung Kee, Jong-Hoon Lee, Suhyun Jung, Byoungwook Park, Seok Kim, Hyungcheol Back, Kilho Yu, and Kwanghee Lee. 2018. "High-efficiency large-area perovskite photovoltaic modules achieved via electrochemically assembled metal-filamentary nanoelectrodes". *Sci. Adv.* no:4(8):eaat3604.

Kaelin, M., D. Rudmann, and A. N. Tiwari. 2004. "Low cost processing of CIGS thin film solar cells". *Solar Energy* no:6 (77):749–756.

Kang, Donghyeon, Dongho Lee, and Kyoung-Shin Choi. 2016. "Electrochemical synthesis of highly oriented, transparent and pinhole-free ZnO and Al-doped ZnO films and their use in heterojunction solar cells". *Langmuir* no:32(41):10459–10466.

Kim, Yong-Tae, Jung Hee Han, Byung Hee Hong, and Young-Uk Kwon. 2010. "Electrochemical synthesis of CdSe quantum-dot arrays on a graphene basal plane using mesoporous silica thin-film templates". *Adv. Mater.* no:22:515–518.

Lahmar, Halla, Fatima Setifi, Amor Azizi, Guy Schmerber, and Aziz Dina. 2017. "On the electrochemical synthesis and characetrization of p-Cu_2O/n-ZnO heterojunction". *J. Alloys Compd.* no:718:36–45.

Lei, Bin-Xin, Jin-Yun Liao, Ran Zhang, Jing Wang, Cheng-Yong Su, and Dai-Bin Kuang. 2010. "Ordered crystalline TiO_2 nanotube arrays on transparent FTO glass for efficient dye-sensitized solar cell". *J. Phys. Chem. C* no:114:15228–15233.

Li, Meixiu, Xuqiang Ji, Liang Cui, and Jingquan Liu. 2017. "In situ preparation of graphene/polypyrrole nanocomposite via electrochemical co-deposition methodology for anti-corrosion application". *J. Mater. Sci.* no:52(20):12251–12265.

Lincot D. 2005. "Electrodeposition of semiconductors". *Thin Solid Films* no:487:40–48.

Liu, Wantao, Yanyan Fang, Peng Xu, Yuan Lin, Xiong Yin, Guangshi Tang, and Meng He. 2014. "Two-step electrochemical synthesis of polypyrrole/reduced graphene oxide composites as efficient Pt-free counter electrode for plastic dye-sensitized solar cells". *ACS Appl. Mater. Inter.* no:6(18):16249–16256.

Macek, J. M., H. Tsuchiya, A. Gicov, R. Hahn, S. Bauer, and P. Schmuki. 2007. "TiO_2 nanotubes: self-organized electrochemical formation, properties and applications". *Curr. Opin. Solid State Mater. Sci.* no:11(1–2):3–18.

Makhlouf, Houssin, Matthieu Weber, Olfa Messaoudi, Sophie Tingry, Matthieu Moret, Olivier Briot, Radhouane Chtoutou, and Mikhael Bechelany. 2017. "Study of Cu_2O/ZnO nanowires heterojunction designed by combining electrodeposition and atomic layer deposition". *Appl. Surf. Sci.* no:426:301–306.

Mor, Gopal K., Karthick Shankar, Maggie Paulose, Oomman K. Varghese, and Craig A. Grimes. 2006. "Use of highly-ordered TiO_2 nanotube arrays in dye-sensitized solar cells". *Nano Lett.* no:6(2):215–218.

Musselman, Kevin P., Andreas Wisnet, Diana C. Iza, Holger C. Hesse, Christina Scheu, Judith L. MacManus-Driscoll, and Lukas Schmidt-Mende. 2010a. "Strong efficiency improvements in ultra-low-cost inorganic nanowire solar cells". *Adv. Mater.* no:22(35):E254–E258.

Musselman, Kevin P., Andrew Marin, Andreas Wisnet, Christina Scheu, Judith L. MacManus-Driscoll, and Lukas Schmidt-Mende. 2010b. "A novel buffering technique for aqueous processing of zinc oxide nanostructures and interfaces and corresponding improvement of electrodeposited $ZnO-Cu_2O$ photovoltaics". *Adv. Fun. Mater.* no:21(3):573–582.

Musselman, Kevin P., Andrew Marin, Lukas Schmidt-Mende, and Judith L. MacManus-Driscoll. 2012. "Incompatible length scales in nanostructured Cu_2O solar cells". *Adv. Fun. Mater.* no:22(10):2202–2208.

Nozik, A. J. 2010. "Nanoscience and nanostructures for photovoltaics and solar fuels". *Nano Lett.* no:10(8):2735–2741.

Pagare, Pavan, K. G. Kanade, and A. P. Torane. 2017. "Effect of air and nitrogen annealing on TiO_2/Cu_2O heterojunction photoelectrochemical solar cells". *Mater. Res. Express* no:10 (4):1–23.

Park, Ik Jae, Gyeongho Kang, Min Ah Park, Ju Seong Kim, Se Won Seo, Dong Hoe Kim, Kai Zhu, Taiho Park, and Jin Young Kim. 2017. "Highly efficient and uniform 1 cm^2 perovskite solar cells with an electrochemically deposited NiO_x hole-extraction layer". *ChemSusChem* no:10(12):2660–2667.

Ramasamy, Madeshwaran Sekkarapatti, Archontoula Nikolakapoulou, Dimitrios Raptis, Vassilios Dracopoulos, Georgios Paterakis, and Pengiotis Lianos. 2015. "Reduced graphene oxide/ploypyrrole/PEDOT composite films as efficient Pt-free counter electrode for dye-sensitized solar cells". *Electrochimica Acta* no:173:276–281.

Reginald M. Penner. 2000. "Hybrid electrochemical/chemical synthesis of quantum dots". *Acc. Chem. Res.* no:33(2):78–86.

Shankar, Karthick, Gopal K. Mor, Maggie Paulose, Omman K. Varghese, and Craig A. Griemes. 2008. "Effect of device geometry on the performance of TiO_2 nanotube array-organic semiconductor double heterojunction solar cells". *J. Non-crystalline Solids* no:354 (19–25):2767–2771.

Wang, Jun, Lei Zhao, Victor S.-Y. Lin, and Zhiqun Lin. 2009. "Formation of various TiO_2 nanostructures from electrochemically anodized titanium". *J. Mater. Chem.* no:19:3682–3687.

Wang, Xiaoyan, Zhen Li, Wenjing Xu, Sneha A. Kulkarni, Sudip K. Batabyal, Sam Zhang, Anyuan Cao, and Lydia Helena Wong. 2015. "TiO_2 nanotube arrays based flexible perovskite solar cells with transparent carbon nanotube electrode". *Nano Energy* no:11:728–735.

Wei, D., Pieres Andrew, and Tapani Ryhanen. 2010. "Electrochemical photovoltaic cells-review of recent developments". *J. Chem. Tech. Biotech.* no:85(12):1547–1552.

Wei, H. M., H. B. Gong, L. Chen, M. Zi, and B. Q. Cao. 2012. "Photovoltaic efficiency enhancement of Cu_2O solar cells achieved by controlling homojunction orientation and surface microstructure". *J. Phys. Chem. C* no:116(9):10510–10515.

Xi, Dongjuan, Han Zhang, Stephen Furst, Bin Chen, and Qibing Pei. 2008. "Electrochemical synthesis and photovoltaic property of cadmium sulfide-polybithiophene intergigitatednanohybrid thin films". *J. Phys. Chem. C* no:112(49):19765–19769.

Xu, Xiabao, Dekang Huang, Kun Cao, Mingkui Wang, Shaik M. Zakeeruddin, and Michael Gratzel. 2013. "Electrochemically reduced graphene oxide multilayer films as efficient counter electrode for dye-sensitized solar cells". *Sci. Rep.* no:3:1489.

Yan, Weibo, Yunlong Li, Weihai Sun, Haitao Peng, Senyun Ye, Zhiwei Liu, Zuqiang Bian, and Chunhui Huang. 2014. "High-performance hybrid perovskite solar cells with polythiophene as hole-transporting layer via electrochemical deposition". *RSC Adv.* no:4:33039–33046.

Yoon, Sanghwa, Jae-Hong Lim, and Bongyoung Yoo. 2016. "Electrochemical synthesis of cuprous oxide on highly conducting metal micro-pillar arrays for water splitting". *J. Alloys Compd.* no:677:66–71.

Yu, Rui, Qingfeng Lin, Siu-Fung Leung, and Zhiyoung Fan. 2012. "Nanomaterials and nanostructures for efficient light absorption and photovoltaics". *Nano Energy* no:1(1):57–72.

Yu, Xiao-Yun, Jin-Yun Liao, Kang-Qiang Qiu, Dai-Bin Kuang, and Cheng-Yong Su. 2011. "Dynamic study of highly efficient CdS/CdSe quantum dot-sensitized solar cells fabricated by electrodeposition". *ACS Nano* no:5(12):9494–9500.

Zaban, Arie and Yishay Diamant. 2000. "Electrochemical deposition of organic semiconductors on high surface area electrodes for solar cells". *J. Phys. Chem. B* no:104(43):10043–10046.

Zhang, Jie, Philippe Barboux, and Thierry Pauporte. 2014. "Electrochemical design of nanostructured ZnO charge carrier layers for efficient solid-state perovskite-sensitized solar cells". *Adv. Ener. Mater.* no:4(18):1400932.

4 Role of Advanced Materials in Electrochemical Supercapacitors

Joyita Banerjee, Kingshuk Dutta, and M. Abdul Kader

CONTENTS

4.1 Introduction	63
4.2 General Discussion of Electrochemical Supercapacitors	64
4.2.1 Electric Double-Layer Capacitors	65
4.2.2 Pseudocapacitors	66
4.3 Application of Materials in the Field of Electrochemical Supercapacitors	68
4.3.1 Advances in Materials for EDLC-Type Supercapacitors	68
4.3.2 Advances in Materials for Pseudocapacitors	69
4.4 Recent Advancements in Electrolytes for Electrochemical Reactions	72
4.4.1 Aqueous Medium	73
4.4.2 Organic Solvent Medium	74
4.4.3 Ionic Liquids	74
4.5 Conclusion	75
References	75

4.1 INTRODUCTION

The fast depletion of fossil fuels and its adverse effect on the environment have led researchers to focus on alternative and clean energy (Das, Dutta, and Rana 2018; Dutta, Das, and Kundu 2015; Banerjee and Dutta 2017; Dutta et al. 2017; Dutta, Kundu, and Kundu 2014; Dutta 2017; Kundu and Dutta 2018; Banerjee, Dutta, and Rana 2019). Renewable sources of energy are abundantly available in nature; however, their intermittent nature of supply necessitated the emergence of more advanced energy storage materials and devices to store the energy for future use. Electrical energy storage devices help in storing electrical energy generated by different means for the purpose of future utilization. Electrochemical supercapacitors have widespread applications, starting from consumer electronics, memory backup systems, fuel cells, and low emission hybrid electric vehicles to industrial power and energy management. The most recent applications include emergency doors on Airbus 380 and to serve as temporary energy storage devices mainly coupled with fuel cells or primary high-energy batteries.

Herein, an attempt has been done to give a basic idea about the working principle, design, and application of electrochemical capacitors. In this book chapter, we tried to collate information regarding various types of materials that have been used as electrode material for application in supercapacitors. Moreover, an attempt has been made to discuss the present status and advancements made in the area of electrolytes utilized in electrochemical reactions.

4.2 GENERAL DISCUSSION OF ELECTROCHEMICAL SUPERCAPACITORS

Unlike capacitors, supercapacitors possess two electrodes—cathode and anode—with an electrolyte between the two plates instead of a solid dielectric. In the case of supercapacitors, as the ion transfer takes place through the electrolytes, it offers high-power density, high-energy efficiency, and a long charge–discharge cycle life. These electrochemical supercapacitors can store electrical charges in two ways: (1) thorough reversible faradaic reaction or (2) by storing opposite electrical charges apart electrostatically. The Ragone plot in Figure 4.1 shows the plot for specific energy versus specific power for various energy storage devices. It has been observed that the electrochemical capacitors can deliver much higher power compared to batteries and fuel cells but lower than that of capacitors. These supercapacitors have comparatively higher specific energy than batteries. In this context, it is important to mention two parameters that dictate the performance of supercapacitors: energy density and power density. Energy and power densities are defined as the amount of energy stored and the rate of energy transfer per unit volume of the active materials used as electrodes for supercapacitors.

In general, a supercapacitor consists of two electrodes, viz. cathode and anode, kept isolated from each other by a semipermeable membrane in an ionic electrolyte placed between the electrodes, and current collectors that collect current from the electrodes (Stoller et al. 2008; Chen, Song, and Xue 2014; Al-Osta et al. 2015). The semipermeable membrane, known as separator, is porous in nature so that it can allow easy shuttling of ions between the oppositely charged electrodes. It also helps in keeping the oppositely charged electrodes apart to avoid short-circuit.

Figure 4.2 depicts the schematic diagrams of the working principles of an electric double-layer capacitor (EDLC) and a pseudocapacitor. The details regarding the working principles for EDLCs and pseudocapacitors will be discussed in the later sections. The basic difference between pseudocapacitors and EDLCs is in the charge storage mechanism. Pseudocapacitors involve fast and reversible reaction along with charge storage near the surface or in the bulk of the electrode, which infers that these types of capacitors have higher energy density than the EDLC types. However, the kinetics of charge transfer is faster in case of EDLCs.

Based on the type of materials used for the cathode and anode, supercapacitors can be further divided into two groups, namely, symmetric and asymmetric. Symmetric supercapacitors consist of electrodes of the same materials ($C_1 = C_2$), such that the total capacitance becomes half of either electrode. On the other hand, asymmetric supercapacitors consist of electrodes of different materials

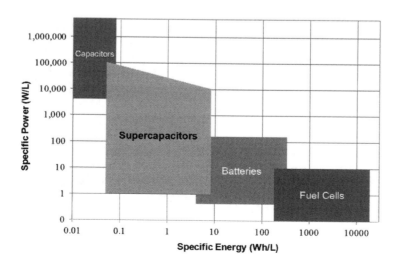

FIGURE 4.1 Specific energy versus specific power plot for various energy storage devices. (Reproduced from www.cap-xx.com/resource/energy-storage-technologies/, © CAP-XX 2019, accessed on March 9, 2019.)

Role of Advanced Materials in Electrochemical Supercapacitors

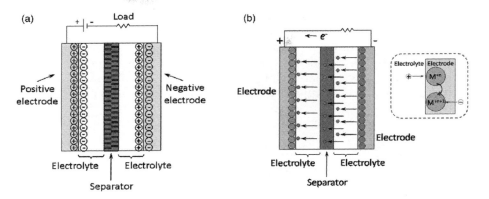

FIGURE 4.2 Schematic diagram of (a) EDLC type and (b) pseudocapacitor type. (Reproduced with permission from Vangari, Pryor, and Jiang 2013, © American Society of Civil Engineers.)

with different capacitance ($C_1 > C_2$), exhibiting lower capacitance. Usually, the unique design of an asymmetric supercapacitor comprises a battery-like faradaic electrode and a carbon-based capacitive electrode. By merely adjusting the operating voltage window of the electrodes constituting the device, the capacitance of the asymmetric supercapacitors can be improved.

4.2.1 Electric Double-Layer Capacitors

Conventional capacitors are able to store little energy due to geometrical constraints (limited surface area) imposed by a dielectric material between the two electrodes. In this case, the opposite charges are stored on the surface of the two electrodes (separated at a comparatively large distance by a solid dielectric) with minuscule capacity to store the charge. EDLC stores electrical charge by electrical double layer mechanism on the two oppositely charged electrodes, which is similar to the mechanism of capacitors (Banerjee, Dutta, Kader et al. 2019). However, the major difference is that, instead of dielectric solid material, it consists of an electrolyte between the two electrodes. The crucial parameters that determine the performance of the EDLCs are the type of electrolytes, the chemical affinity between the adsorbed ions and electrode surface, electrical field applied across the electrodes, pore size, and specific surface area of the electrodes.

The capacitance (C_{DL}) of the EDLC-based supercapacitors can be presented as

$$C_{DL} = \varepsilon_0 \varepsilon_r * A/D \qquad (4.1)$$

where ε_0 and ε_r represent permittivity of free space and dielectric constant (or relative permittivity) of the material between the plates, respectively; A represents the area of each electrode; and D represents the distance between the electrodes. In brief, EDLC stores charge through the "electrical double layer" model, which involves an adsorbed layer on the metal electrode adjacent to the electrolyte known as the inner Helmholtz plane (IHP), followed by an outer Helmholtz plane (OHP) with loosely attached oppositely charged ions. These types of supercapacitors resemble conventional capacitors and this model is valid for 1 nm from the electrode with a very high electric field. According to this model, there exists a constant differential capacitance across the double layer adjacent to the surface of the electrode. The outer diffuse layer of the ions in the electrolyte extends from the inner adsorbed layer to the bulk solution (You et al. 2018). On the other hand, the Gouy–Chapman model is applicable for a distance of 20 nm from the electrode, and in this case the electric potential decreases exponentially from the surface of the electrode. Last, the Stern model is applicable for highly charged double layer systems (Stern 1924), and basically it is a combination of the electrical double layer model and the Gouy–Chapman model. For the Stern model, the adsorbed ion layer defines the inner Stern layer (as in the Helmholtz model) and the diffuse ion layer forms the Gouy-Chapman diffuse layer. Apart from these models, there exists several other models that can explicitly describe the behavior of ions at the electrode–electrolyte interface.

As in this case, the charge is accumulated and stored on the surface of the plates. It is considered to be a purely physical process; and thus, EDLCs exhibit high rate capability but limited capacitance compared to a battery (Wang et al. 2017a). Hence, at this point, a new concept of storage of charge through redox reaction comes into existence.

4.2.2 Pseudocapacitors

The very concept of charge storage through redox reaction between the electrodes and electrolyte was discovered in 1971 (Conway 1991). The value of capacitance of the pseudocapacitors is basically proportional to the charge transferred in the whole process, and it is given as

$$C_{pc} = q(d\theta/dv) \qquad (4.2)$$

where $d\theta$ represents the change in fractional coverage of the surface, dv represents the potential change, and q represents the faradaic charge. However, it has been observed that for pseudocapacitors, the term $d\theta/dv$ does not vary linearly with its specific capacitance. The faradaic reaction is responsible for the pseudocapacitors to achieve much higher specific capacitance than EDLCs.

The faradaic process for pseudocapacitors mainly stems from three mechanisms: (1) underpotential deposition (Figure 4.3a), (2) redox pseudocapacitance (Figure 4.3b), and (3) intercalation pseudocapacitance (Figure 4.3c). In the first case, the monolayer deposition of one metal ion from solution on another metal surface and the redox pseudocapacitance involves storage of charge in the surface through electrochemical redox reaction via adsorption/desorption of ions between the solution and the electrode. In intercalation pseudocapacitance, the pseudocapacitance arises from the intercalation of active ions between the crystallographic layers of the electrode without changing the crystal structure of the electrode material. For a long time, the word "pseudocapacitor" has been used for electrode materials like Ni(II) hydroxide [Ni(OH)$_2$] and several other materials (Kandalkar, Gunjakar, and Lokhande 2008; Senthilkumar et al. 2014; Gaikar et al. 2017), which had exhibited purely battery-like behavior (Hu et al. 2011a; Wang et al. 2013). However, there exists several electrode materials like transition metal oxides (TMOs), sulfides, hydroxides, carbides,

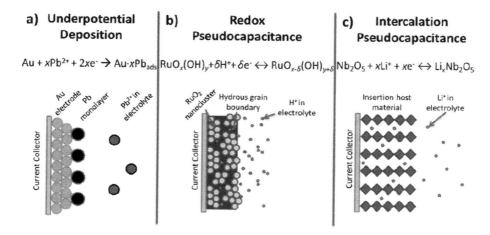

FIGURE 4.3 Different types of reversible redox mechanisms that give rise to pseudocapacitance (a) underpotential deposition, (b) redox pseudocapacitance, and (c) intercalation pseudocapacitance. (Reproduced from Augustyn, Simon, and Dunn 2014 with permission from The Royal Society of Chemistry.)

nitrides, and even conducting polymers that show properties of both a purely capacitive system as well as of a faradaic system (Gogotsi and Penner 2018). These particular class of electrode materials has been increasingly popular for their extraordinary electrochemical properties. Simon, Gogotsi, and Dunn (2014) have given a new identity to the word "pseudocapacitor" signifying the electrode material that can exhibit behavior tending to capacitors. In the case of EDLC, the capacitance arises from the potential-dependence of the electrostatically stored charges on the surface of the electrode (i.e., non-faradaically). However, a completely different charge storage mechanism is involved for pseudocapacitors. Pseudocapacitors involve charge storage at the interface between the electrode and the electrolyte, but the mechanism is different from EDLCs (Brousse, Bélanger, and Long 2015). The charge storage takes place through faradaic reaction (like battery) and the charge transfer takes place through the double layer (like EDLC). On the other hand, electrode materials like $Ni(OH)_2$ and cobalt oxides in potassium hydroxide (KOH) electrolyte have shown pronounced faradaic peaks in cyclic voltammetric (CV) curves, which is typically a battery behavior (Costentin, Porter, and Savéant 2017). Further, Shukla, Sampath, and Vijayamohanan (2000) postulated that the EDLC operates through charge–discharge of the interfacial double layer; and in the case of pseudocapacitors, the transfer of electrons occurs without any bulk phase transformation. It was further speculated that the electrons involved in electrical double layer are contributed from the conduction band and the electrons from the valence band are involved in Faradaic reaction. In fact, the pseudocapacitance is given by

$$O_{ad} + ne^- \to R_{ad} \qquad (4.3)$$

where O_{ad} and R_{ad} represent the adsorbed oxidants and reductants, respectively; and n represents the number of electrons involved in the faradaic reaction. The main characteristic of pseudocapacitance stems from absorption and desorption of O_{ad}/R_{ad} on the surface of the electrode, which can be explained by the Frumkin model.

It is evident from the preceding discussion that EDLC possesses high power density but lacks energy density. Therefore, in order to improve the charge density of the electrode, researchers have focused on hybrid supercapacitors. The hybrid supercapacitors have been found to possess a better combination of electrochemical properties compared to either EDLCs or pseudocapacitors (Zhou et al. 2014). These hybrid supercapacitors are the amalgamation of two completely diverse electrodes

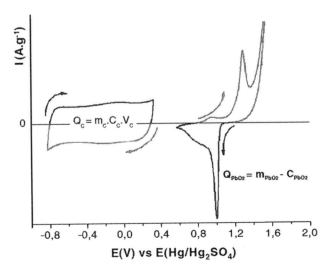

FIGURE 4.4 Cyclic voltammogram of PbO_2 and carbon-based hybrid (asymmetric) supercapacitor. (Reproduced from Brousse, Bélanger, and Long 2015, © the authors.)

of a battery-type material and a capacitor-type material (Li, Cheng, and Xia 2005). The battery-type materials undergo faradaic reaction for charge transfer, whereas the capacitor-type materials form electrical double layers for electron storage (Lota, Fic, and Frackowiak 2011). Figure 4.4 shows the cyclic voltammogram of a hybrid asymmetric supercapacitor composed of carbon electrode as the cathode and PbO_2 as the anode (Pell and Conway 2004; Zheng 2003). It was observed that a full cell supercapacitor with two electrodes working in complimentary potential windows have shown much capacitance due to wide operating window potential (Béguin and Frackowiak 2013).

4.3 APPLICATION OF MATERIALS IN THE FIELD OF ELECTROCHEMICAL SUPERCAPACITORS

4.3.1 Advances in Materials for EDLC-Type Supercapacitors

Unlike conventional capacitors, EDLCs possess two electrodes—cathode and anode— separated by an electrolyte with a very small gap. In this case, no faradaic reaction takes place, and the ions or the electrons shuttle between the electrodes through the electrolyte. EDLCs can be charged or discharged within a short time span with high power density. In order to fulfill the requirement for an EDLC electrode, the active material is supposed to have good electrical conductivity, porosity, and high surface area.

Electrodes based on activated carbon fiber cloth obtained from polyacrylonitrile (PAN) by optimizing the carbonization had been found to possess a high capacitance value of 208 F g^{-1}, upon using a 6 M KOH aqueous solution (Xu et al. 2007). Zhao et al. (2008) had also studied the excellent electrochemical properties of mesoporous or macroporous carbon formed by using Pluronic F127 and SiO_2 opal as templates. Nanostructured graphene nanoparticles prepared by ball-milling at four different conditions exhibited specific capacitance values of 3 μF cm^{-2} (per surface area) for 4 h and 12 F g^{-1} (per weight) for 8 h of milling (Gomibuchi et al. 2006). Electrode material based on freestanding mats of entangled multiwalled carbon nanotubes (MWCNTs) along with sulfuric acid (H_2SO_4) electrolyte has exhibited high capacitance (Niu et al. 1997). Earlier, the effect of vertically aligned and random carbon nanotubes (CNTs) on the specific capacitance of the supercapacitor electrodes have also been extensively investigated (An et al. 2001; Futaba et al. 2006). Gold (Au)–manganese dioxide (MnO_2)/CNT coaxial electrodes, prepared by a complex process, exhibited an energy density value of 4.5 Wh kg^{-1} and a specific capacitance value of 68 F g^{-1} in an aqueous electrolyte (Reddy et al. 2010). Yi et al. (2011) had reported CNTs-implanted mesoporous carbon spheres as a potential electrode material for supercapacitors with additional porosity and improved electrical conductivity. These mesoporous carbon spheres possessed a diameter of approximately 0.5 μm to 1.0 μm, an average pore size of 3.9 nm, and a specific surface area of 284 m^2 g^{-1}. This electrode produced an equivalent series resistance (ESR) value of 0.83 V cm^{-2}, a maximum specific capacitance value of 189 F g^{-1} and a power density value of 8.7 kW kg^{-1} at an energy density value of 6.6 Wh kg^{-1}. Electrodes based on random single-walled carbon nanotubes (SWCNTs) networks exhibited a specific capacitance of 180 F g^{-1} in presence of KOH electrolyte, a power density of 20 kW kg^{-1}, and an energy density of 10 Wh kg^{-1} (Chen and Dai 2013). Jin et al. (2017) had reported that phosphorus and nitrogen codoped exfoliated carbon nanosheets exhibited a considerably high capacitance of 265 F g^{-1} at a current density of 0.5 A g^{-1} with excellent capacity retention (84% after 5000 cycles).

Recently, flexible electrodes have accrued immense attention due to their low cost, fast charging and discharging capabilities, high power per volume, portability, and good electrochemical properties. These flexible electrodes find application from health care and flexible display to wearable energy storage devices (Jost et al. 2015; Tang et al. 2016; Vlad et al. 2015; Xue et al. 2017). So, it is clear that flexible supercapacitors have opened a completely new plethora of advanced materials as emergent energy storage devices. Hence, the increasing demand for flexible electronics boosted the development of supercapacitors, both yarn/fiber-shaped and planar ones, with excellent electrochemical performances (Choi et al. 2016; Wang et al. 2017b). Also, 3D-porous flexible graphene foam-based supercapacitors have shown magnificent electrochemical properties (Manjakkal et al. 2018). The graphene–Ag conductive epoxy-graphene foam exhibited an aerial capacitance of 38 mF cm^{-2} at a current density of 0.67 mA cm^{-2}, upon using phosphoric acid (H$_3$PO$_4$) as the electrolyte. Purkait et al. (2018) had reported the performance of porous electrochemically grown reduced graphene oxide (pErGO) networks on copper wire as a potential electrode for flexible supercapacitor applications. The pErGO-based supercapacitor showed a specific capacitance of 81 ± 3 F g^{-1} at a current density of 0.5 A g^{-1}, using polyvinyl alcohol (PVA)/H$_3$PO$_4$ as the electrolyte. Flexible supercapacitors based on carbon fibers are also studied. Electrodes based on vertically aligned CNTs forest and carbon nanopetals grown on unidirectional carbon fibers by chemical vapor deposition have shown excellent electrochemical performances (Cherusseri and Kar 2015a; Cherusseri and Kar 2015b).

4.3.2 Advances in Materials for Pseudocapacitors

TMOs and conducting polymers are the most investigated electrode material for pseudocapacitors. Metal oxides possess several attracting properties that help them in securing a position for their use as electrode material for supercapacitors. These metal oxides show pseudocapacitance along with double layer capacitance, with wide charge/discharge potential range. However, they possess limited surface area with low conductivity and stability, high cost, and toxicity. Many TMOs have been investigated for their capabilities as supercapacitor electrodes. However, the high cost of the ruthenium oxide that exhibits the highest capacitance values restricts its commercialization. This paved the way for combinations of ruthenium oxide with other metal oxides, as well as, mixed metal oxides. Stannic oxide (SnO$_2$)/vanadium pentoxide (V$_2$O$_5$), formed by hydrothermal synthesis, in combination with CNTs have shown a specific capacitance value of 121.4 F g^{-1}, at a scan rate of 100 mV s^{-1} in 0.1 M potassium chloride (KCl) solution (Jayalakshmi et al. 2007).

Recently, mixed transition metal oxides (MTMOs) have been highly investigated as they have superior properties than that of the TMOs. MTMOs have been denoted as A$_x$B$_{3-x}$O$_4$ (where A, B = Co, Ni, Zn, Mn, Fe, and so on) with stoichiometric or even nonstoichiometric compositions (Yuan et al. 2014). These MTMOs are composed of single phases of two ternary transition metal oxides with two different metal cations but not a binary mixture of two metal oxides (Teh et al. 2011; Zhang et al. 2012a; Zhou, Zhao, and Lou 2012). Due to their complex spinel structure, they exhibit multiple oxidation states possessed by different transition metals in a single compound. This significantly increases their capacitance due to a synergistic effect as compared to single TMOs. Moreover, MTMOs have higher electrical conductivity in comparison to that of TMOs, due to lower activation energies of electron transfers between cations. Typical examples of MTMOs are CuV$_2$O$_6$ (Ma et al. 2008; Xiao et al. 2009), FeVO$_4$ (Sim et al. 2012), Li$_3$VO$_4$ (Li et al. 2013; Mei et al. 2014), ZnV$_2$O$_6$ (Sun et al. 2012), Zn$_3$V$_2$O$_8$ (Gan et al. 2014), Co$_3$V$_2$O$_8$·nH$_2$O (Wu et al. 2015b), and Co$_2$V$_2$O$_7$ hexagonal microplatelets (Wu et al. 2015a). These compounds have already found profound application in lithium-ion batteries. However, in recent times, these MTMOs have also shown good pseudocapacitive behavior.

It was reported by Dar, Moonoosawmy, and Es-Souni (2013) that the pseudocapacitance of the metal oxides depends on the morphology developed during the synthesis of metal oxide nanoparticles. It was observed that the electrodes of nickel(II) oxide (NiO) nanotubes produced much higher capacitance value (i.e., 2093 F g^{-1}) compared to that produced by NiO nanorods (i.e., 797 F g^{-1}). It has been shown in another study that cobalt tetraoxide (Co$_3$O$_4$) possessed a very high theoretical capacitance of 3560 F g^{-1} and low environmental toxicity. Hence, it has been used extensively as an electrode material for electrochemical capacitors. As a reason behind this high capacitance value, it was said that both the pseudocapacitive and battery-type behavior leads to such high capacitance (Yuan et al. 2011; Lee, Chin, and Sow 2014).

$$\text{Battery-type behavior}: Co_3O_4 + OH^- + H_2O \leftrightarrow 3MOOH + e^- \quad (4.4)$$

$$\text{Pseudocapacitive-type behavior}: CoOOH + OH^- \leftrightarrow CoO_2 + H_2O + e^- \quad (4.5)$$

Vanadium oxides are also examples of such important transition metal oxides that have been studied extensively for their use as electrode materials for supercapacitors. They exhibit multiple valences, such as +5, +4, +3, and +2. Among these the most investigated forms of vanadium oxides are V_2O_5, V_2O_3, and VO_2. Lee and Goodenough (1999) reported the electrochemical properties of the amorphous $V_2O_5 \cdot nH_2O$, which showed a capacitance value of 350 F g^{-1} along with good cyclability.

A MnO_2-based nanocomposite, formed by electrophoretic deposition of exfoliated CNTs on nickel foam followed by deposition of nanoparticles like platinum (Pt), palladium (Pd), or Au, has been observed to exhibit a specific capacitance value of 559.1 F g^{-1} (Zhao et al. 2013). This 3D porous structure electrode showed extremely good conducting due to ease in electrolyte penetration and ion delivery in the presence of ionic liquid (IL) (i.e., 1-butyl-3-methylimidazolium hexafluorophosphate/N, N-dimethylformamide) electrolyte.

Supercapacitors based on functionalized freestanding CNTs as the positive electrode and freestanding CNT/MoO_{3-x} as the negative electrode have shown a high volumetric capacitance value of 3.0 F cm^{-3} and high power and energy densities of 4.2 W cm^{-3} and 1.5 mWh cm^{-3} (Xiao et al. 2014), respectively. Highly ordered V_2O_5–TiO_2 prepared by self-organization of Ti–V alloys, with vanadium (up to 18%), had shown specific capacitance of 220 F g^{-1} along with energy density of 19.56 Wh kg^{-1} (Yang et al. 2011). An asymmetric supercapacitor based on α-Fe_2O_3 on Fe foil as the negative electrode and $ZnCo_2O_4$@MnO_2 core-shell nanotube arrays on Ni foam as the positive electrode exhibited a specific capacitance of 161 F g^{-1} along with an energy density of 37.8 Wh kg^{-1} (Ma et al. 2015). It was realized that most of the metal oxides, like $NiCo_2O_4$, NiO, and Co_3O_4, are considered to behave like p-type semiconductors with band gap varying between 2.2 eV and 3.6 eV (Choi and Im 2005; Hu et al. 2011b; Kumar, Diamant, and Gedanken 2000).

Figure 4.5 illustrates a schematic diagram of the synthetic procedure of mesoporous NiO/$NiCo_2O_4$/Co_3O_4 composites by the sol–gel process, followed by calcination at 250°C (Liu et al. 2012). These NiO/$NiCo_2O_4$/Co_3O_4 composites showed a specific capacitance value of 1717 F g^{-1} with 94.9% retention after 1000 cycles. Hierarchical $MnMoO_4$–$CoMoO_4$ nanowires disclosed a capacitance value of 187.1 F g^{-1} at a current density value of 1 A g^{-1}, along with higher stability, which was higher than that of the electrodes composed of $MnMoO_4$ and $CoMoO_4$ individually (Mai et al. 2011). $CoMoO_4$–$NiMoO_4 \cdot xH_2O$ bundles, synthesized by chemical coprecipitation, have manifested 1039 F g^{-1} capacitance at a current density of 1 A g^{-1} along with excellent rate capability (Senthilkumar et al. 2013). Hierarchical nanosheet-based $CoMoO_4$–$NiMoO_4$ nanotubes, formed by the hydrothermal process, have also shown excellent pseudocapacitance (Yin et al. 2015). Zhang et al. (2014b) reported that $Co_3V_2O_8$ thin nanoplates prepared by hydrothermal reaction have shown good electrochemical properties. These electrodes exhibited a high capacitance of 739 F g^{-1} at

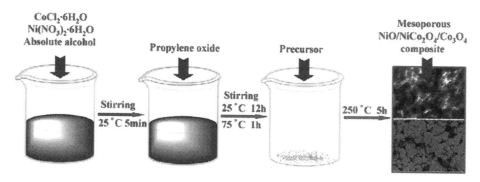

FIGURE 4.5 A schematic diagram of the synthetic process involved in the production of mesoporous $NiCo_2O_4$/NiO/Co_3O_4 composite. (Reprinted with permission from Liu et al. 2012. © 2012 American Chemical Society.)

a current density of 0.5 A g^{-1}, along with retention of the capacitance value of 704 F g^{-1} after 2000 cycles in the presence of aqueous 3 M KOH solution. Table 4.1 provides a short description of several other metal oxides that find potential application as electrodes for supercapacitors.

As mentioned earlier, flexible electronics are promising materials that have several advantages. Pseudocapacitive materials have also been used in flexible-materials-based electronics. Huang et al. (2018) had recently reported the capacitance of a cathode based on aluminum-doped cobalt sulfide nanosheet anchored on nickel foam along with carbon cloth to be 2434 F g^{-1} at a current density of

TABLE 4.1
List of Metal Oxides That Find Potential Application as Electrodes for Supercapacitors

Electrode Materials	Method of Synthesis	Specific Capacitance (F g^{-1})	Reference
Manganese oxide nanostructure grown on nanoporous gold film	Anodic deposition	432	Shi et al. 2017
Mn$_3$O$_4$/graphene	Hydrothermal	367	Lee et al. 2015
RuO$_2$ thin film	Chemical bath deposition	73	Patil et al. 2011
Ni(OH)$_2$/rGO	Solvothermal	1886	Zang et al. 2017
Nitrogen-doped mesoporous carbon/nickel cobalt layered double hydroxide	Microwave	2498	Xu et al. 2016
Carbon nanofiber/MnO$_2$	Electrospinning	311	Zhi et al. 2012
NiO/rGO	Electrodeposition	950	Zhao et al. 2016
NiO nanoflake/carbon cloth	Chemical bath deposition	660	Zhang et al. 2014a
Nickel oxide coated graphene/PAni	Hydrothermal and *in situ* chemical oxidative polymerization	1409	Wu et al. 2016
NiO nanomaterial	Hydrothermal	1337	Pei et al. 2016
Manganese oxide film	Sol–gel method	360	Sarkar et al. 2015
Co$_3$O$_4$–MnO$_2$–NiO nanotubes	Electrodeposition	2525	Singh et al. 2016
Cobalt–manganese layered double hydroxide	Electrodeposition	1062	Jagadale et al. 2016
Porous nickel oxide/mesoporous carbon	Chemical precipitation method	2570	Zhang et al. 2010
NiO nanosheets on Ni foam	Hydrothermal	674	Tahmasebi et al. 2016
Nickel oxide quantum dots embedded with graphene	Alternating voltage approach	1181	Zhao et al. 2016
Manganese oxide decorated graphene nanosheets	Sacrificial reaction	280	Wang et al. 2012b
PAni–RuO$_2$	Successive ionic layer adsorption and reaction (SILAR)	664	Chen et al. 2016
Ni(OH)$_2$/MnO$_2$/rGO	Hydrothermal	1985	Wang et al. 2012a
Mn–Ni oxide	Anodic deposition	250	Yang et al. 2012
NiCo$_2$O$_4$ single-wall carbon nanotubes (SWCNT)	Controlled hydrolysis process	1642	Wang et al. 2012b
Manganese oxide/multiwalled carbon nanotubes (MWCNT)	Sol–gel method	339	Chen et al. 2016
Co$_3$O$_4$	Sol–gel method followed by freeze-drying	742.3	Wang et al. 2012a
NiO	Hydrothermal method	1700	Yang et al. 2012
CoMoO4	Hydrothermal method	286	Liu et al. 2013
CoMoO$_4$ nanoplate arrays	Hydrothermal method	1558	Guo et al. 2013

1 A g^{-1}. The reported value was quite high and the asymmetric supercapacitor showed an energy density of 65.7 Wh kg^{-1} with good cyclability. For flexible supercapacitors, the current collector plays a pivotal role, and it needs to be both flexible and conductive in nature. However, the quantity of the active material that needs to be used is restricted by the current collector; thus, it is important to develop a new advanced current collector. Recently, Ko et al. (2017) studied a novel approach that involved the conversion of cellulose paper using a ligand-mediated layer-by-layer assembly of metal nanoparticles to a highly porous and flexible current collector.

4.4 RECENT ADVANCEMENTS IN ELECTROLYTES FOR ELECTROCHEMICAL REACTIONS

Electrolytes are integral parts of electrochemical supercapacitors, which help in shuttling of ions between the two electrodes. Mainly, the electrolytes for supercapacitors can be divided into three types: aqueous medium, organic solvents, and ILs. However, there are other types of electrolytes, like redox-type electrolytes and solid or semisolid electrolytes, that have been explored in the past. For an efficient electrochemical supercapacitor, the electrolyte needs to be nontoxic, possess a wide voltage window, exhibit high conductivity, be environmentally friendly, possess self-discharge capability, have low cost, and be electrochemically stable with a wide temperature range of applicability. Each category of electrolyte has its own pros and con. Besides the electrolyte operating window, the performance of electrochemical capacitors also depends on the interaction between the electrodes and the electrolytes. In this context, it is important to mention here that the matching of pore size of the carbon-based electrode and ion size of the electrolyte helps in improving the performance of the supercapacitors. It has further been realized that the internal resistance of the supercapacitors also gets influenced by the ionic conductivity of the electrolyte, mainly for ILs and organic electrolytes. In addition, parameters like viscosity, freezing point, boiling point, and degradation temperature greatly influence the thermal stability as well as the operating potential window of the electrolyte.

Figure 4.6 illustrates the effect rendered by electrolytes on several performance parameters of electrochemical supercapacitors. It has been observed that decomposition of electrolytes may lead

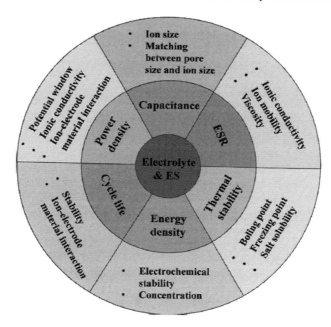

FIGURE 4.6 Effects of electrolytes on electrochemical supercapacitor performance. (Reproduced from Zhong et al. 2015 with permission from The Royal Society of Chemistry.)

Role of Advanced Materials in Electrochemical Supercapacitors

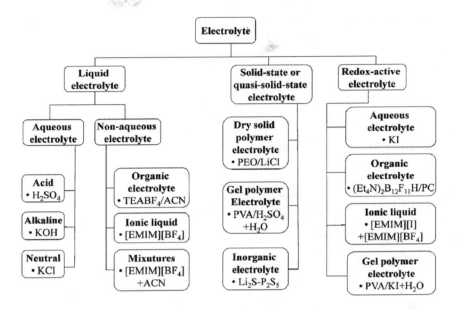

FIGURE 4.7 Classification of electrolytes for electrochemical supercapacitors. (Reproduced from Zhong et al. 2015 with permission from The Royal Society of Chemistry.)

to severe damage of electrochemical supercapacitors. Figure 4.7 depicts the broad classification of electrolytes that are used in supercapacitors.

Recently, redox-type electrolytes have gained focus owing to their good performance as electrolytes for supercapacitors. These redox-type electrolytes also possess additional pseudocapacitance from redox reactions taking place at the electrode–electrolyte interface.

The specific energy density (in Wh kg^{-1}) is given by

$$\text{Specific energy density} = \tfrac{1}{2} C_s (\Delta V)^2 = \tfrac{1}{2} C/m (\Delta V)^2 \tag{4.6}$$

where m represents the active mass, and C_s and C represent the specific capacitance and total capacitance of the supercapacitor, respectively. It has been observed that the energy density of supercapacitors is directly proportional to the potential window range. Hence, the working potential window of the electrolytes plays a pivotal role in dictating the performance of the supercapacitors. It has been reported in various studies that organic electrolytes and ILs possess wider potential window (higher energy density) than aqueous electrolytes. In this respect it should be kept in mind that although organic solvents and ILs can operate at a wider potential window, they exhibit high resistances toward ion transport. This factor limits the power density of the supercapacitors.

4.4.1 Aqueous Medium

Although aqueous medium electrolytes enjoy low cost fabrication as well as rapid ion transfer with demonstration of high power and energy densities, they exhibit a narrower voltage window. It has been reported that if the voltage of the supercapacitors is increased above 1.23 V, it evokes water decomposition that leads to oxygen/hydrogen evolution. The evolution of gases builds enormous pressure on the electrodes, resulting in tampering of electrodes. This affects the usage of aqueous media for commercial applications in aqueous-based supercapacitors. Nevertheless, among all the types of electrolytes that are used for electrochemical supercapacitors, the aqueous-based supercapacitors have demonstrated high conductivity, which is much higher than that of other electrolytes (approximately a conductivity value of 0.8 S cm^{-1} for 1 M H_2SO_4 at 25°C). Furthermore, the aqueous-based electrolytes can be further subdivided into three categories based on its pH value: acidic, basic, or neutral. In order to select the electrolyte for electrochemical capacitors, it is important to take certain parameters into account, such as size of bare and hydrated ions and even mobility of the ions. Moreover, the degree of corrosion and the electrochemical potential window play an important role in selecting the pH of the electrolyte. The high conductivity of the aqueous medium helps in decreasing the ESR of the electrolyte, and thus, improves the power density of the electrochemical capacitors. It has been observed that for an aqueous solution of H_2SO_4, the conductivity depends on the concentration of H_2SO_4. Low and high concentration of H_2SO_4 in aqueous medium might not have the same conductivity. It was observed that 1 M H_2SO_4 in aqueous medium possesses the highest conductivity value. Therefore, in most cases for aqueous-based electrochemical capacitors, it is being used. Torchała, Kierzek, and Machnikowski (2012) reported that the specific capacitance of activated carbon increased with an increase in electrolyte concentration and conductivity. Moreover, it was postulated that the acid-based aqueous medium showed much higher conductivity than the neutral medium (Wu et al. 2013; Zhang et al. 2012b). Recent reports on carbon-based electrodes have revealed that EDLC-based carbon electrodes also undergo redox reaction, which added some sort of pseudocapacitance to the total capacitance of the electrode (Conway 1999). This redox reaction was attributed to the presence of oxygenated carbon on the surface of the electrode. The interaction between the electrolyte and the electrode was observed to be very specific. It has also been observed that metal sulfides (like CoS_x and NiS) exhibited low pseudocapacitance value as compared to $Co(OH)_2$ and $Ni(OH)_2$ in KOH electrolyte, as well as higher conductivity than any other bases (Hou et al. 2011; Yuan et al. 2009). In this context, it is also necessary to mention that the size of the ions plays a crucial role in dictating the electrochemical properties of the supercapacitors. Researchers have reported that MnO_2 showed higher capacitance in lithium hydroxide (LiOH) as compared to KOH or sodium hydroxide (NaOH) electrolyte, which is due to the smaller size of Li^+ ions that help in easy intercalation (Misnon et al. 2014; Yuan and Zhang 2006). A similar observation was reported for NiO (Inamdar et al. 2011), which demonstrated higher capacitance value in NaOH as compared to KOH.

4.4.2 Organic Solvent Medium

Numerous research works have been focused on aqueous electrolyte in the last few decades. However, due to the inherent drawbacks of aqueous-based electrolytes, recently more attention has been given to organic electrolytes for their versatile properties. Supercapacitors with organic electrolytes possess wide operating voltage potential, and thus, help in improving the power and energy densities of the supercapacitors. However, the decomposition of organic electrolytes blocks the pores of the electrode, which affects the capacitance and cycling behavior of the electrochemical capacitors. Moreover, organic electrolytes exhibit environmental hazards and are also toxic in nature. Nevertheless, in the case of organic electrolytes, cheaper materials can be used for current collectors and packages compared to aqueous electrolytes. Generally, it is considered that organic electrolytes consist of conductive salts, for example, tetraethylammonium tetrafluoroborate, dissolved in acetonitrile or propylene carbonate solvents.

4.4.3 Ionic Liquids

ILs are low melting salts that do not crystallize at room temperature. They have accrued immense attention because of their versatile properties. ILs are liquid at temperatures below 100°C. These types of electrolytes have electrostatic interactions at long ranges, which contribute to Debye screening and also to charge ordering effects (Hansen and McDonald 2013). On the other hand, fluids only consist of neutral particles that have distinguished characteristics from that of ILs. In fact, in the case of ILs, the two ions (cations and anions) have different ionic sizes with low degree of symmetry, leading to reduced lattice energy that results in their lower melting points. The outstanding properties possessed by ILs include:

- Excellent ionic conductivity at room temperature
- Ease and flexibility in processing
- Environmentally benign
- Possess ions for conductivity
- Acceptable viscosity
- Wide working potential window
- Molten salts that are liquids at low temperatures

ILs consist of highly charged organic molecules; therefore, it is important to purify them before use, since they are prone to contamination. In fact, ILs need to be used in pure form, otherwise a small quantity of moisture may reduce the potential window. Usually, ILs may consist of only one component; however, in some cases, they may consist of multiple components.

4.5 CONCLUSION

Electrochemical capacitors possess much higher charge storage capability compared to conventional capacitors. Although their working principle is somewhat different from that of the battery, they are advantageous in advanced applications. The quest for higher energy and power capabilities for advanced energy storage systems led to improvements in supercapacitors. The performance of electrochemical capacitors is mainly dictated by the type and size of nanoparticles, along with the method of fabrication of the electrodes. Innovation of new functional nanomaterials, along with their intricate design, helps in developing advanced electrode materials. In this chapter, we have summarized the several advanced materials that have been used as electrode materials in the last few years. As mentioned earlier, the fabrication of electrode nanomaterials plays an important role in improving the performance of electrochemical capacitors. Amongst all the fabrication methods, electrospinning has gained immense attention for the development of electrodes due to better control over the fiber morphology developed during the processing. Electrospinning is a facile process producing nanofibers with good heterogeneity and porosity, which are beneficial for application in supercapacitors. Electrochemical capacitors are mostly used in case of applications where high power density and charge storage are required, like memory backup power systems, portable electronic fuses, and pulse power sources for smart weapons. These devices are likely to help society to indulge in production of new and more advanced energy storage devices on a global scale.

REFERENCES

Al-Osta, A., V. V. Jadhav, N. A. Saad, R. S. Mane, M. Naushad, K. N. Hui, S. H. Han. 2015. "Diameter-dependent electrochemical supercapacitive properties of anodized titanium oxide nanotubes." *Scripta Materialia* no. 104, 60–63. doi: 10.1016/j.scriptamat.2015.03.025

An, K. H., W. S. Kim, Y. S. Park, J.-M. Moon, D. J. Bae, S. C. Lim, Y. S. Lee, and Y. H. Lee. 2001. "Electrochemical properties of high-power supercapacitors using single-walled carbon nanotube electrodes." *Advanced Functional Materials* no. 11 (5):387–392.

Augustyn, Veronica, Patrice Simon, and Bruce Dunn. 2014. "Pseudocapacitive oxide materials for high-rate electrochemical energy storage." *Energy and Environmental Science* no. 7 (5):1597–1614.

Banerjee, Joyita, and Kingshuk Dutta. 2017. "Materials for electrodes of Li-ion batteries: issues related to stress development." *Critical Reviews in Solid State and Materials Sciences* no. 42 (3):218–238.

Banerjee, Joyita, Kingshuk Dutta, and Dipak Rana. 2019. "Carbon nanomaterials in renewable energy production and storage applications." In *Emerging Nanostructured Materials for Energy and Environmental Science*, eds.: R. Saravanan, Mu. Naushad, K. Raju, and R. Boukherroub, doi: 10.1007/978-3-030-04474-9_2, pp 51–104, Springer Nature: Cham.

Banerjee, Joyita, Kingshuk Dutta, M. Abdul Kader, and Sanjay K. Nayak. 2019. "An overview on the recent developments in polyaniline-based supercapacitors." *Polymers for Advanced Technologies* no. 30 (8):1902–1921. doi: 10.1002/pat.4624

Béguin, François, and Elzbieta Frackowiak. 2013. *Supercapacitors: Materials, Systems, and Applications.* Wiley-VCH: Weinheim.

Brousse, Thierry, Daniel Bélanger, and Jeffrey W. Long. 2015. "To be or not to be pseudocapacitive?" *Journal of the Electrochemical Society* no. 162 (5):A5185–A5189.

Chen, Kunfeng, Shuyan Song, and Dongfeng Xue. 2014. "An ionic aqueous pseudocapacitor system: electroactive ions in both a salt electrode and redox electrolyte." *RSC Advances* no. 4 (44):23338–23343.

Chen, Shih-Hsun, Cheng-HsuanWu, Alex Fang, and Chung-Kwei Lin. 2016. "Effects of adding different morphological carbon nanomaterials on supercapacitive performance of sol-gel manganese oxide films." *Ceramics International* no. 42 (4):4797–4805.

Chen, Tao, and Liming Dai. 2013. "Carbon nanomaterials for high-performance supercapacitors." *Materials Today* no. 16 (78):272–280.

Cherusseri, Jayesh, and Kamal K. Kar. 2015a. "Hierarchically mesoporous carbon nanopetal based electrodes for flexible supercapacitors with super-long cyclic stability." *Journal of Materials Chemistry A* no. 3 (43):21586–21598.

Cherusseri, Jayesh, and Kamal K. Kar. 2015b. "Self-standing carbon nanotube forest electrodes for flexible supercapacitors." *RSC Advances* no. 5 (43):34335–34341.

Choi, Changsoon, Hyeon Jun Sim, Geoffrey M. Spinks, Xavier Lepró, Ray H. Baughman, and Seon Jeong Kim. 2016. "Elastomeric and dynamic MnO_2/CNT core-shell structure coiled yarn supercapacitor." *Advanced Energy Materials* no. 6 (5):1502119.

Choi, Jeong-M, and Seongil Im. 2005. "Ultraviolet enhanced Si-photodetector using p-NiO films." *Applied Surface Science* no. 244 (1–4):435–438.

Conway, B. E. 1991. "Transition from 'supercapacitor' to 'battery' behavior in electrochemical energy storage." *Journal of the Electrochemical Society* no. 138 (6):1539–1548.

Conway, B. E. 1999. *Electrochemical supercapacitors: scientific fundamentals and technological applications.* Kluwer Academic/Plenum: New York.

Costentin, Cyrille, Thomas R. Porter, and Jean-Michel Savéant. 2017. "How do pseudocapacitors store energy? Theoretical analysis and experimental illustration." *ACS Applied Materials and Interfaces* no. 9 (10):8649–8658.

Dar, Farrukh Iqbal, Kevin Radakishna Moonoosawmy, and Mohammed Es-Souni. 2013. "Morphology and property control of NiO nanostructures for supercapacitor applications." *Nanoscale Research Letters* no. 8:363.

Das, Suparna, Kingshuk Dutta, and Dipak Rana. 2018. "Polymer electrolyte membranes for microbial fuel cells: a review." *Polymer Reviews* no. 58 (4):610–629.

Dutta, Kingshuk. 2017. "Polymer-inorganic nanocomposites for polymer electrolyte membrane fuel cells." In *Polymer-Engineered Nanostructures for Advanced Energy Applications*, eds.: Zhiqun Lin, Yingkui Yang, and Aiqing Zhang, doi: 10.1007/978-3-319-57003-7_15, pp. 577–606, Springer Nature: Cham.

Dutta, K., D. Rana, H. S. Han, and P. P. Kundu. 2017. "Unitized regenerative fuel cells: a review on developed catalyst systems and bipolar plates." *Fuel Cells* no. 17 (6):736–751.

Dutta, K., P. P. Kundu, and A. Kundu. 2014. "Fuel cells – exploratory fuel cells | micro-fuel cells." In *Reference Module in Chemistry, Molecular Sciences and Chemical Engineering*, ed.: Jan Reedijk, doi: 10.1016/B978-0-12-409547-2.10975-8, Elsevier: Amsterdam.

Dutta, Kingshuk, Suparna Das, and Patit P. Kundu. 2015. "Synthesis, preparation, and performance of blends and composites of π-conjugated polymers and their copolymers in DMFCs." *Polymer Reviews* no. 55 (4):630–677.

Futaba, Don N., Kenji Hata, Takeo Yamada, Tatsuki Hiraoka, Yuhei Hayamizu, Yozo Kakudate, Osamu Tanaike, Hiroaki Hatori, Motoo Yumura, and Sumio Iijima. 2006. "Shape-engineerable and highly densely packed single-walled carbon nanotubes and their application as super-capacitor electrodes." *Nature Materials* no. 5:987–994.

Gaikar, P. S., S. T. Navale, S. L. Gaikwad, A. Al-Osta, V. V. Jadhav, P. R. Arjunwadkar, M. Naushad, R. S. Mane. 2017. "Pseudocapacitive performance of a solution-processed β-Co(OH)2 electrode monitored through its surface morphology and area." *Dalton Transactions* no. 46:33933399. doi: 10.1039/c6dt04581b

Gan, Li-Hua, Dingrong Deng, Yanjun Zhang, Gen Li, Xueyun Wang, Li Jiang, and Chun-Ru Wang. 2014. "$Zn_3V_2O_8$ hexagon nanosheets: a high-performance anode material for lithium-ion batteries." *Journal of Materials Chemistry A* no. 2 (8):2461–2466.

Gogotsi, Yury, and Reginald M. Penner. 2018. "Energy storage in nanomaterials – capacitive, pseudocapacitive, or battery-like?" *ACS Nano* no. 12 (3):2081–2083.

Gomibuchi, Emi, Takayuki Ichikawa, Koichi Kimura, Shigehito Isobe, Koji Nabeta, and Hironobu Fujii. 2006. "Electrode properties of a double layer capacitor of nano-structured graphite produced by ball milling under a hydrogen atmosphere." *Carbon* no. 44 (5):983–988.

Guo, Di, Haiming Zhang, Xinzhi Yu, Ming Zhang, Ping Zhang, Qiuhong Li, and Taihong Wang. 2013. "Facile synthesis and excellent electrochemical properties of CoMoO$_4$ nanoplate arrays as supercapacitors." *Journal of Materials Chemistry A* no. 1 (24):7247–7254.

Hansen, Jean-Pierre, and Ian R. McDonald. 2013. *Theory of Simple Liquids*. doi: 10.1016/C2010-0-66723-X, Elsevier: Amsterdam.

Hou, Linrui, Changzhou Yuan, Diankai Li, Long Yang, Laifa Shen, Fang Zhang, and Xiaogang Zhang. 2011. "Electrochemically induced transformation of NiS nanoparticles into Ni(OH)$_2$ in KOH aqueous solution toward electrochemical capacitors." *Electrochimica Acta* no. 56 (22):7454–7459.

https://www.cap-xx.com/resource/energy-storage-technologies/, Accessed on 09/03/2019.

Hu, Guangxia, Chunhua Tang, Chunxiang Li, Huimin Li, Yu Wang, and Hao Gong. 2011a. "The sol-gel-derived nickel-cobalt oxides with high supercapacitor performances." *Journal of the Electrochemical Society* no. 158 (6):A695–A699.

Hu, Linfeng, Limin Wu, Meiyong Liao, and Xiaosheng Fang. 2011b. "High-performance NiCo2O4 nano-film photodetectors fabricated by an interfacial self-assembly strategy." *Advanced Materials* no. 23 (17):1988–1992.

Huang, Jun, Junchao Wei, Yingbo Xiao, Yazhou Xu, Yujuan Xiao, Ying Wang, Licheng Tan, Kai Yuan, and Yiwang Chen. 2018. "When Al-doped cobalt sulfide nanosheets meet nickel nanotube arrays: a highly efficient and stable cathode for asymmetric supercapacitors." *ACS Nano* no. 12 (3):3030–3041.

Inamdar, A. I., YoungSam Kim, S. M. Pawar, J. H. Kim, Hyunsik Im, and Hyungsang Kim. 2011. "Chemically grown, porous, nickel oxide thin-film for electrochemical supercapacitors." *Journal of Power Sources* no. 196 (4):2393–2397.

Jagadale, Ajay D., Guoqing Guan, Xiumin Li, Xiao Du, Xuli Ma, Xiaogang Hao, and Abuliti Abudula. 2016. "Ultrathin nanoflakes of cobalt-manganese layered double hydroxide with high reversibility for asymmetric supercapacitor." *Journal of Power Sources* no. 306:526–534.

Jayalakshmi, M., M. Mohan Rao, N. Venugopal, and K.-B. Kim. 2007. "Hydrothermal synthesis of SnO$_2$-V$_2$O$_5$ mixed oxide and electrochemical screening of carbon nano-tubes (CNT), V$_2$O$_5$, V$_2$O$_5$-CNT, and SnO$_2$-V$_2$O$_5$-CNT electrodes for supercapacitor applications." *Journal of Power Sources* no. 166 (2):578–583.

Jin, Jutao, Xiaochang Qiao, Feng Zhou, Zhong-Shuai Wu, Lifeng Cui, and Hongbo Fan. 2017. "Interconnected phosphorus and nitrogen codoped porous exfoliated carbon nanosheets for high-rate supercapacitors." *ACS Applied Materials and Interfaces* no. 9 (20):17317–17325.

Jost, Kristy, David P. Durkin, Luke M. Haverhals, E. Kathryn Brown, Matthew Langenstein, Hugh C. De Long, Paul C. Trulove, Yury Gogotsi, and Genevieve Dion. 2015. "Natural fiber welded electrode yarns for knittable textile supercapacitors." *Advanced Energy Materials* no. 5 (4):1401286.

Kandalkar, S. G., J. L. Gunjakar, and C. D. Lokhande. 2008. "Preparation of cobalt oxide thin films and its use in supercapacitor application." *Applied Surface Science* 254 (17):5540–5544.

Ko, Yongmin, Minseong Kwon, Wan Ki Bae, Byeongyong Lee, Seung Woo Lee, and Jinhan Cho. 2017. "Flexible supercapacitor electrodes based on real metal-like cellulose papers." *Nature Communications* no. 8:536.

Kumar, R. Vijay, Y. Diamant, and A. Gedanken. 2000. "Sonochemical synthesis and characterization of nanometer-size transition metal oxides from metal acetates." *Chemistry of Materials* no. 12 (8):2301–2305.

Kundu, Patit P., and Kingshuk Dutta. 2018. *Progress and Recent Trends in Microbial Fuel Cells*. doi: 10.1016/C2016-0-04695-8, Elsevier: Amsterdam.

Lee, Hae-Min, Gyoung Hwa Jeong, Doo Won Kang, Sang-Wook Kim, and Chang-Koo Kim. 2015. "Direct and environmentally benign synthesis of manganese oxide/graphene composites from graphite for electrochemical capacitors." *Journal of Power Sources* no. 281:44–48.

Lee, Hee Y., and J. B. Goodenough. 1999. "Ideal supercapacitor behavior of amorphous V$_2$O$_5 \cdot n$H$_2$O in potassium chloride (KCl) aqueous solution." *Journal of Solid State Electrochemistry* no. 148 (1):81–84.

Lee, Kian Keat, Wee Shong Chin, and Chorng Haur Sow. 2014. "Cobalt-based compounds and composites as electrode materials for high-performance electrochemical capacitors." *Journal of Materials Chemistry A* no. 2 (41):17212–17248.

Li, Huiqiao, Liang Cheng, and Yongyao Xia. 2005. "A hybrid electrochemical supercapacitor based on a 5 V Li-ion battery cathode and active carbon." *Electrochemical and Solid-State Letters* no. 8 (9):A433–A436.

Li, Huiqiao, Xizheng Liu, Tianyou Zhai, De Li, and Haoshen Zhou. 2013. "Li$_3$VO$_4$: a promising insertion anode material for lithium-ion batteries." *Advanced Energy Materials* no. 3 (4):428–432.

Liu, Mao-Cheng, Ling-Bin Kong, Chao Lu, Xiao-Ming Li, Yong-Chun Luo, and Long Kang. 2012. "A sol-gel process for fabrication of NiO/NiCo$_2$O$_4$/Co$_3$O$_4$ composite with improved electrochemical behavior for electrochemical capacitors." *ACS Applied Materials and Interfaces* no. 4 (9):4631–4636.

Liu, Mao-Cheng, Ling-Bin Kong, Chao Lu, Xiao-Ming Li, Yong-Chun Luo, and Long Kang. 2013. "Facile fabrication of CoMoO4 nanorods as electrode material for electrochemical capacitors." *Materials Letters* no. 94:197–200.

Lota, Grzegorz, Krzysztof Fic, and Elzbieta Frackowiak. 2011. "Alkali metal iodide/carbon interface as a source of pseudocapacitance." *Electrochemistry Communications* no. 13 (1):38–41.

Ma, Hua, Shaoyan Zhang, Weiqiang Ji, Zhanliang Tao, and Jun Chen. 2008. "α-CuV$_2$O$_6$ nanowires: hydrothermal synthesis and primary lithium battery application." *Journal of the American Chemical Society* no. 130 (15):5361–5367.

Ma, Wenqin, Honghong Nan, Zhengqiang Gu, Baoyou Geng, and Xiaojun Zhang. 2015. "Superior performance asymmetric supercapacitors based on ZnCo$_2$O$_4$@MnO$_2$ core-shell electrode." *Journal of Materials Chemistry A* no. 3 (10):5442–5448.

Mai, Li-Qiang, Fan Yang, Yun-Long Zhao, Xu Xu, Lin Xu, and Yan-Zhu Luo. 2011. "Hierarchical MnMoO4/CoMoO4 heterostructured nanowires with enhanced supercapacitor performance." *Nature Communications* no. 2:381.

Manjakkal, Libu, Carlos García Núñez, Wenting Dang, and Ravinder Dahiya. 2018. "Flexible self-charging supercapacitor based on graphene-Ag-3D graphene foam electrodes." *Nano Energy* no. 51:604–612.

Mei, Peng, Xing-Long Wu, Haiming Xie, Liqun Sun, Yanping Zeng, Jingping Zhang, Linghua Tai, Xin Guo, Lina Cong, Shunchao Ma, Cen Yao, and Rongshun Wang. 2014. "LiV$_3$O$_8$ nanorods as cathode materials for high-power and long-life rechargeable lithium-ion batteries." *RSC Advances* no. 4 (49):25494–25501.

Misnon, Izan Izwan, Radhiyah Abd Aziz, Nurul Khairiyyah Mohd Zain, Baiju Vidhyadharan, Syam G Krishnan, and Rajan Jose. 2014. "High performance MnO$_2$ nanoflower electrode and the relationship between solvated ion size and specific capacitance in highly conductive electrolytes." *Materials Research Bulletin* no. 57:221–230.

Niu, Chunming, Enid K. Sichel, Robert Hoch, David Moy, and Howard Tennent. 1997. "High power electrochemical capacitors based on carbon nanotube electrodes." *Applied Physics Letters* no. 70 (11):1480.

Patil, U. M., S. B. Kulkarni, V. S. Jamadade, and C. D. Lokhande. 2011. "Chemically synthesized hydrous RuO$_2$ thin films for supercapacitor application." *Journal of Alloys and Compounds* no. 509 (5):1677–1682.

Pei, Lei, Xia Zhang, Lingling Zhang, Yujun Zhang, and Yuandong Xu. 2016. "Solvent influence on the morphology and supercapacitor performance of the nickel oxide." *Materials Letters* no. 162:238–241.

Pell, Wendy G., and Brian E. Conway. 2004. "Peculiarities and requirements of asymmetric capacitor devices based on combination of capacitor and battery-type electrodes." *Journal of Power Sources* no. 136 (2):334–345.

Purkait, Taniya, Guneet Singh, Dinesh Kumar, Mandeep Singh, and Ramendra Sundar Dey. 2018. "High-performance flexible supercapacitors based on electrochemically tailored three-dimensional reduced graphene oxide networks." *Scientific Reports* no. 8:640.

Reddy, Arava Leela Mohana, Manikoth M. Shaijumon, Sanketh R. Gowda, and Pulickel M. Ajayan. 2010. "Multisegmented Au-MnO$_2$/carbon nanotube hybrid coaxial arrays for high-power supercapacitor applications." *The Journal of Physical Chemistry C* no. 114 (1):658–663.

Sarkar, Abhimanyu, Ashis Kumar Satpati, Vikram Kumar, and Sanjiv Kumar. 2015. "Sol-gel synthesis of manganese oxide films and their predominant electrochemical properties." *Electrochimica Acta* no. 167:126–131.

Senthilkumar, Baskar, Danielle Meyrick, Yun-Sung Lee, and Ramakrishnan Kalai Selvan. 2013. "Synthesis and improved electrochemical performances of nano β-NiMoO$_4$-CoMoO$_4$.xH$_2$O composites for asymmetric supercapacitors." *RSC Advances* no. 3 (37):16542–16548.

Senthilkumar, Baskar, Ramakrishnan Kalai Selvan, Leonid Vasylechko, and Manickam Minakshi. 2014. "Synthesis, crystal structure and pseudocapacitor electrode properties of γ-Bi$_2$MoO$_6$ nanoplates." *Solid State Sciences* no. 35:18–27.

Shi, Xiaobo, Zhigang Zeng, Erjuan Guo, Xiao Long, Haijun Zhou, and Xiaohong Wang. 2017. "A growth mechanism investigation on the anodic deposition of nanoporous gold supported manganese oxide nanostructures for high performance flexible supercapacitors." *Journal of Alloys and Compounds* no. 690:791–798.

Shukla, A. K., S. Sampath, and K. Vijayamohanan. 2000. "Electrochemical supercapacitors: energy storage beyond batteries." *Current Science* no. 79 (12):1656–1661.

Sim, Dao Hao, Xianhong Rui, Jing Chen, Huiteng Tan, Tuti Mariana Lim, Rachid Yazami, Huey Hoon Hng, and Qingyu Yan. 2012. "Direct growth of FeVO$_4$ nanosheet arrays on stainless steel foil as high-performance binder-free Li ion battery anode." *RSC Advances* no. 2 (9):3630–3633.

Simon, Patrice, Yury Gogotsi, and Bruce Dunn. 2014. "Where do batteries end and supercapacitors begin?" *Science* no. 343 (6176):1210–1211.

Singh, Ashutosh K., Debasish Sarkar, Keshab Karmakar, Kalyan Mandal, and Gobinda Gopal Khan. 2016. "High-performance supercapacitor electrode based on cobalt oxide-manganese dioxide-nickel oxide ternary 1D hybrid nanotubes." *ACS Applied Materials and Interfaces* no. 8 (32):20786–20792.

Stern, O. 1924. "The theory of the electric double layer." *Z. Electrochemistry* no. 30:508.

Stoller, Meryl D., Sungjin Park, Yanwu Zhu, Jinho An, and Rodney S. Ruoff. 2008. "Graphene-based ultracapacitors." *Nano Letters* no. 8 (10):3498–3502.

Sun, Yan, Chunsheng Li, Lina Wang, Yaozu Wang, Xuegang Ma, Peijuan Ma, and Mingyang Song. 2012. "Ultralong monoclinic ZnV_2O_6 nanowires: their shape-controlled synthesis, new growth mechanism, and highly reversible lithium storage in lithium-ion batteries." *RSC Advances* no. 2 (21):8110–8115.

Tahmasebi, Mohammad H., Antonello Vicenzo, Mazdak Hashempour, Massimiliano Bestetti, Mohammad A. Golozar, and Keyvan Raeissi. 2016. "Nanosized Mn-Ni oxide thin films via anodic electrodeposition: a study of the correlations between morphology, structure and capacitive behaviour." *Electrochimica Acta* no. 206:143–154.

Tang, Jiayong, Ping Yuan, Chuanlin Cai, Yanbao Fu, and Xiaohua Ma. 2016. "Combining nature-inspired, graphene-wrapped flexible electrodes with nanocomposite polymer electrolyte for asymmetric capacitive energy storage." *Advanced Energy Materials* no. 6 (19):1600813.

Teh, Pei Fen, Yogesh Sharma, Stevin Snellius Pramana, and Madhavi Srinivasan. 2011. "Nanoweb anodes composed of one-dimensional high aspect ratio, size tunable electrospun $ZnFe_2O_4$ nanofibers for lithium ion batteries." *Journal of Materials Chemistry* no. 21 (38):14999–15008.

Torchała, Kamila, Krzysztof Kierzek, and Jacek Machnikowski. 2012. "Capacitance behavior of KOH activated mesocarbon microbeads in different aqueous electrolytes." *Electrochimica Acta* no. 86:260–267.

Vangari, Manisha, Tonya Pryor, and Li Jiang. 2013. "Supercapacitors: review of materials and fabrication methods." *Journal of Energy Engineering* no. 139 (2):72–79.

Vlad, Alexandru, Neelam Singh, Charudatta Galande, and Pulickel M. Ajayan. 2015. "Design considerations for unconventional electrochemical energy storage architectures." *Advanced Energy Materials* no. 5 (19): 1402115.

Wang, Huanwen, Huan Yi, Xiao Chen, and Xuefeng Wang. 2013. "Facile synthesis of a nano-structured nickel oxide electrode with outstanding pseudocapacitive properties." *Electrochimica Acta* no. 105:353–361.

Wang, Jie, Shengyang Dong, Bing Ding, Ya Wang, Xiaodong Hao, Hui Dou, Yongyao Xia, and Xiaogang Zhang. 2017a. "Pseudocapacitive materials for electrochemical capacitors: from rational synthesis to capacitance optimization." *National Science Review* no. 4 (1):71–90.

Wang, Libin, Huiling Yang, Xiaoxiao Liu, Rui Zeng, Ming Li, Yunhui Huang, and Xianluo Hu. 2017b. "Constructing hierarchical tectorum-like α-Fe_2O_3/PPy nanoarrays on carbon cloth for solid-state asymmetric supercapacitors." *Angewandte Chemie International Edition* no. 56 (4):1105–1110.

Wang, Xu, Afriyanti Sumboja, Eugene Khoo, Chaoyi Yan, and Pooi See Lee. 2012a. "Cryogel synthesis of hierarchical interconnected marco-/mesoporous Co_3O_4 with superb electrochemical energy storage." *The Journal of Physical Chemistry C* no. 116 (7):4930–4935.

Wang, Xu, Xuanding Han, Mengfang Lim, Nandan Singh, Chee Lip Gan, Ma Jan, and Pooi See Lee. 2012b. "Nickel cobalt oxide-single wall carbon nanotube composite material for superior cycling stability and high-performance supercapacitor application." *The Journal of Physical Chemistry C* no. 116 (23):12448–12454.

Wu, Fangfang, Chunhui Yu, Wenxiu Liu, Ting Wang, Jinkui Feng, and Shenglin Xiong. 2015a. "Large-scale synthesis of $Co_2V_2O_7$ hexagonal microplatelets under ambient conditions for highly reversible lithium storage." *Journal of Materials Chemistry A* no. 3 (32):16728–16736.

Wu, Fangfang, Shenglin Xiong, Yitai Qian, and Shu-Hong Yu. 2015b. "Hydrothermal synthesis of unique hollow hexagonal prismatic pencils of $Co_3V_2O_8 \cdot nH_2O$: a new anode material for lithium-ion batteries." *Angewandte Chemie International Edition* no. 54 (37):10787–10791.

Wu, Hao, Xianyou Wang, Lanlan Jiang, Chun Wu, Qinglan Zhao, Xue Liu, Ben'an Hu, and Lanhua Yi. 2013. "The effects of electrolyte on the supercapacitive performance of activated calcium carbide-derived carbon." *Journal of Power Sources* no. 226:202–209.

Wu, Xinming, Qiguan Wang, Wenzhi Zhang, Yan Wang, and Weixing Chen. 2016. "Nano nickel oxide coated graphene/polyaniline composite film with high electrochemical performance for flexible supercapacitor." *Electrochimica Acta* no. 211:1066–1075.

Xiao, Lifen, Yanqiang Zhao, Jia Yin, and Lizhi Zhang. 2009. "Chewlike ZnV_2O_4 hollow spheres: nonaqueous sol-gel synthesis, formation mechanism, and lithium storage properties." *Chemistry – A European Journal* no. 15 (37):9442–9450.

Xiao, Xu, Tianqi Li, Zehua Peng, Huanyu Jin, Qize Zhong, Qiyi Hu, Bin Yao, Qiuping Luo, Chuanfang Zhang, Li Gong, Jian Chen, Yury Gogotsi, and Jun Zhou. 2014. "Freestanding functionalized carbon nanotube-based electrode for solid-state asymmetric supercapacitors." *Nano Energy* no. 6:1–9.

Xu, Bin, Feng Wu, Shi Chen, Cunzhong Zhang, Gaoping Cao, and Yusheng Yang. 2007. "Activated carbon fiber cloths as electrodes for high performance electric double layer capacitors." *Electrochimica Acta* no. 52 (13):4595–4598.

Xu, Juan, Zhengwei Ju, Jianyu Cao, Wenchang Wang, Cheng Wang, and Zhidong Chen. 2016. "Microwave synthesis of nitrogen-doped mesoporous carbon/nickel-cobalt hydroxide microspheres for high-performance supercapacitors." *Journal of Alloys and Compounds* no. 689:489–499.

Xue, Qi, Jinfeng Sun, Yan Huang, Minshen Zhu, Zengxia Pei, Hongfei Li, Yukun Wang, Na Li, Haiyan Zhang, and Chunyi Zhi. 2017. "Recent progress on flexible and wearable supercapacitors." *Small* no. 13 (45):1701827.

Yang, M., J. X. Li, H. H. Li, L. W. Su, J. P. Wei, and Z. Zhou. 2012. "Mesoporous slit-structured NiO for high-performance pseudocapacitors." *Physical Chemistry Chemical Physics* no. 14 (31):11048–11052.

Yang, Yang, Doohun Kim, Min Yang, and Patrik Schmuki. 2011. "Vertically aligned mixed V_2O_5-TiO_2 nanotube arrays for supercapacitor applications." *Chemical Communications* no. 47 (27):7746–7748.

Yi, Bin, Xiaohua Chen, Kaimin Guo, Longshan Xu, Chuansheng Chen, Haimei Yan, and Jianghua Chen. 2011. "High-performance carbon nanotube-implanted mesoporous carbon spheres for supercapacitors with low series resistance." *Materials Research Bulletin* no. 46 (11):2168–2172.

Yin, Zhuoxun, Yujin Chen, Yang Zhao, Chunyan Li, Chunling Zhu, and Xitian Zhang. 2015. "Hierarchical nanosheet-based $CoMoO_4$-$NiMoO_4$ nanotubes for applications in asymmetric supercapacitors and the oxygen evolution reaction." *Journal of Materials Chemistry A* no. 3 (45):22750–22758.

You, Xiangyu, Manjusri Misra, Stefano Gregori, and Amar K. Mohanty. 2018. "Preparation of an electric double layer capacitor (EDLC) using *Miscanthus*-derived biocarbon." *ACS Sustainable Chemistry and Engineering* no. 6 (1):318–324.

Yuan, Anbao, and Qinglin Zhang. 2006. "A novel hybrid manganese dioxide/activated carbon supercapacitor using lithium hydroxide electrolyte." *Electrochemistry Communications* no. 8 (7):1173–1178.

Yuan, Changzhou, Bo Gao, Linhao Su, Li Chen, and Xiaogang Zhang. 2009. "Electrochemically induced phase transformation and charge-storage mechanism of amorphous CoS_x nanoparticles prepared by interface-hydrothermal method." *Journal of the Electrochemical Society* no. 156 (3):A199–A203.

Yuan, Changzhou, Hao Bin Wu, Yi Xie, and Xiong Wen Lou. 2014. "Mixed transition-metal oxides: design, synthesis, and energy-related applications." *Angewandte Chemie International Edition* no. 53 (6):1488–1504.

Yuan, Changzhou, Long Yang, Linrui Hou, Laifa Shen, Fang Zhang, Diankai Li, and Xiaogang Zhang. 2011. "Large-scale Co_3O_4 nanoparticles growing on nickel sheets *via* a one-step strategy and their ultra-highly reversible redox reaction toward supercapacitors." *Journal of Materials Chemistry* no. 21 (45):18183–18185.

Zang, Xiaoxian, Chencheng Sun, Ziyang Dai, Jun Yang, and Xiaochen Dong. 2017. "Nickel hydroxide nanosheets supported on reduced graphene oxide for high-performance supercapacitors." *Journal of Alloys and Compounds* no. 691:144–150.

Zhang, Genqiang, Le Yu, Hao Bin Wu, Harry E. Hoster, and Xiong Wen Lou. 2012a. "Formation of $ZnMn_2O_4$ ball-in-ball hollow microspheres as a high-performance anode for lithium-ion batteries." *Advanced Materials* no. 24 (34):4609–4613.

Zhang, Jing, Ling-Bin Kong, Jian-Jun Cai, Heng Li, Yong-Chun Luo, and Long Kang. 2010. "Hierarchically porous nickel hydroxide/mesoporous carbon composite materials for electrochemical capacitors." *Microporous and Mesoporous Materials* no. 132 (1–2):154–162.

Zhang, Ruiping, Jun Liu, Hongge Guo, and Xili Tong. 2014a. "Hierarchically porous nickel oxide nanoflake arrays grown on carbon cloth by chemical bath deposition as superior flexible electrode for supercapacitors." *Materials Letters* no. 136:198–201.

Zhang, Xiaoyan, Xianyou Wang, Lanlan Jiang, Hao Wu, Chun Wu, and Jingcang Su. 2012b. "Effect of aqueous electrolytes on the electrochemical behaviors of supercapacitors based on hierarchically porous carbons." *Journal of Power Sources* no. 216:290–296.

Zhang, Youjuan, Yuanying Liu, Jing Chen, Qifei Guo, Ting Wang, and Huan Pang. 2014b. "Cobalt vanadium oxide thin nanoplates: primary electrochemical capacitor application." *Scientific Reports* no. 4:5687.

Zhao, Bo, Tao Wang, Li Jiang, Kai Zhang, Matthew M. F. Yuen, Jian-Bin Xu, Xian-Zhu Fu, Rong Sun, and Ching-Ping Wong. 2016. "NiO mesoporous nanowalls grown on RGO coated nickel foam as high performance electrodes for supercapacitors and biosensors." *Electrochimica Acta* no. 192:205–215.

Zhao, Dandan, Yongqing Zhao, Xuan Zhang, Cailing Xu, Yong Peng, Hulin Li, and Zhi Yang. 2013. "Application of high-performance MnO_2 nanocomposite electrodes in ionic liquid hybrid supercapacitors." *Materials Letters* no. 107:115–118.

Zhao, Yu, Ming-bo Zheng, Jie-ming Cao, Xing-fei Ke, Jin-song Liu, Yong-ping Chen, and Jie Tao. 2008. "Easy synthesis of ordered meso/macroporous carbon monolith for use as electrode in electrochemical capacitors." *Materials Letters* no. 62 (3):548–551.

Zheng, Jim P. 2003. "The limitations of energy density of battery/double-layer capacitor asymmetric cells." *Journal of the Electrochemical Society* no. 150 (4):A484–A492.

Zhi, Mingjia, Ayyakkannu Manivannan, Fanke Meng, and Nianqiang Wu. 2012. "Highly conductive electrospun carbon nanofiber/MnO2 coaxial nano-cables for high energy and power density supercapacitors." *Journal of Power Sources* no. 208:345–353.

Zhong, Cheng, Yida Deng, Wenbin Hu, Jinli Qiao, Lei Zhang, and Jiujun Zhang. 2015. "A review of electrolyte materials and compositions for electrochemical supercapacitors." *Chemical Society Reviews* no. 44 (21):7484–7539.

Zhou, Liang, Dongyuan Zhao, and Xiong Wen Lou. 2012. "Double-shelled $CoMn_2O_4$ hollow microcubes as high-capacity anodes for lithium-ion batteries." *Advanced Materials* no. 24 (6):745–748.

Zhou, Zhou, Yirong Zhu, Zhibin Wu, Fang Lu, Mingjun Jing, and Xiaobo Ji. 2014. "Amorphous RuO_2 coated on carbon spheres as excellent electrode materials for supercapacitors." *RSC Advances* no. 4 (14):6927–6932.

5 Microbial Electrochemical Technologies for Fuel Cell Devices

S. V. Sheen Mers, K. Sathish-Kumar, L. A. Sánchez-Olmos, M. Sánchez-Cardenas, and Felipe Caballero-Briones

CONTENTS

5.1	Introduction to Microbial Electrochemical Technologies	84
	5.1.1 Components of MFCs	86
	5.1.1.1 Electrodes	86
	5.1.1.2 Electrolyte	86
	5.1.1.3 Separators	86
	5.1.1.4 Microorganisms	87
	5.1.2 Various Parameters Used in MFC Treatment	87
	5.1.2.1 Power	87
	5.1.2.2 Hydraulic Retention Time (HRT)	87
	5.1.2.3 Specific Energy Consumption (μ)	88
	5.1.2.4 Chemical Oxygen Demand	88
	5.1.2.5 Coulombic Efficiency (CE)	88
	5.1.2.6 Normalized Energy Recovery (NER)	88
5.2	Opportunities and Thermodynamic/Electromotive Force Constraint of Microbial Fuel Cell Technologies	88
	5.2.1 Thermodynamics of the MFC	89
	5.2.2 Potential Losses	91
	5.2.3 Activation Losses	92
	5.2.4 Concentration Losses	93
	5.2.5 Biocatalyst Metabolic Losses	93
	5.2.6 Ohmic Losses	93
5.3	Microbial Conversion of Organic Material into Electricity Production Prototypes	94
5.4	Microbial Conversion of Organic Matters into Fuel	97
5.5	Microbial Electrochemical Technology Limitations	97
	5.5.1 Biochemical Constraints	98
	5.5.2 Functional Materials Constraints	99
	5.5.3 Economic Constraints	99
5.6	Future Direction of Microbial Electrochemical Technologies	100
5.7	Conclusions	100
References		101

5.1 INTRODUCTION TO MICROBIAL ELECTROCHEMICAL TECHNOLOGIES

The depletion of fossil fuels alarms the increased demand for seeking alternate technologies that utilize renewable sources (solar energy, wind, hydro energy, geothermal energy, and biomass) to meet the growing demand for energy. Among them, energy production through fuel cells seems promising since it is clean, meaning that it is free of contaminating by-products. Various types of fuel cells are in practice, and distinguished among them is the microbial fuel cell (MFC) that can work collaboratively as an electricity generator, a sensor, and a wastewater/bioremediation unit. The knowledge about the basic structure and the further advancement of MFC design that influence the performance of MFCs should be analyzed and brought together for determining the barriers to develop new prototypes. It is a promising field that has more scope for research and will be more innovative when chemists, material scientists, engineers, and microbiologists collaborate. Due to the lack of studies for a deeper understanding of the mechanism behind the microbial activity and drop in researching MFC parameter values (chemical oxygen demand [COD], biochemical oxygen demand [BOD], power and current densities, voltage, or coulombic efficiency) to the expected level, there are still miles to go for taking MFCs to the next level of practical applications. In this chapter, we discuss the basic principles of MFCs, various prototypes proposed for the conversion of waste materials either to fuel or energy, and their future prospects. This chapter will provide readers access to information in a simplified form for better understanding.

A microbial fuel cell comprises a simple electrochemical setup whereby biocatalyst (microbes) functions lead to reactions at the anode. An MFC serves the dual purpose as a power generator and as a wastewater/bioremediation unit through the removal of organic residues/contamination from the environment. MFC technology is a century-old technique, and the number of publications and patents in this field has shown rapid growth during the last few decades. (Schröder 2012; Figure 5.1). The advancement in modern technology ensures the increasing demand for energy and utilization of alternative renewable energy, thus methods like MFC technology are in high need. It should be noted that we still depend on nonrenewable fossil fuels for 70% of our energy needs (Ieropoulos et al., 2013). However, fossil fuels are depleting year by year, and we are aware that it will take millions of years for them to replace themselves. Suitable alternative, renewable clean energy production can be made real through fuel cells. Although various forms of fuel cells like polymer electrolyte membrane fuel cells (PEMs), solid oxide fuel cells (SOFCs), alkaline fuel cells (AFCs), phosphoric acid fuel cells (PAFCs), and molten carbonate fuel cells (MCFCs) are available, MFCs are advantageous since they require a simple setup, comparatively low-cost components, and are more convenient to use (Wang et al. 2011). A major limitation of this technique is that it still lacks commercialization due to the inadequacy in performance and high fabrication cost and design when applied for practical use.

The MFC setup is composed of minimal components usually possessing an anodic chamber containing microorganisms (biocatalyst) and wastewater, while an oxidant at the cathode is connected

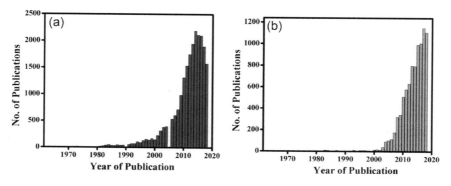

FIGURE 5.1 Number of (a) publications and (b) patents from 1962 to 2018 using the search word "microbial fuel cell." (Data taken from Scopus, March 2019.)

Microbial Electrochemical Technologies for Fuel Cell Devices

to an external circuit to generate electricity. Different cell configurations were developed to eliminate certain charge transfer constraints like voltage drop due to internal resistance (arising due to highly populated organic residues in wastewater), interelectrode distance, separator problems, and undesirable side reactions. The various prototypes recently proposed and the comparison of their efficiencies are given in Sections 5.3 and 5.4.

Now let us have a look into different conditions under which MFCs operate and the resultant changes in terms of chemical products. MFCs can be operated in two ways:

1. Mediated
2. Nonmediated

In MFCs, the organic residues are digested by microorganisms to produce protons and electrons. The protons diffuse through the separator and get reduced at the cathode. The electrons produced during this process can be transferred directly to an anodic current collector if exoelectrogenic bacteria are used, and the process is said to be nonmediated. It facilitates direct electron transfer from the bacterial cell wall through the cytochromes to the electrode surface (Adebule et al. 2018; Schröder 2012; Inamuddin et al. 2016). In the other case, mediators like potassium ferric cyanide, thionine, or neutral red are used to receive an electron (redox shuttle) from the bacteria to the electrode surface. Moreover, the use of a mediator is not advisable for the sustainable operation of MFCs. The atmospheric condition at which the MFC operates, either in aerobic or anaerobic condition, or the impact of oxygen present in the air could adaptably create better performances. For a more straightforward understanding, this can be explained by taking glucose and acetate as model analytes in the MFC and varying the atmospheric conditions (Moqsud et al. 2013). In the anode compartment, glucose is oxidized to CO_2 and H_2O under aerobic conditions, while in anaerobic conditions it is oxidized to CO_2, protons, and electrons. On the other hand, the cathode consumes atmospheric oxygen and converts them into the water. In this way, electricity can be produced in MFCs only under anaerobic conditions. The corresponding anodic and cathodic reactions are given as follows (Santiago et al. 2016):

Anodic reaction:

$$C_6H_{12}O_6 + 6H_2O \rightarrow 6CO_2 + 24H^+ + 24e^- \text{ (anaerobic condition)}$$

Cathode reaction:

$$O_2 + 4H^+ + 4e^- \rightarrow 2H_2O$$

Hence, it is an environmentally friendly approach since bacteria and wastewater form the electrolyte and carbon-based materials are used as electrodes (both for anode and cathode). The productivity of MFCs can be enhanced through the choice of suitable electrolytes and approaches such as the integrated active sludge method and by the addition of buffers. Buffers (>7 mS cm^{-1}) possess better conductivity than that of wastewater (~1 mS cm^{-1}) and hence the improved activity (Ahn et al. 2014). However, in terms of practical applications, the addition of buffers is not a good idea as it is economically less plausible and active sludge being an independent technique will be an additional step during MFC operation. Hence, the main focus should be in reducing internal resistance to obtain the increased electrical output. In this aspect, miniaturization of MFC is a feasible design generally known as mini microbial electrolysis cell (mini-MEC) where the geometry considerably influences the power output in the cell having a cross-sectional area of 2 cm^2 and volume 1.2 cm^3 (Ringeisen et al. 2006). Before going into the existing technologies for efficient power production using MFCs, the basic structure of MFCs needs to be understood. A microbial electrolytic cell (MEC) is one of the bioelectrochemical technologies where an external voltage is applied for the reduction of protons to form hydrogen or the formation of any other gaseous products that can be able to produce an electric current.

5.1.1 Components of MFCs

The schematic diagram of a typical MFC is shown in Figure 5.1. It is composed of three primary components as seen in a normal electrochemical cell: anode, cathode, and electrolyte along with a separator (Figure 5.2).

5.1.1.1 Electrodes

Carbon-based materials are usually used as the anode and cathode. Some of the materials include graphite rod, carbon cloth, and graphite brush (Ahn et al. 2014; Nam et al. 2010). In MECs, a noble metal catalyst like platinum is coated on the cathode to enhance the oxygen reduction reaction. A few alternatives, like activated carbon and transition metal oxides, can be used for enhancing the power output. Another way to enhance the cathodic activity is to increase the surface area twice that of the anode. Among various anodes, carbon brushes are found to show supreme behavior since they possess high surface area, high conductivity, and its peculiar bristles prevent the electrode from fouling (Lanas et al. 2014). The maximum power density of an MFC is proportional to the logarithm of the surface area of the electrodes (Oliveir et al. 2013):

$$\text{Power density} = -0.0369 \times \ln\left(\text{surface area}\right) + 0.3371)$$

5.1.1.2 Electrolyte

Waste effluents like food wastes, artificial wastes and industrial wastes are used as the anolyte. MFCs operating in an acidic medium are found to be more active. In the cathode side, the electrodes will be exposed to air, or an electron acceptor containing solution is taken as the catholyte. To enhance the MFC performance, sometimes phosphate buffers are added. This approach has its limitations due to economic feasibility, which adds to the operating cost when used for real-time applications (Nam et al. 2010). More recently, it was shown that CO_2 can also be used as the catholyte, as it forms carbonic acid in alkaline solution and acts as a buffer (Fornero et al. 2010).

5.1.1.3 Separators

The role of the separator is to keep the anode and cathode close without coming into contact with each other and transfer the specific ions (anion or cation). Even though separators offer internal ionic

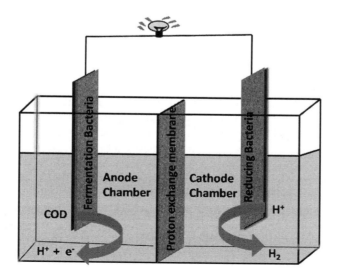

FIGURE 5.2 Schematic diagram showing a simple microbial fuel cell system.

Microbial Electrochemical Technologies for Fuel Cell Devices

resistance toward the electrolyte, a systematic study comparing MFCs operated using refinery waste with and without operators showed that the power output of a separator electrode assembly (SEA) is found to be ~30 mW/m² more than a spaced electrode assembly (SPA) and the separators offer only very low resistance (Zhang et al. 2014a). In some cases, it was found that if separators are not used, the evolved H⁺ will interfere with anode reactions and undesirable side products will be formed.

5.1.1.4 Microorganisms

Bacteria like Oscillibacter, Chitinophagaceae, and Acidobacteria are composed of organic constituents such as cellulose and chitin, and exoelectrogenic species like *Clostridium butyricum* EG3, *Arcobacter*, *Desulfobulbus*, *Desulfovibrio*, and *Geobacter* are some of the known bacteria used for MFC applications (Wang et al. 2013). The role of bacteria in energy production is very crucial. It digests the organic residues and converts them to electricity. Also, this process will remove inorganic N, P, and K constituents from water, as they are essential for the new cell formation (Ieropoulos et al. 2013). Some bacteria possess specially designed pili, which are forms of a certain type of cytochrome that transfer the electrons to the electrode surface (Kumar et al. 2016). In such cases, there is no need for mediators for the metabolically produced electrons to reach the anode surface. The cathode surface can also be modified with certain types of the bacterial consortium to accept electrons. The microorganisms will either reduce oxygen or act as redox mediators in the oxygen reduction reaction (Schröder 2012). Recently Cahoon et al. (2018) found that an extracellular electron transfer (EET) mechanism is responsible for electron transfer to metals after proceeding through a series of electron transfer phenomenon through NADH oxidation, quinone reduction in lipids, and proteins containing heme groups.

5.1.2 VARIOUS PARAMETERS USED IN MFC TREATMENT

5.1.2.1 Power

The output of MFC is measured in terms of voltage concerning time, which is then converted to current (I) using Ohm's law:

$$I = V/R$$

where V is the voltage and R is the resistance.

The required value for power output is then obtained by taking the product of current and voltage:

$$P = I \times V$$

Combining the preceding equations we get

$$P = V^2/R \tag{5.1}$$

Power is measured in W/m² (Umachagi et al. 2017).

5.1.2.2 Hydraulic Retention Time (HRT)

HRT is the duration at which the wastewater remains within the MFC cell. The unit of HRT is in hours and can be calculated using the formula (Umachagi et al. 2017)

$$\text{HRT} = \frac{V_1}{Q} \tag{5.2}$$

where
  V_1 is the volume of the cell in m³
  Q is the low rate in m³/h.

5.1.2.3 Specific Energy Consumption (μ)

Specific energy consumption (μ) is defined as the amount of current produced per unit of COD removal. Further, it measured in kWh/kg COD and can be calculated using the following equation (Yan et al. 2016):

$$\mu = \frac{E_w}{1000 M_o} \qquad (5.3)$$

where E_w is the electricity consumed for water treatment (kWh) and M_o is the mass of organic matter removed (ton).

5.1.2.4 Chemical Oxygen Demand

Chemical oxygen demand (COD) is the parameter used to measure the amount of organic matter removed from wastewater. It defines the amount of oxygen required to convert organic constituents in wastewater into CO_2 and H_2O. It is measured in mg/L (McCarty et al. 2011).

5.1.2.5 Coulombic Efficiency (CE)

Coulombic efficiency is defined as the ratio of the total coulombs of current transferred to the anode through the digestion of organic residue to the maximum coulombs that can be produced through the process. It is measured using the formula (Umachagi et al. 2017)

$$CE = \frac{C_t}{C_p} \times 100 \qquad (5.3)$$

C_t = Average current (A) × Time (s)

$$C_p = \frac{F \times n \times \Delta s(g) \times v(L)}{M}$$

where F is Faraday's constant in coulombs, n is the number of electrons exchanged per mole of oxygen, M is the molecular weight of oxygen, Δs is the change in COD concerning time, and v is the volume of electrolyte taken in the anode chamber.

5.1.2.6 Normalized Energy Recovery (NER)

The normalized energy density is a newly introduced term to quantify energy and is given by the equation (He 2017)

$$NER = \frac{\text{Power} \times \text{Time}(t)}{\text{Volume of waste water in anode compartment}}$$

$$= \frac{\text{Power}}{\text{Waste water flow rate}} \qquad (5.4)$$

The unit of NER is kWh m^{-3}.

5.2 OPPORTUNITIES AND THERMODYNAMIC/ELECTROMOTIVE FORCE CONSTRAINT OF MICROBIAL FUEL CELL TECHNOLOGIES

The overall reaction for producing electrochemical energy is thermodynamically favorable, which can be evaluated in terms of Gibbs free energy. Gibbs free energy is a measure of the maximum work that can be derived from the chemical reactions (Bard et al. 1985; Newman 1973) calculated as

$$\Delta G_r = \Delta G_r^0 + RT \ln(\Pi) \tag{5.5}$$

where ΔG_r (J mol^{-1}) is the Gibbs free energy for the specific conditions, ΔG_r^0 (J mol^{-1}) is the Gibbs free energy under standard conditions (T = 298.15 K, concentration = 1 mol L^{-1}, and P = 1 atm), R is the universal gas constant (8.31447 J mol^{-1} K^{-1}), T(K) is the temperature, and (Π) is the reaction quotient calculated as the activities of the products divided by those of the reactants, each raised to the power of its stoichiometric coefficient. The standard Gibbs free energy reaction can be calculated from the tabulated energy formation of organic compounds in water.

Nevertheless, thermodynamics in electrochemical systems are evaluated in terms of the electromotive force (E_{emf}). In the case of negative Gibbs free energy, the reaction is spontaneous and occurs without external energy input. An MFC is an energy-producing bioelectrochemical system and has negative Gibbs free energy ($E_{emf} > 0$), whereas an MEC system is a nonspontaneous reaction ($E_{emf} < 0$) and it requires an external energy input to drive the reactions. The relation between ΔG and E_{emf} can be expressed by the following equation:

$$E_{emf} = -\frac{\Delta G_r}{nF} \tag{5.6}$$

where E_{emf} is the electron motive force, n is the number of moles of electrons transferred in the reaction (per mole of reactant or product), and F is the Faraday's constant (96485 C/mol e$^-$). Furthermore, E_{emf} is calculated as the difference between the theoretical cathodic and anodic potentials. Thus, the cathodic potential is higher than the anodic potential in the case of a spontaneous reaction.

$$E_{emf} = E_{Cathode} - E_{Anode} \tag{5.7}$$

where $E_{Cathode}$ (V) and E_{Anode} (V) are the theoretical equilibrium potential of cathodic and anodic potentials at specific conditions (concentration, pH, and temperature).

5.2.1 Thermodynamics of the MFC

MFC is appropriate to evaluate the reaction by the electromotive force (E_{emf}) of overall cell reaction, as mentioned in Equation 5.7. E_{emf} is related to the work, W (joule, J), done by the fuel cell:

$$W = E_{emf} Q = -\Delta G_r \tag{5.8}$$

where

$$Q = nF \tag{5.9}$$

It represents the charge transferred in the reaction, expressed in coulombs (C), which is determined by the number of electrons exchanged in the reaction, n is number of electrons per reaction mol, and F is Faraday's constant (96485 C/mole). Merging these two equations (Equation 5.8 and Equation 5.9) yields

$$E_{emf} = -\frac{\Delta G_r}{nF} \tag{5.10}$$

If all the reactions are evaluated at standard conditions, $\Pi = 1$, then

$$E_{emf}^0 = -\frac{\Delta G_r^0}{nF} \tag{5.11}$$

where E_{emf}^0 (V) is the standard cell electromotive force.

Therefore, we can use the preceding equations to express the overall reaction in terms of potential as

$$E_{emf} = E_{emf}^0 - \frac{RT}{nF} \ln \Pi \tag{5.12}$$

This equation is positive for a favorable reaction and directly produces a value of the emf for the reaction. The Nernst equation expresses the emf of a cell in terms of activities of products and reactants taking place in the cell reaction. The Nernst equation relates the potential of an electrochemical cell (half-cell or full-cell reaction) containing a reversible system with fast kinetics, and it is valid only at equilibrium and at the surface of the electrode. These would help to evaluate the potential of the cell under nonstandard conditions (i.e., Equation 5.12) and can be used to calculate the reduction potential at various pHs, temperatures, and concentrations of product/reactant. The reaction quotient (Π) is the ratio of the activities of the products divided by the reactants raised to their respective stoichiometric coefficients:

$$\Pi = \frac{[\text{products}]^p}{[\text{reactants}]^r} \tag{5.13}$$

In the IUPAC convention, all the chemical reactions are written in the chemical reduction direction. Thus, the products are always the reduced species, and the reactants are the oxidized species (i.e., oxidized species + e$^-$ → reduced species). Likewise, as a standard condition, a temperature of 298 K and chemical concentrations of 1 M for liquids and 1 bar for gases (1 bar = 0.9869 atm = 100 kPa) by the IUPAC convention. The values of E^0 are calculated concerning that of hydrogen under standard conditions. These are defined to be $E^0(H_2) = 0$, referred to as the "normal hydrogen electrode" (NHE). Hence, the standard potentials for all the chemical reactions are obtained with $\Pi = 1$ relative to a hydrogen electrode.

The half-cell reactions at the anode and cathode of an MFC with acetate as an electron donor and the air as an electron acceptor are

Anode:

$$CH_3COOO^- + 4H_2O \rightarrow 2HCO_3^- + 9H^+ + 8e^- \qquad E^0 = 0.187 \text{ V vs SHE}$$

Cathode:

$$O_2 + 4H^+ + 4e^- \rightarrow 2H_2O \qquad E^0 = 1.229 \text{ V vs SHE}$$

In anodic theoretical reduction potential of acetate is oxidized at the specific conditions and can be calculated as follows:

$$E_{An} = E_{An}^0 - \frac{RT}{8F} \ln \frac{[CH_3COO^-]}{[HCO_3^-]^2 [H^+]^9} \tag{5.14}$$

where the standard potential of acetate, $E^0 = 0.187$ V, with the concentration of 1 g/L (16.9 mM) and under conditions of neutral pH = 7 and alkalinity set by the bicarbonate concentration of $HCO_3^- = 5$ mM,

$$E_{An} = 0.187 - \frac{\left(8.31 \frac{J}{\text{mol K}}\right)(298.15 \text{ K})}{(8)\left(9.65 \times 10^4 \frac{C}{\text{mol}}\right)} \ln \frac{[0.0169]}{[0.005]^2 [10^{-7} \text{ M}]^9}$$

$$E_{An} = -0.300 \text{ V}$$

Microbial Electrochemical Technologies for Fuel Cell Devices

The cathodic reduction potential of air cathode at the specific conditions can be calculated as

$$E_{Cat} = E_{Cat}^0 - \frac{RT}{2F} \ln \frac{1}{[O_2]^{\frac{1}{2}}[H^+]^2} \tag{5.15}$$

where the standard potential of oxygen reduction $E^0 = 1.229$ V, pH = 7, and $pO_2 = 0.2$

$$E_{Cat} = 1.229 - \frac{\left(8.31 \frac{J}{mol\,K}\right)(298.15\,K)}{(2)\left(9.65 \times 10^4 \frac{C}{mol}\right)} \ln \frac{1}{[0.2]^{\frac{1}{2}}[10^7\,M]^2}$$

$$E_{Cat} = 0.805\,V$$

However, the potential with oxygen is much less in practical application than the predicted.
Thus, for an air-cathode MFC maximum cell potential:

$$E_{MFC} = E_{Cat} - E_{An}$$

$$E_{MFC} = 0.805 - (-0.300)\,V$$

$$E_{MFC} = 1.105\,V$$

This calculated cell potential of MFC provides a maximum limit for the whole cell voltage. However, the actual potential derived from MFC would be lower (≤0.600 to 0.800 V), due to various potential losses. This potential loss is often referred to as overpotential or the difference between the potential under equilibrium conditions and the actual potential.

With the aid of thermodynamic calculations, it can identify the nature of energy losses and range of size.

5.2.2 Potential Losses

The experimental voltage in an MFC can be expressed as follows (Logan et al., 2006):

$$E_{Cell} = E_{emf} - \eta_{Cathode} - \eta_{Anode} - \eta_{Ohm} \tag{5.16}$$

where E_{Cell} (V) is the experimental voltage of an MFC; η_{Anode} (V) and $\eta_{Cathode}$ (V) are the overpotential of the anode and cathode, respectively; and η_{Ohm} (V) is the ohmic overpotential.

The overpotential of both the anode and cathode can be calculated as the difference between the experimental electrode potential and the theoretical reduction potential at specific conditions:

$$\eta_{Anode} = E_{Anode} - E_{Anode}^{theory} \tag{5.17}$$

$$\eta_{Cathode} = E_{Cathode} - E_{Cathode}^{theory} \tag{5.18}$$

where E_{Anode}^{theory} and $E_{Cathode}^{theory}$ are the theoretical electrode potential of the anode and cathode respectively.

Moreover, overpotentials are the sum of activation, concentration, and biocatalyst metabolic losses. Thus, the following can be stated:

$$\eta_{Anode} = \eta_{Act,An} + \eta_{Con,An} + \eta_{Met,Biocat} \tag{5.19}$$

$$\eta_{Cathode} = \eta_{Act,Ca} + \eta_{Con,Ca} \tag{5.20}$$

where $\eta_{Act,An}$ and $\eta_{Act,Ca}$ are the activation losses for the anode and cathode, respectively; $\eta_{Con,An}$ and $\eta_{Con,Ca}$ are the concentration losses for the anode and cathode, respectively; and $\eta_{Met,Biocat}$ is the metabolic loss of the biocatalyst.

5.2.3 Activation Losses

The activation losses represent the overpotential required to overcome the activation energy of an electrochemical reaction on a catalyst surface, i.e., the voltage loss required to initiate a reaction. It depends on the catalyst used. The better catalyst reduces the activation energy, and as a result, lowers the activation losses. The electrochemical kinetic equation of Butler-Volmer describes the relationship between the activation losses and the current intensity. The oxygen reduction reaction of the Butler-Volmer equation is

$$I = I_{O,c} \left[\left(\frac{C^*_{O_2}}{C^B_{O_2}}\right)\left(\frac{C^*_{H+}}{C^B_{H+}}\right)^4 \exp\left(\frac{-ab_{O_2}F\eta_{Act,C}}{RT}\right) - \exp\left(\frac{(1-a)b_{O_2}F\eta_{Act,C}}{RT}\right) \right] \tag{5.21}$$

where I (A) is the current intensity, I_{O_2} (A) is the current exchange intensity (background current or current intensity at zero overpotential), C^* (mol L^{-1}) stands for the concentration at the catalyst surface, C^B (mol L^{-1}) is the concentration at the bulk solution, and α is the transfer coefficient. Equally $I_{O,c}$ and α are the parameters related to the catalyst activity and should be determined for each specific case. High activation energy exhibits at the low exchange current intensity (Freguia et al. 2007). In MFC cathodes, a concentration gradient of oxygen from the bulk solution to the catalyst surface can be neglected. Hence, limitations from oxygen transport occur at very high intensities. This condition is not yet achieved in MFCs (Popat et al., 2012; Wang et al., 2001). Thus, the term $\left(\frac{C^*_{O_2}}{C^B_{O_2}}\right)$ in Equation 5.21 could be eliminated.

Biocatalysts serve as catalysts in the anode of MFCs, therefore to reduce the activation losses of the oxidation reaction, the biocatalyst could effectively transfer the electron to the electrode by the adaptation/stress conditions. Moreover, the anode reaction is governed by both the enzymatic kinetics (i.e., conversion of the substrate [organic] into the product [carbon dioxide, protons, and electrons]) and the electron transfer kinetics (i.e., direct/indirect electron transfer from biocatalyst to anode).

The Nernst-Monod model considers the anode as a final electron acceptor and is a modified version of the Monod model (Torres et al., 2008). Therefore, it describes the bio-anode kinetics as a function of the substrate concentration and the anode potential. The Nernst-Monod model is as follows:

$$I = I_{max}\left(\frac{1}{1+e^{-\frac{F}{RT}(E_{Anode}-E_{KA})}}\right) \tag{5.22}$$

where I_{max} (A) is the maximum current intensity determined by the enzymatic reaction, E_{KA} (V) and K_s (mol L^{-1}) are the anode potential and the substrate concentration at which the current intensity is half the I_{max}, respectively; and C_s is the substrate concentration (mol L^{-1}).

Microbial Electrochemical Technologies for Fuel Cell Devices

$$I = I_{\max}\left(\frac{1-e^{-\frac{F}{RT}\eta_{\text{Anode}}}}{K_1.e^{-(1-a)\frac{F}{RT}\eta_{\text{Anode}}} + K_2.e^{-\frac{F}{RT}\eta_{\text{Anode}}+\left(\frac{K_s}{C_s}+1\right)}}\right) \quad (5.23)$$

where η_{Anode} is the anodic overpotential (V), K_1 is a lumped parameter describing the ratio between the biochemical and the electrochemical reaction rates, and K_2 is a lumped parameter describing the ratio between the forward and the backward biochemical rate constants. The anodic overpotential (η_{Anode}) is used in the Butler-Volmer-Monod equation instead of the anodic activation losses ($\eta_{\text{Act,An}}$), which means that the different anodic overpotentials are considered in this model. The experiments for fitting the model were conducted by avoiding mass transfer limitations and therefore the anodic concentration losses ($\eta_{\text{Con,An}}$) could be neglected.

5.2.4 Concentration Losses

The concentration losses (η_{Con}) affect the thermodynamics of a reaction. It is caused by the reactant or product diffusion limitation between the bulk solution and the electrode surface. η_{Con} can be estimated as the difference between the theoretical reduction potential calculated at the bulk solution concentrations (E^{eq}) and at the local electrode concentrations (note by the superscript asterisk [*]).

$$\eta_{\text{Con,An}} = E_{\text{Anode}}^{eq} - E_{\text{Anode}}^* \quad (5.24)$$

$$\eta_{\text{Con,Ca}} = E_{\text{Cathode}}^{eq} - E_{\text{Cathode}}^* \quad (5.25)$$

As previously mentioned, in the cathodes of MFC concentration, overpotentials related to oxygen can be neglected. Hence, the concentration losses are associated with protons.

$$\eta_{\text{Con,Ca}} = \frac{RT}{b_{O_2}F}\ln\left(\frac{C_{H^+}^B}{C_{H^+}^*}\right)^4 \quad (5.26)$$

The maximum attainable voltage was hindered by an increase in the cathodic surface area of MFC. Further, in the anodic reaction $\eta_{\text{Con,An}}$ can be associated with either reactant (glucose or acetate) or proton (products) mass transfer limitations. In the case of proton transfer, limitations are reflected in the pH decrease in the biofilm that affects the bioanodes kinetics using thermodynamics reaction.

5.2.5 Biocatalyst Metabolic Losses

Biocatalyst metabolic energy (ATP and NADH) is generated by transporting electrons from a substrate (glucose, acetate, etc.,) to the final terminal electron acceptors (oxygen, sulfites, nitrates, etc.). The metabolic energy gain of the biocatalyst will be determined by the difference between the potential of the anode and the substrate redox potential.

5.2.6 Ohmic Losses

The resistance to the flow of ions in the electrolyte and through the separators (membrane) and the resistance to flow of electrons through electrodes (anode and cathode) and interconnections exhibit the ohmic overpotential. Ohm's law can be used to calculate this loss:

$$\eta_{\text{Ohm}} = I.R_{\text{int}}$$

where R_{int} (Ω) is the internal resistance of MFCs.

5.3 MICROBIAL CONVERSION OF ORGANIC MATERIAL INTO ELECTRICITY PRODUCTION PROTOTYPES

The following are some of the prototypes fabricated in our laboratory.

1. An efficient electrochemically active sodic–saline inoculum was enriched through electrical stress, and it exhibited superior performance in the single chamber microbial fuel cells (Sathish-Kumar et al. 2012a; Sathish-Kumar et al. 2012b). The prototype thus developed was used to power an LED and a calculator (Sathish-Kumar 2014) (Figure 5.3).
2. Our group has developed the novel vertical integrated MFC design exploitation of natural composite materials such as lignocellulose and clay (Figure 5.4a to g) (Sathish-Kumar et al. 2015a). This type of new design will diminish the cumbersome design of the MFC scale-up process by inserting the vertical integrated MFC in the existing wastewater treatment process that will accelerate the removal of DQO and meanwhile generate electrical energy to achieve the sustainable treatment. Moreover, with the use of natural materials like clay and softwood, we can replace the cost-effective commercial membranes (Nafion, etc.). Further, we have extracted the MnO_x catalyst from used alkaline batteries since we aimed at avoiding the usage of noble metals in the cathodes (Angel Rodrigo et al. 2015).

FIGURE 5.3 Electrical stress directed enriched inoculum seeded in a single-chamber MFC prototype powered LED and calculator.

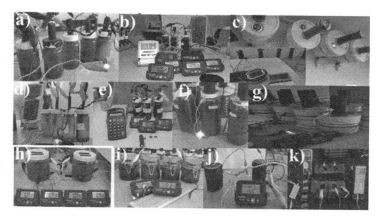

FIGURE 5.4 (a–c) Novel clay tube integrated MFCs prototypes lab scale-up (120 ml to 20 L). (d–g) Novel Softwood integrated MFC prototypes lab scale-up (120 ml to 1.5 L). (h, i) Clay-cup-modified dual-chamber MFC prototypes. (j) Clay-cup-modified air cathode MFC prototypes. (k) Brick-modified MFC prototypes.

Microbial Electrochemical Technologies for Fuel Cell Devices

The surface of the clay was modified with the aid of acrylic varnish, which will improve the performance of MFCs. This setup was then used to digest the biomass effluent from a Nopal biogas plant. The control experiment without applied varnish did not show a stable potential, supporting the idea that the acryloyl group in varnish could favor the performance. Finally, a four-digit clock was powered with the two serial connected MFCs (Figure 5.4h) (Sathish-Kumar et al. 2015b; Sathish-Kumar et al. 2019).

3. Reuse of materials is attractive for their sustainable environmental and economic benefits. At this context, we have extracted the graphite rod (GR) from a spent battery and used it as electrode materials in a dual-chamber MFC for treating wastewater. Interestingly, the surface of the graphite rod was fractured/irradiated by laser ablation to increase the superficial of the electrode, and as a result this improved the volumetric power production (2210.55 mW/m^3) 15 times higher than the control electrode. The power generated from this system was used to power the LED (Figure 5.4i) (Ochoa et al. 2017).

4. Another prototype, a clay cup modified MFC, was designed that used the MnOx from the spent battery as the air catalyst. Two of the clay cup MFCs connected in serial were able to power the digital clock (Figure 5.4j) (Bárcenas Durón et al. 2017). The brick was modified into MFC by placing the electrode on the two sides (on the internal side the anode, and on the external side the cathode). It was inoculated with the wastewater acting as the medium of treatment and inoculum. This type of MFC was easily scalable and would have potential for real-world application. We demonstrated the brick-MFCs prototype to power digital clocks (Figure 5.4k) (Sathish-Kumar et al. 2018).

5. A paper-cup-based integrated microbial fuel cell (PC-MFC) was produced for instant power production (Figure 5.5). In this, a cellulose-based paper material can act as a separator to facilitate the air cathodic reaction by a Pt/carbon cloth wrapped on the paper cup outside as the standard cell. Standard PC-MFC shows the maximum volumetric power density of 231.56 mW/m^3. Further, we applied the used blattery material as a cathode for the different catalyst loading rates of 1, 2 and 3 grams per cm^2 to avoid the usage of Pt and to prepare the cost-effective real-time electronic applications. Remarkably, the used battery material cathode of 3-PC-MFC exhibited 3 times higher volumetric power density than the Pt cathode. Moreover increasing the catalyst loading could significantly improve the power in PC-MFCs. This work demonstrates the potential realization of practical applications where we can generate instant electricity directly from the wastewater with the aid of used/dead batteries (Figure 5.4) (María Catalina et al. 2015).

6. We extended our novel vertical integrated MFCs intermediated by the plants to generate electrical energy from the growing plants. We applied different plants such as *Philodendron scandens*, green grass, and cactus plants. These would have potential applications in the

FIGURE 5.5 Biodegradable paper-cup-based MFC prototype.

green roof—to reduce the temperature of the buildings and give lighting to the surrounding and vertical garden hanging on the roadside bridges—to reduce the emission of greenhouse gases, for lighting roads, and in sustainable agriculture systems (Sathish-Kumar et al. 2017).

The maximum voltage that can be obtained using a single stack MFC is approximately 1 V, and in most of the cases, it will be reduced to half of its value and even lower due to poor bacterial metabolism, weak electrode activity, and high internal resistance. (Kim et al. 2012). Several prototypes have been used during the last decades for improving the efficiencies of MFC toward organic matter

FIGURE 5.6 Plant-intermitted MFC prototypes.

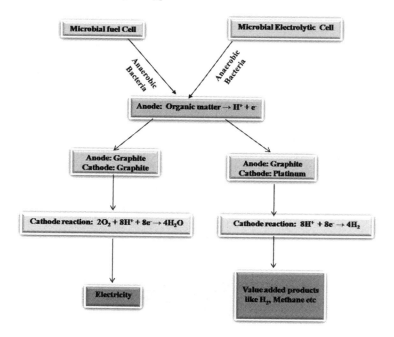

FIGURE 5.7 Pathways for current production and gaseous products in microbial fuel cell.

conversion and for the production of electricity. The main requirements for improving MFC performance are (1) increasing the surface area of the anode for bacterial inoculation, (2) identification of anaerobic microbes responsible for electron transfer, and (3) establishing favorable cell conditions for bacterial growth. One of the noteworthy innovations in MFC is the utilization of urine samples for energy production. Since urine possesses the major organic component, urea, and contains N, P, and K, it works as an ideal setup for bacterial growth, electricity generation, and water treatment. The usage of urine as the anolyte is a suitable preference for usage in remote areas and a preferable solution for the wastage of water in urinals and lavatories. A Samsung mobile phone was charged for 2 h using the power developed from this urine containing MFC setup, producing 5 min of talk time (Walter et al. 2016). Tender et al. (2008) implemented an MFC setup in the marine environment where MFC-friendly microorganisms like the Geobacteraceae family of bacteria (eliminated the use of mediators), *Clostridium* (production of acetate from organic matter), and *Desulfobulbus* or *Desulfocapsa* genera (conversion of elemental sulfur to sulfate) are present. Arrays of graphite anodes were used, and it is found that the power production was stable for 2 years and can be used for operating marine-deployed scientific instruments. MFCs have been executed in other environments like river sediments and soil substrates, typically near the roots of plants (Ewing et al. 2014; Kouzuma et al. 2014).

Cha et al. (2010) combined the single-chambered MFC technique with the active sludge method and found that the cell performance increased linearly with the amount of oxygen purging. This particular MFC prototype functioning using graphite felt anode had shown a power density of 16.7 Wm^{-3} and minimal internal resistance of 17 Ω. A comparison of the power outputs using the various techniques discussed above is given in Table 5.1.

5.4 MICROBIAL CONVERSION OF ORGANIC MATTERS INTO FUEL

A microbial electrolysis cell (MEC) is a modified form of MFCs where the main motive is the conversion of organic matter into value-added products like hydrogen, methane, acetate, and hydrogen peroxide from which electricity can be produced indirectly (Kumar et al. 2017). The difference between a microbial fuel cell and a microbial electrolytic cell is depicted graphically in Figure 5.7. Of these value-added products, hydrogen evolution is more desirable as it can be directly used for electricity production. Usually, in MFCs, the protons produced due to anode reaction are reduced in the cathodic compartment to produce hydrogen gas. For the reduction of protons, platinum is found to be highly active and is generally used as the cathode. Several challenges are there regarding the production of hydrogen from such a setup. The methanogenic bacteria compete with the electroactive bacteria that are present within the wastewater. The methanogenic bacteria consume the produced hydrogen from the cathode and convert them into methane. Loss of hydrogen yield will also happen when no membrane is used during the cell operation. The reduced hydrogen gas reaches the anode, and again, it will be converted to protons (Rozendal et al. 2008; Patil et al. 2015). Moreover, various studies show that activated carbon is a suitable alternative for the high-cost cathode material, Pt (Ge et al. 2015).

Another form of an MFC, a microbial desalination cell (MDC), works with the aim of desalinating water to emanate hydrogen gas (Aghababaie et al. 2015). Interestingly, high-temperature operation of the MECs (at 30°C) using Pt-decorated carbon cloth enhances the evolution of hydrogen (Cusick et al. 2011). Heat shock is applied to retard the growth of methanogenesis that restricts the evolution of hydrogen. The heat treatment was between 75°C and 121°C, and some of the hydrogen evolution-friendly bacteria are *Bacillus* sp., *Clostridium* sp., and *Desulfitobacterium hafniense* (Butti et al. 2016).

5.5 MICROBIAL ELECTROCHEMICAL TECHNOLOGY LIMITATIONS

As previously stated, MFCs struggle to meet the essential energy requirement due to increasing demand as a result of increased usage of electronic devices. It is successful in providing power to operate small devices, which requires low power values. Some of the constraints that will retard

TABLE 5.1
Comparison of Power Outputs Using Various Techniques

	Method of Operation	Cathode	Anode	Separator	Reference	Type of Wastewater Used	Maximum Power Output (mW/m²)
1	Dual cathode mode	Graphite fiber	Activated carbon on stainless steel	Plastic, polycarbonate, painted steel wire	He et al. (2015)	Raw domestic wastewater	1100 (steel)
2	Multiple unit cells connected in series	Graphite felt, (A)	Platinum	Nafion N424	Kim et al. (2012)	Brewery wastewater	1184.80 (unshared) 282.63 (shared anolyte)
3	MFC in series	Carbon cloth	Platinum grid	—	Oliot et al. (2017)	Acetate	0.9
4	SEA, SPA	Graphite fiber	Platinum	Textile separator (46% cellulose and 54% polyester)	Ahn et al. (2014)	Domestic wastewater	328 (SEA) 282 (SPA)
5	Fed-batch method	Non-wet-proofed carbon cloth	Pt on wet-proofed carbon cloth	—	Nam et al. (2010)	Activated sludge	410
6	Single-chamber MFC	NH₃-treated graphite fiber	Pt on carbon cloth	—	Cheng et al. (2011)	Acetate	1260
7	Stacked microbial electrodeionization cell	Graphite fiber	Pt with Nafion	5 CEM and 6 AEM	Shehab et al. (2014)	Brackish water, seawater	650
8	PF-MFC	Carbon fiber stack	Pt/C	—	Karra et al. (2013)	University wastewater	281.74
	CM-MFC	Granular activated carbon					239.56
9	Single-chamber MFC	Graphite	Graphite	Nafion 117	Goud et al. (2013)	Synthetic wastewater	309
10	Single-chamber MFC	Graphite	Pt/C	Nafion 117	Liu et al. (2004)	Domestic wastewater	26

SEA, separator electrode assembly; SPA, closely spaced electrode assembly; CEM, cation exchange membrane; AEM, anion exchange membrane; PF-MFC, plug flow MFC; CM-MFC, complete mixing MFC.

better implementation of MFCs for practical applications are microorganisms, electrode materials, and separator cost. How it affects the overall performance and feasibility of this technique is in the following sections.

5.5.1 BIOCHEMICAL CONSTRAINTS

The efficiency of an MFC is determined in terms of COD removal and current production. There are situations where the COD removal is found to be superior while the current efficiency is poor. The reason is that in real samples, various types of bacteria are present in which the nonelectrogenic bacteria process the organic matters and supply the electrons for the metabolic activities of microorganisms (Karra et al. 2013). Some of the exogenous compounds used are potassium ferric cyanide, thionine, and neutral red, which produce a toxic environment to bacterial growth (Adebule et al. 2018), whereas bacteria *Shewanella putrefaciens*, *Geobacter sulfurreducens*, *Geobacter metallireducens* and *Rhodoferax ferrireducens* act as self mediators and improve the efficiency of MFCs (Min et al. 2004).

It is advisable to keep the pH stable without fluctuation, and one such effort is through the addition of buffers. These will favor better growth of microorganisms. In real samples, it is difficult to buffer the wastewater, and the growth of microorganisms is thus constrained (Nam et al. 2010).

One of the known hydrogen-producing bacteria, acidophilic bacteria, prefers to inoculate at acidic pH (~6). Simultaneously, the acidic pH condition inhibits the growth of methanogenic bacteria, thus improving efficiency (Mohan et al. 2008). Since it is well known that the bacteria will produce current only under anaerobic conditions, the usage of substrates containing dissolved oxygen is restricted (Min et al. 2004). It is also likely possible that dissolved oxygen will diffuse from the cathode side to the anode compartment where bacteria consume it and reduce the efficiency (Ahn et al. 2014). However, genetic engineering of bacteria is sure to increase productivity (Choudery et al. 2014). Still, the problem of additional labor requirement and the cost for programming bacterial genes will be a concern. Another constraint regarding bacterial growth is the temperature. Bacteria require optimum temperature for its inoculation. Usually, bacteria requires an optimum temperature for carrying out anodic microbial activity. Michie et al. (2010) showed that only psychotolerant bacteria can be used for MFCs operating at 10°C and mesophilic bacteria can alone be used at 35°C.

5.5.2 Functional Materials Constraints

In MFCs, the electrodes are predominantly made of carbon-based materials such as carbon felt, (Wang et al. 2013), carbon cloth, and activated carbon (Karra et al. 2013; Zhang et al. 2014b). Regarding the activity of MFCs, the voltage loss occurs due to the emergence of internal resistance, which is associated with solution resistance and resistance offered by electrode materials (Karra et al. 2013; Wang et al. 2013). This ohmic loss can be compensated by designing morphologically significant electrode materials and through regulating the flux of substrates near the electrode surface (Karra et al. 2013). So far, two types of MFC configurations were followed: (1) connecting MFCs in series and (2) insertion of multiple electrodes in a single unit cell (Kim et al. 2012). Considering the whole MFC configuration, the voltage drop will predominantly occur in the cathode side. A cathode with low surface area cannot compensate with the electron flow from the anode.

To accelerate the electron accepting phenomena, either the cathode material should possess higher surface area than the anode, or the cathode should be modified with a kinetically favorable high-electron-accepting bacteria. Usually, graphene felts are used as cathodes and anodes in the manufacture of MFCs as they provide high surface area and a constructive environment for bacterial inoculation. Another important component of MFC is the proton exchange membrane, which is usually made of Nafion. One of the drawbacks of Nafion is that it will peel off easily when the electrolyte flows in high pressure. Similarly, the salinity of the anolyte is also known to affect bacterial growth due to the diffusion of Cl$^-$ ions from the cathode to neutralize the charge (Shehab et al. 2014).

5.5.3 Economic Constraints

While using MFCs for electricity production, the setup is simple and cost-effective since graphite rods are used as anodes and cathodes. But in case of the microbial electrolytic cell, for the reduction of protons to form hydrogen, the cathode should be made of platinum. It will increase the cost of the whole MFC setup. Another scenario is the large-scale production of electricity or fuel using real-time wastewater. Most of the prototypes proposed are in the lab scale, and when trying to explore for practical use, they suffer from economic feasibility and technical issues. Generally, in lab-scale studies, the electrode area taken is 35–40 cm^2, but to be applied in the practical scenario it requires hundreds of meters. Usage of platinum for such large areas is highly impossible in terms of cost.

Another constraint is the use of selective loading of genetically modified bacteria, which requires an additional process that requires huge time and expense (Choudery et al. 2017). Activated carbon, carbon fiber, and cobalt-based materials are alternatives for platinum to carry out reduction reaction but the cost is still high for activated carbon and cobalt is a toxic material (He et al. 2015; Logan et al. 2014; Ghasemi et al. 2011). Also, since the resistivity of carbon fiber increases with surface area, it cannot confer the efficiency provided by platinum (Ieropoulos et al. 2010). The pH of the solution between 6 to 9 is also known to have a huge impact on the growth of microorganisms typically and hence buffering is a proven way to improve the MFC performance. But buffering such a large pool of wastewater is not cost effective and not a feasible method for real applications. In certain prototypes, to enhance bioelectricity, air should be purged in the cathode compartment, which will further increase the cost of operating MFCs (Cha et al. 2010).

The maintenance cost of electric generators is further costly, and the evolved biogas from the cathode side requires a purification step for the removal of hydrogen sulfide. The energy losses happening due to microbial and functional material constraints also contribute to the high cost of the setup (Ge et al. 2013). When connected in series, separate batteries should be used for each MFC and that too adds to the cost (Kim et al. 2012). The prominent constraint of practical applications is that the domestic water possesses poor electronic conductivity, which is less than 5 mS cm^{-1}, and the neutral pH provides constraint for microorganisms toward efficiency (Stoll et al. 2016).

5.6 FUTURE DIRECTION OF MICROBIAL ELECTROCHEMICAL TECHNOLOGIES

Future development of microbial electrochemical technologies should rely on ideas that can scale up MFCs for production. Serially connected MFCs are expected to satisfy this requirement, but in practice, the theoretical output cannot be obtained due to a phenomenon called ion-short circuit within the electrolyte (Oliota et al. 2017). People have come up with the microbe sprinkle technique to enhance COD removal, and after that the performance was found to be better. The COD removal, which is directly related to power output, can be achieved by improving the treatment efficiency and proper cell configuration, which in turn reduces the cost of MFC operation. Concerning practical use, MFCs are by far the best method of obtaining the essential energy output. In short, design of electrode materials possessing essential characteristics like high surface area, a nontoxic environment for bacterial growth, good conductivity, and a mechanically strong membrane separator that will allow only proton transfer and the better electrode alignment that will prevent ohmic losses are the primary elements where alterations can be made for enhancing the operation of an MFC. The voltage loss in MFCs connected in series is due to ionic short-circuiting, which can be decreased by keeping the anodes and cathodes closer and keeping MFCs apart. In these ways, it is possible to obtain at least 80% of the theoretical voltage output value (Kim et al. 2012). Even though MFCs lag in producing desirable power output, it is growing in terms of technical aspects. The growing technical aspects and the research for practical implementations of MFCs show that this field is expected to be improved in the coming years in terms of excellent power output without fluctuation and low capital investment with cheap raw materials for cell fabrication.

5.7 CONCLUSIONS

This chapter explains the whole aspect of MFCs from the cell design to the recent advancements in the field. MFCs are a promising technology for energy production since it can be operated without any external voltage. The focus on the design of suitable prototypes is the primary requirement for the enhancement of productivity. Apart from this, there are still voids in the understanding of bacterial activity especially regarding their metabolism. Thus, the collaborative analysis of biological and electrochemical influences will definitely help in the prediction of MFC output. Another issue to be noted is that current technologies are in the laboratory scale and practical use is far beyond the required level, which requires a lot of understanding regarding basic chemistry and finding the limitations offered by substrates, electrode materials, separators and cell design to tune up the voltage/power output. Moreover, the data about the various constraints related to microbial activity, functional materials, and the economy will help researchers to resolve the issues. In the practical sense, this can be achieved through the modification of the electrode surface and inoculation of specific bacterial types. In this way, we hope this chapter provides a review of core concepts in condensed form that will support further innovation.

REFERENCES

Adebule, A. P., Aderiye, B. I., and Adebayo, A. A. 2018. "Improving bioelectricity generation of microbial fuel cell (mfc) with mediators using kitchen waste as substrate". *Annals of Applied Microbiology and Biotechnology Journal* no. 2:1–5.

Aghababaie, Marzieh, Mehrdad Farhadian, Azam Jeihanipour, and David Biria. 2015. "Effective factors on the performance of microbial fuel cells in wastewater treatment – A review". *Environmental Technology Reviews* no. 4:71–89.

Ahn, Yongtae, Marta C. Hatzell, Fang Zhang, and Bruce E. Logan. 2014. "Different electrode configurations to optimize performance of multielectrode microbial fuel cells for generating power or treating domestic wastewater". *Journal of Power Sources* no. 249:440–445.

Bard, A. J., R. Parsons, and J. Jordan, Eds. 1985. *Standard Potentials in Aqueous Solution*; Marcel Dekker: New York.

Butti, Sai Kishore, G. Velvizhi, Mira L. K. Sulonen, Johanna M. Haavisto, Emre Oguz Koroglu, Afsin Yusuf Cetinkaya, Surya Singh, Divyanshu Arya, J. Annie Modestra, K. Vamsi Krishna, Anil Verma, Bestami Ozkaya, Aino-Maija Lakaniemi, Jaakko A. Puhakka, and S. Venkata Mohan. 2016. "Microbial electrochemical technologies with the perspective of harnessing bioenergy: maneuvering towards upscaling". *Renewable and Sustainable Energy Reviews* no. 53:462–476.

Cahoon, Latya, and Nancy E. Freitag. 2018. "The electrifying energy of gut microbes". *Nature* no. 562:1–2.

Catalina Vega Reyes, María, Angel Rodrigo M. Ochoa, K. Sathish-Kumar, O. Solorza-Feria, J. Tapia-Ramírez. 2015. "Instant disposable microbial-battery from paper cup for electronic applications". *XV International Congress of the Mexican Hydrogen Society* (Abstract-Pg No. 106; Chapter 5.6-Pg No. 384–393).

Cha, Jaehwan, Soojung Choi, Hana Yu, and Hyosoo Kim, Changwon Kim. 2010. "Directly applicable microbial fuel cells in aeration tank for wastewater treatment". *Bioelectrochemistry* no. 78:72–79.

Cheng, Shaoan, Defeng Xing, and Bruce E. Logan. 2011. "Electricity generation of single-chamber microbial fuel cells at low temperatures". *Biosensors and Bioelectronics* no. 26:1913–1917.

Choudhury, Payel, Uma Shankar Prasad Uday, Tarun Kanti Bandyopadhyay, Rup Narayan Ray, and Biswanath Bhunia. 2017. "Performance improvement of microbial fuel cell (mfc) using suitable electrode and bioengineered organisms: a review". *Bioengineered* no. 8:471–487.

Cusick, Roland, D., Bill Bryan, Denny S. Parker, Matthew D. Merrill, Maha Mehanna, Patrick D. Kiely, Guangli Liu and Bruce E. Logan. 2011. "Performance of a pilot-scale continuous flow microbial electrolysis cell fed winery wastewater". *Applied Microbiology Biotechnology* no. 89, 2053–2063.

Durón, G. J. Bárcenas, K. Gunaseelan, S. Gajalakshmi, F. Caballero Briones, K. Sathish-Kumar, and F. Aguirre Sámano. 2017. "Surface modified clay based air cathode microbial fuel cells for wastewater treatment", *European Fuel Cell Technology & Applications Conference - Piero Lunghi Conference, (EFC17024)*, 53–54. ISBN 978-88-8286-356-2.

Ewing, Timothy, Phuc Thi Ha, Jerome T. Babauta, Nghia Trong Tang, Deukhyoun Heo, and Haluk Beyenal. 2014. "Scale-up of sediment microbial fuel cells". *Journal of Power Sources* no. 272:311–319.

Fornero, Jeffrey, J., Miriam Rosenbaum, Michael A. Cotta, and Largus T. Angenet. 2010. "Carbon di oxide addition to microbial fuel cell cathodes maintains sustainable catholyte ph and improves anolyte ph, alkalinity, and conductivity". *Environmental Science and Technology* no. 44:2728–2734.

Freguia, Stefano, Korneel Rabaey, Zhiguo Yuan, and Jurg Keller, J. 2007. "Non-catalyzed cathodic oxygen reduction at graphite granules in microbial fuel cells". *Electrochimica Acta* no. 53 (2):598–603.

Ge, Zheng, Fei Zhang, Julien Grimaud, Jim Hurst, and Zhen He. 2013. "Long-term investigation of microbial fuel cells treating primary sludge or digested sludge". *Bioresource Technology* no. 136:509–514.

Ge, Zheng, and Zhen He. 2015. "An effective dipping method for coating activated carbon catalyst on the cathode electrodes of microbial fuel cells". *RSC Advances* no. 5:36933–36937.

Ghasemi, Mostafa, Samaneh Shahgaldi, Manal Ismail, Byung Hong Kim, Zahira Yaakob, and Wan Ramli Wan Daud. 2011. "Activated carbon nanofibers as an alternative cathode catalyst to platinum in a two-chamber microbial fuel cell". *International Journal of Hydrogen Energy* no. 36:13746–13752.

Goud, R., Kannaiah, and S. Venkata Mohan. 2013. "Prolonged applied potential to anode facilitate selective enrichment of bio-electrochemically active proteobacteria for mediating electron transfer: microbial dynamics and bio-catalytic analysis". *Bioresource Technology* no. 137:160–170.

He, Weihua, Xiaoyuan Zhang, Jia Liu, Xiuping Zhu, Yujie Feng, and Bruce E. Logan. 2015. "Microbial fuel cells with an integrated spacer and separate anode and cathode modules". *Environmental Science Water Research & Technology* no. 2:186–195.

He, Zhen. 2017. "Development of microbial fuel cells needs to go beyond power density". *ACS Energy Letters* no. 2:700–702.

Ieropoulos, I., I. Gajda, J. You, and J. Greenman. 2013. "Urine–waste or resource?". *The Economic and Social Aspects* no. 2:192–199.

Ieropoulos, I., J. Greenman, and C. Melhuish. 2010. "Improved energy output levels from small-scale microbial fuel cells". *Bioelectrochemistry* no. 78:44–50.

Inamuddin, Beenish, A. A. Ahmed, and M. Naushad. 2016. "Electrochemical study of single wall carbon nanotubes/graphene/ferritin composite for biofuel cell applications". *Russian Journal of Electrochemistry* no. 52, 245–250. doi:10.1134/S1023193516030058

Karra, Udayarka, Elizabeth Troop, Michael Curtis, Karl Scheible, Christopher Tenaglier, Nirav Patel, and Baikun Li. 2013. "Performance of plug flow microbial fuel cell (PF-MFC) and complete mixing microbial fuel cell (CM-MFC) for wastewater treatment and power generation". *International Journal of Hydrogen Energy* no. 38:5383–5388.

Karra, Udayarka, Seetha S. Manickam, Jeffrey R. McCutcheon, Nirav Patel, and Baikun Li. 2013. "Power generation and organics removal from wastewater using activated carbon nanofiber (ACNF) microbial fuel cells (MFCs)". *International Journal of Hydrogen Energy* no. 38:1588–1597.

Kim, Daehee, Junyeong An, Bongkyu Kim, Jae Kyung Jang, Byung Hong Kim, and In Seop Chang. 2012. "Scaling-up microbial fuel cells: Configuration and potential drop phenomenon at series connection of unit cells in shared anolyte". *ChemSusChem* no. 5:1086–1091.

Kouzuma, Atsushi, Nobuo Kaku, and Kazuya Watanabe. 2014. "Microbial electricity generation in rice paddy fields: recent advances and perspectives in rhizosphere microbial fuel cells". *Applied Microbiology and Biotechnology* no. 98:9521–9526.

Kumar, Gopalakrishnan, Rijuta Ganesh Saratale, Abudukeremu Kadier, Periyasamy Sivagurunathan, Guangyin Zhen, Sang-Hyoun Kim, and Ganesh Dattatraya Saratale. 2017. "A review on bioelectrochemical systems (BES) for the syngas and value added biochemicals production". *Chemosphere* no. 177:84–92.

Kumar, Ravinder, Lakhveer Singh, and A. W. Zularisam. 2016. "Exoelectrogens: recent advances in molecular drivers involved in extracellular electron transfer and strategies used to improve it for microbial fuel cell applications". *Renewable and Sustainable Energy Reviews* no. 56:1322–1336.

Lanas, Vanessa, Yongtae Ahn and Bruce E. Logan. 2014. "Effects of carbon brush anode size and loading on microbial fuel cell performance in batch and continuous mode". *Journal of Power Sources*, no. 247:228–234.

Liu, Hong, Ramanathan Ramanarayanan, and Bruce E. Logan. 2004. "Production of electricity during wastewater treatment using a single chamber microbial fuel cell". *Environmental Science and Technology* no. 38:2281–2285.

Logan, Bruce, E., Bert Hamelers, Rene Rozendal, Uwe Schröder, Jurg Keller, Stefano Freguia, and Korneel Rabaey, K. 2006. "Microbial fuel cells: methodology and technology". *Environmental Science & Technology* no. 40 (17):5181–5192.

McCarty, Perry, L., Jaeho Bae, and Jeonghwan Kim. 2011. "Domestic wastewater treatment as a net energy producer can this be achieved?". *Environmental Science and Technology* no. 45:7100–7106.

Melhado, J. Tyce, R. C., Flynn, D., Petrecca, R., and Dobarro, J. 2008. "The first demonstration of a microbial fuel cell as a viable power supply: powering a meteorological buoy". *Journal of Power Sources* no. 179:571–575.

Michie, Iain S., Jung Rae Kim, Richard M. Dinsdale, Alan J. Guwy, and Giuliano C., Premier. 2011. "The influence of psychrophilic and mesophilic start-up temperature on microbial fuel cell system performance". *Energy & Environmental Science* no. 4:1011–1019.

Min, Booki, and Bruce E. Logan. 2004. "Continuous electricity generation from domestic wastewater and organic substrates in a flat plate microbial fuel cell". *Environmental Science and Technology* no. 38:5809–5814.

Mohan, Venkat, S., R. Saravanan, S. Veer Raghavulu, G. Mohanakrishna, and P. N. Sarma. 2008. "Bioelectricity production from wastewater treatment in dual chambered microbial fuel cell (MFC) using selectively enriched mixed microflora: effect of catholyte". *Bioresource Technology* no. 99:596–603.

Moqsud, M. Azizul, Kiyoshi Omine, Noriyuki Yasufuku, Masayuki Hyodo, and Yukio Nakata. 2013. "Microbial fuel cell (MFC) for bioelectricity generation from organic wastes". *Waste Management* no. 33:2465–2469.

Nam, Joo-Youn, Hyun-Woo Kim, Kyeong-Ho Lim, Hang-Sik Shin, and Bruce E. Logan. 2010. "Variation of power generation at different buffer types and conductivities in single chamber microbial fuel cells". *Biosensors and Bioelectronics* no. 25:1155–1159.

Newman, J. S. 1973. *Electrochemical Systems*; Prentice Hall: Englewood Cliffs, NJ.

Ochoa, Angel Rodrigo Montes, K. Sathish-Kumar, O. Solarza-Feria, Manuel Sanchez Cardenas, Fernando Trejo Zarraga, and J. Tapia-Ramírez. 2015. "European fuel cell technology and applications conference". *Proceeding of the 6th European fuel cell Piero Lunghi Conference, facile microbial fuel cell from pencil utilizing sustainable cathode as of used battery.* (EFC15154), 185–186 (Poster) ISBN 978-88-8286-324-1.

Oliota, Manon, Luc Etcheverry, Renaut Mosdale, and Alain Bergel. 2017. "Microbial fuel cells connected in series in a common electrolyte underperform: understanding why and in what context such a set-up can be applied". *Electrochimica Acta* no. 246:879–889.

Oliveira, V. B., M. Simões, L. F. Melo, and A. M. F. R. Pinto. 2013. "Overview on the developments of microbial fuel cells", *Biochemical Engineering Journal* no. 73:53–64.

Patil, Sunil A., Sylvia Gildemyn, Deepak Pant, Karsten Zengler, Bruce E. Logan, and Korneel Rabaey. 2015. "A logical data representation framework for electricity-driven bioproduction processes". *Biotechnology Advances* no. 33:736–744.

Popat, Sudeep, C., DongwonKi, Bruce E. Rittmann, and Cesar Torres. 2012. "Importance of OH– transport from cathodes in microbial fuel cells". *ChemSusChem* no. 5 (6):1071–1079.

Ringeisen, B. R., Emily Henderson, Peter K. Wu, Jeremy Pietron, Ricky Ray, Brenda Little, Justin C. Biffinger, and Joanne M. Jones- Meeehan. 2006. "High power density from a miniature microbial fuel cell using *Shewanella oneidensis* DSP10". *Environmental Science and Technology* no. 40:2629–2634.

Rozendal, Rene´ A., Hubertus V. M. Hamelers, Korneel Rabaey, Jurg Keller, and Cees J. N. Buisman. 2008. "Towards practical implementation of bioelectrochemical wastewater treatment". *Trends in Biotechnology* no. 26:450–459.

Santiago, Oscar, Emilio Navarro, Miguel A. Raso, and Teresa J. Leo. 2016. "Review of implantable and external abiotically catalysed glucose fuel cells and the differences between their membranes and catalysts". *Applied Energy* no. 179:497–522.

Sathish Kumar, Kamaraj, O. Solorza-Feria, G. Vázquez-Huerta, J. P. Luna-Arias, and H. M. Poggi-Varaldo. 2012a. "Electrical stress-directed evolution of biocatalysts community sampled from a sodic-saline soil for microbial fuel cells". *Journal of New Materials for Electrochemical Systems* no. 15:181–186.

Sathish-Kumar, Kamaraj, O. Solorza-Feria, R. Hernández-Vera, G. Vazquez-Huerta, and H. M. Poggi-Varaldo. 2012b. "Comparison of various techniques to characterize a single chamber microbial fuel cell loaded with sulfate reducing biocatalysts". *Journal of New Materials for Electrochemical Systems* no. 15:195–201.

Sathish-Kumar, Kamaraj. 2014. "Decorated nano carbon electrodes for current generation from the sodic-saline microbial community for wastewater treatment process in microbial fuel cell". Doctoral thesis, CINVESTAV, Mexico.

Sathish-Kumar, Kamaraj, Angel Rontrigo Montes Ochoa, O. Solorza Feria and Jose Isabel Tapia Ramirez. 2015a. "Microbial fuel cells with a wood electrode with graphite core for wastewater treatment applications". Mexican patent: MX/a/2015/001573.

Sathish-Kumar, Kamaraj, Alejandro Esqueda Rivera, Selvasankar Murugesan, Jaime García-Mena, Claudio Frausto Reyes, and José Tapia-Ramírez. 2015b. "Electricity generation from nopal biogas waste biomass using clay cup (cantarito) modified microbial fuel cell", *European Fuel Cell Technology & Applications Conference, Proceeding of the 6th European fuel cell Piero Lunghi Conference.* (EFC15153), 183–184, (Oral) ISBN 978-88-8286-324-1.

Sathish-Kumar, Kamaraj, Venkatasamy Vignesh, and Felipe Caballerero-Briones. 2017. "Sustainable power production from plant-mediated microbial fuel cells", in A. Dhanarajan, Ed., *Sustainable Agriculture towards Food Security*, Springer Nature Publisher Book (ISBN 978-981-10-6646-7).

Sathish-Kumar, Kamaraj, Alejandro Esqueda Rivera, Selvasankar Murugesan, Otoniel Maya, Jaime García-Mena, Claudio Frausto-Reyes, José Tapia-Ramírez, Hector Silos Espino, and Felibe Caballero. 2019. "Electricity generation from nopal biogas effluent using a surface modified clay cup (cantarito) microbial fuel cell", *Heliyon* no. 5:e01506.

Sathish-Kumar Kamaraj, Felipe Caballero Briones, Fabio Felipe Chale Lara, Sandra Edith Benito Santiago, Cesia Guarneros Aguilar, Pedro Nava Diguero and Elia Esther Hoz Zavala. 2018. "Scalable electrochemical device to generate electrical energy and treat wastewater based on solidclay support". Mexican patent: MX/a/2018/012847.

Schröder, Uwe. 2012. "Microbial fuel cells and microbial electrochemistry: into the next century!", *ChemSusChem* no. 5:959–961.

Shehab, Noura, A., Gary L. Amy, Bruce E. Logan, and Pascal E. Saikaly. 2014. "Enhanced water desalination efficiency in an air-cathode stacked microbial electrode ionization cell (SMEDIC)". *Journal of Membrane Science* no. 469:364–370.

Stoll, Zachary A., Zhaokun Ma, Christopher B. Trivedi, John R. Spear, and Pei Xu. 2016. "Sacrificing power for more cost-effective treatment: a techno-economic approach for engineering microbial fuel cells". *Chemosphere* no. 161:10–18.

Tender, Leonard, M., Sam A. Gray, Ethan Groveman, Daniel A. Lowy, Peter Kauffmand, Julio Melhado, Robert C. Tyce, Darren Flynn, Rose Petrecca, and Joe Dobarro. 2008. "The first demonstration of a microbial fuel cell as a viable power supply: powering a meteorological buoy". *Journal of Power Sources* no. 179:571–575.

Torres, C. I., Marcus, A. K., Parameswaran, P., and Rittmann, B. E. 2008. "Kinetic experiments for evaluating the nernst–monod model for anode-respiring bacteria (ARB) in a biofilm anode". *Environmental Science & Technology* no. 42 (17):6593–6597.

Umachagi, Sunil, and G. M. Hiremath. 2017. "Impact of electronic configurations on hydraulic retention time (hrt) in treatment of sugar mill waste water using microbial fuel cell". *International Journal of Innovative Research in Science Engineering and Technology* no. 6:13539–13549.

Walter, Xavier Alexis, Andrew Stinchcombe, John Greenman, and Ioannis Ieropoulos, I. 2016. "Urine transduction to usable energy: a modular mfc approach for smartphone and remote system charging". *Applied Energy* no. 192:575–581.

Wang, Yun, Ken S. Chen, Jeffrey Mishler, Sung Chan Cho, and Xavier Cordobes Adroher. 2011. "A review of polymer electrolyte fuel cells: technology, applications, and needs on fundamental research". *Applied Energy* no. 88:981–1007.

Wang, Zhiwei, Jinxing Ma, Yinlun Xu, Hongguang Yu, and Zhichao Wu. 2013. "Power production from different types of sewage sludge using microbial fuel cells: a comparative study with energetic and microbiological perspectives". *Journal of Power Sources* no. 235:280–288.

Wang, Z. H., C. Y. Wang, and K. S. Chen. 2001. "Two-phase flow and transport in the air cathode of proton exchange membrane fuel cells". *Journal of Power Sources* no. 94 (1):40–50.

Yan, Peng, Rongcong Qin, Jinsong Guo, Qiang Yu, Zhe Li, You-Peng Chen, Yu Shen, and Fang Fang. 2016. "A net-zero energy model for sustainable wastewater treatment". *Environmental Science and Technology* no. 51:1017–1023.

Zhang, Fang, Yongtae Ahn, and Bruce E. Logan. 2014. "Treating refinery wastewaters in microbial fuel cells using separator electrode assembly or spaced electrode configurations". *Bioresource Technology* no. 152:46–52.

Zhang, Xiaoyuan, Xue Xia, Ivan Ivanov, Xia Huang, and Bruce E. Logan. 2014. "Enhanced activated carbon cathode performance for microbial fuel cell by blending carbon black". *Environmental Science and Technology* no. 48:2075–2081.

6 Photoelectrochemical Process for Hydrogen Production

S. Devi and V. Tharmaraj

CONTENTS

6.1	Introduction	105
6.2	Hydrogen (H$_2$) and Solar Energy	106
6.3	Photoelectrochemical Process	106
	6.3.1 Experimental Setup	107
	6.3.2 General Mechanism of PEC	107
	6.3.3 Calculations in PEC Process	108
6.4	Metal Oxide as Photoelectrode in PEC Water Splitting	109
	6.4.1 Titanium-Dioxide-Based Electrodes	109
	6.4.2 Zinc-Oxide-Based Electrodes	109
	6.4.3 Iron-Oxide-Based Electrodes	110
	6.4.4 Copper-Oxide-Based Electrodes	110
6.5	Nonmetal Oxide as Photoelectrode in PEC Water Splitting	110
	6.5.1 Nitride in PEC	110
	6.5.2 Phosphide in PEC	112
	6.5.3 Oxynitride in PEC	113
	6.5.4 Sulfide in PEC	113
6.6	Quantum-Dot-Based Photoelectrode	113
6.7	Hydrogen Production from Biomass and Water	113
6.8	Conclusions	116
References		116

6.1 INTRODUCTION

Recently, increases in both population and standards of living have caused an increase in energy demand worldwide. The demand for energy has been increasing continuously in highly populated countries (e.g., India and China). Around 85% of the energy has been supplied by fossil fuels for transportation and electricity generation. However, the use of fossil fuels for energy leads to environmental pollution due to the emission of greenhouse gases (CO$_2$) during their combustion that is accountable for global warming. Fossil fuels are also considered as a nonrenewable energy source because of their slow formation, nonhomogeneous distribution, rapid depletion, and limited nature (Fulcheri and Schwob 1995). Hence, it is essential to explore an alternative clean energy system for the sustainable development of the environment. The clean energy system is sustainable when it gives (1) better energy security, (2) better design and analysis, (3) better resource utilization, (4) better cost efficiency, (5) better effectiveness, and (6) a non-polluted environment.

6.2 HYDROGEN (H$_2$) AND SOLAR ENERGY

In recent years, the development of clean energy systems has attracted great attention for solving the increasing energy demand without polluting the environment and depleting energy sources that already exist in nature. Solar energy is considered a suitable resource for supplying the energy needed for the future that could replace the use of fossil fuels. For example, direct sunlight irradiation of earth surface for 30 min can produce the worldwide consumption of energy for one year. Solar energy is inexhaustible in nature and also the end product is nonpolluting to the environment. However, the quantity of sunlight reaching the Earth's surface can be influenced by the intermittent and fluctuating (day/night) character of solar energy. Hence, a medium is needed to store solar energy for its continuous supply. Conversion of solar energy into easily storable chemical fuel has recently attracted considerable attention as a strategic technology to provide clean energy. The solar energy can be converted into a chemical fuel, namely hydrogen, through the dissociation of water or renewable biomass thus providing a green environment (Turner 2004). Among the reported chemical elements, the lightest and plentiful element in the universe is H$_2$. It mainly exists with other elements, for example, it is combined with oxygen in water and also combined with nitrogen, carbon, and oxygen in living things. In a fuel cell, H$_2$ can be converted into electricity by producing water as a byproduct. H$_2$ can also produce liquid fuels by the reduction of carbon dioxide. The methods involving the production of H$_2$ are classified based on solar energy (Figure 6.1) (Saraswat, Rodene, and Gupta 2018; Dincer and Acar 2015; Veras et al. 2016). Most of the methods are not well recognized and also convert solar energy to H$_2$ with low efficiency. The techno-economic analyses of H$_2$ production by different methods are studied (Shaner et al. 2016).

The reasons for choosing H$_2$ as an energy carrier are (1) it is easily generated from the renewable energy source and water without producing greenhouse gases like CO$_2$ resulting in a green environment; (2) it can be stored in various phase such as solid, liquid, or gas; (3) it is abundant in nature with other elements; (4) it is an easily transportable fuel with minimum loss and also has good energy storage densities.

6.3 PHOTOELECTROCHEMICAL PROCESS

The photoelectrochemical (PEC) process, also known as "artificial photosynthesis," is the more suitable route for the direct conversion of solar energy into H$_2$ through water splitting. The advantages of the PEC process are (a) easy fabrication of chemically modified electrode materials, (b) no post

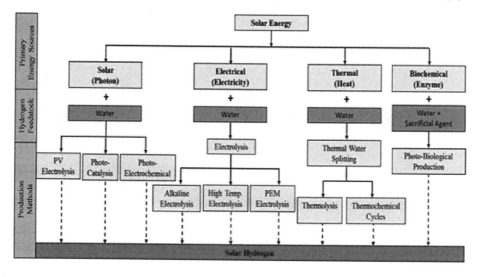

FIGURE 6.1 Types of hydrogen production methods using solar energy. (Adopted from Saraswat 2018 with permission.)

separation of evolved H$_2$ gas, (c) low temperature process, (d) environmentally safe, and (e) potential for combination with conventional photovoltaics (Walter et al. 2010; McKone, Lewis and Gray 2013; Lewis and Nocera 2006). Here, a photon in the visible range of solar spectrum with energies in the range 1 to 3 eV was enough for generating H$_2$ through splitting of water. H$_2$ was also generated from biomass and fossil hydrocarbons via the photoelectrochemical process. In some cases, the by-products obtained via water dissociation were converted into useful industrial products like sodium hydroxide and chlorine (Rabbani, Dincer, and Naterer 2016).

6.3.1 Experimental Setup

A PEC cell consists of three electrodes, namely reference, working or photo, and counter electrodes. All the electrodes were immersed in an electrolyte solution. Mostly, Ag/AgCl or standard calomel electrode (SCE) and Pt were used as reference and counter electrodes. The working electrodes should be an n-type (anode) or p-type (cathode) semiconductor material that plays an important role in the absorption of light, generation of electron–hole pairs, and charge transfer in PEC cell. The applied bias and photocurrent (I_{ph}) density as a function of counter electrode potential (V_{CE}) can be measured from the potential difference between working and counter electrodes. The PEC cell was constructed in such a way to collect the produced gases. Most PEC processes involved simulated sunlight (air mass [AM] 1.5G) as the light source that was obtained from a commercially available solar simulator. An attached potentiostat and gas chromatography analyzer with PEC cell were employed to determine the photoelectrochemical properties and the composition of evolved gases. In some cases, two electrode PEC cells, namely photoelectrode and counter electrode without reference electrode, were used for the photoelectrochemical water splitting. Both working and counter electrodes were connected using a metal wire. A membrane or a divider was used to collect and separate the gases produced in the PEC process and also for preventing the recombination of water. In most of the PECs the electrolyte medium is water but acid/alkaline solution can also be used.

6.3.2 General Mechanism of PEC

At first, an excited electron was generated on the semiconductor (photoanode) by the absorption of photonic energy. The excited electrons were jumped from the valence band (VB) to the conduction band (CB) of the semiconductor thereby creating electron–hole pairs in the photoanode. The holes were used to oxidize water on the photoelectrode surface with subsequent reduction of water into H$_2$ at the counter electrode (cathode) by the migration of electrons toward the cathode side.

The overall water splitting reaction was written as follows.

1. $h\vartheta \rightarrow e^- + \text{hole}^+$
2. At cathode: $4\text{hole}^+ + 2H_2O \rightarrow O_2 \uparrow + 4H^+$
3. At anode: $2H^+ + 2e^- \rightarrow H_2 \uparrow$
4. The overall water splitting reaction is $4h\vartheta + 2H_2O \rightarrow O_2 \uparrow + 2H_2 \uparrow$

Fujishima and Honda first reported the PEC H$_2$ production under irradiation of visible light using TiO$_2$ as the photoelectrode without applied electrical bias (1972). They achieved a PEC device with very low solar to hydrogen (STH) efficiency. Currently, the research is focusing on the development of more cost-effective photoelectrodes with improved light harvesting efficiency. The STH efficiency can be improved by applying external bias. STH conversion efficiency directly depends on the quantity of light absorbed by the semiconductor. A semiconductor with suitable bandgap or photosensitizer or H$_2$-forming catalyst as the photoelectrode was used to improve the STH efficiency by efficient light harvesting and reducing the electron–hole recombination. The minimum voltage required for water splitting is 1.23 V. The ideal semiconductor photoelectrodes have to meet the following requirements: (1) narrow bandgap to absorb the entire solar spectral region, (2) highly durable; (3) easy charge transport, and (4) suitable band alignment that straddle with redox potential of water splitting.

6.3.3 Calculations in PEC Process

The following formulas are used to assess the performance of PEC devices.

1. The bandgap of the new material was obtained from the absorption spectrum using the Tauc formula:

$$a = \frac{1}{d}\ln(1 - R|T) \ ; ah\vartheta = \beta(h\vartheta - E_g)^n$$

where
  α = optical absorption coefficient
  T = transmittance
  d = film thickness
  R = reflectance
  h = Planck's constant
  n – 2 or ½ for the direct and indirect allowed transition

2. STH (η) is the ratio between the power used for splitting of water and the input light power:

$$\eta_{PEC} = i\frac{(E^\circ_{rev} - V_{bias})}{I_0}$$

where
  i = photocurrent density (mA/cm^2)
  E°_{rev} = standard water splitting reaction potential with respect to the normal hydrogen electrode (NHE)
  I_o = light intensity (mW/cm^2)
  V_{bias} = applied external potential with respect to the reversible hydrogen electrode (RHE)

3. The incident photon to current conversion efficiency (IPCE) is also employed for evaluating PEC materials:

$$\text{IPCE} = \frac{(1240 \times I)}{\lambda \times P_{light}}$$

where
  I = measured photocurrent density in mA cm^{-2} at a specific wavelength
  λ = incident light wavelength in nm
  P_{light} = power density of monochromatic light at a specific wavelength in mW cm^{-2}

4. Applied bias photon to current efficiency (ABPE):

$$\text{ABPE}(\%) = i_{ph}(\text{mA/cm}^2) \times \frac{(V_{redox} - V_b)(V)}{P_{light}\left(\frac{mW}{cm^2}\right)AM} 1.5G \times 100(\%)$$

where
  i_{ph} = photocurrent density (mA cm^{-2}) at applied bias V_b
  V_{redox} = redox potential for hydrogen production (0 V)
  V_b = externally applied bias potential
  P_{light} = intensity of the incident light for AM 1.5G condition (~100 mW/cm^2)

5. H_2 generation rate was calculated from the photocurrent as follows:

$$\eta_{H_2} = \frac{Q \times \eta_F}{2F} \; ; \upsilon_{H_2} = \frac{\eta_{H_2}}{t \times S} = \frac{1800 \times I_{ph} \times \eta_F}{e \times N_A}$$

where
- Q = quantity of electricity
- I_{ph} = photocurrent density in mA/cm^2
- η_F = faradic efficiency
- e = electronic charge
- N_A = Avogadro constant

In the following section, we report on the literature on PEC hydrogen production with special emphasis on different metal oxides, sulfides, nitrides, silicon metal, carbides, oxynitrides, phosphides, and nanostructures materials.

6.4 METAL OXIDE AS PHOTOELECTRODE IN PEC WATER SPLITTING

6.4.1 Titanium-Dioxide-Based Electrodes

Since 1972, titanium dioxide (TiO_2) is one of the popular semiconductors for PEC water splitting due to its extraordinary properties such as high stability, high photocatalytic activity, nontoxicity, low cost, resistance to corrosion, and abundance. However, TiO_2 only absorbs light in the ultraviolet (UV) region due to its large bandgap (3.2 eV) and also the fast electron–hole recombination, resulting in low STH efficiency. The bandgap of TiO_2 can be diminished by doping with elements, surface modification, construction of the heterostructure, synthesis in the form of nanostructures, and sensitization with dye.

Even though TiO_2 nanoparticles (NPs) have high surface-to-volume ratio, the use of TiO_2 nanoparticles in PEC was limited due to the reflection of light on the surface of TiO_2 that leads to poor photon absorption. As compared with NPs, one-dimensional (1D) nanostructures such as nanorods (NRs) (Zheng et al. 2015), nanowires (NWs), and nanotubes (NTs) (Zhang, Hossain, and Takahashi 2010) enhanced the charge separation between electron–hole pairs and absorbed the light efficiently while the kinetics of water oxidation was sluggish.

Morita et al. (2017) modified a mesoporous TiO_2 electrode using platinum porphyrin as the catalyst to achieve the H_2 production from water. The mesoporous TiO_2 thin film was fabricated on the conductive fluorine-doped tin oxide (FTO) glass through screen printing and the sintering method followed by modification with porphyrin via an immersion method. The absorption spectra were used to evaluate the amount of porphyrin deposited on TiO_2. PEC measurements was performed in three electrodes containing PEC cell using 0.1 M aqueous phosphate buffer (PBS) containing 0.1 M Na_2SO_4 as the supporting electrolyte. The total amount of H_2 at −0.80 V vs. standard calomel electrode (SCE) generated was found to be 34 μmol. This electrode has high stability and longtime durability and can serve as a potential electrode in dye-sensitized solar cells.

6.4.2 Zinc-Oxide-Based Electrodes

Zinc oxide (ZnO) (bandgap 3.37 eV) has similar characteristic properties to TiO_2 except it possesses high electron mobility. The fast recombination of photogenerated electron and hole pairs in ZnO decreased the efficiency in H_2 production. It can be improved by fabricating a heterostructure with ZnS. Sánchez-Tovar et al. (2016) synthesized a heterostructure based on ZnO/ZnS for studying photoelectrochemical water splitting using solar energy. Nitrogen-doped ZnO nanowire as the photoanode was also used for the generation of H_2 through water splitting (Yang et al. 2009). The photoanode was synthesized via hydrothermal and thermal annealing methods.

6.4.3 IRON-OXIDE-BASED ELECTRODES

Hematite was considered as a promising material in PEC water splitting due to its advantages of easier synthesis; visible light absorption; stability in neutral, acid, and alkaline solutions; nontoxicity; cheapness; narrow bandgap (2 eV); and abundance. However, its drawbacks such as low conductivity, short excited lifetime, low electron mobility due to the small hole diffusion length, and rapid recombination inhibited its applications in PEC. The performance of hematite can be improved by elemental doping. Titanium was doped on hematite (α-Fe_2O_3) by electrodeposition followed by thermal annealing methods (Kalamaras et al. 2016). Ti/hematite was used for generating H_2 in the presence of ethanol as a sacrificial agent. The PEC measurements at 0.9 V vs. RHE indicated that the photocurrent of hematite was increased by increasing the loading of Ti on hematite, which further increased by using ethanol. This was due to trapping of photogenerated holes by ethanol that result in the separation of electron and holes for longer time. At this time, a greater number of hydrogen ions are reduced into hydrogen by the electrons at the cathode, thereby increasing the amount of hydrogen evolved. The obtained photocurrent was stable up to 16 h.

6.4.4 COPPER-OXIDE-BASED ELECTRODES

A p-type semiconductor, Cu_2O, has a narrow bandgap of 1.9 to 2.2 eV. The Cu_2O NP was used to modify the wide bandgap semiconductors such as TiO_2 and ZnO to improve the STH conversion efficiencies. Fan et al. (2016) fabricated a heterostructure based on semiconductor metal oxides and reduced graphene oxide (rGO). Herein, rGO plays a role in optimizing the interfacial area between different semiconductors, which influences the separation between electron and hole pairs. The TiO_2/rGO/Cu_2O was fabricated on FTO conductive glass via hydrothermal, spin coating, and chemical solution deposition. PEC measurements were performed in the three-electrode PEC cell using 0.5 M Na_2SO_4 as the supporting electrolyte and a 150 W xenon lamp with intensity of 47 mW/cm^2 as the light source. The higher H_2 production rate for the heterostructure was attributed to the inhibition of electron–hole pair recombination and improved charge transfer efficiency. The stability test indicated that there was an increase in photocurrent after irradiation for 4 h. This was due to the change in morphology of Cu_2O. During PEC reaction, the photogenerated holes can oxidize Cu (I) to Cu (II) that was suppressed by adding the sacrificial reagent such as methanol. The pure electrodeposited Cu_2O on stainless steel plate was employed for the PEC hydrogen production (Bicer and Dincer 2017). The produced H_2 can be used to synthesize ammonia by attaching the relevant reactor. The same group also investigated the effect of H_2 production using copper oxide deposited on various substances at different experimental conditions (Bicer, Chehade, and Dincer 2017). A Cu_2O/CuO bilayer composite was also used for the production of H_2 (Yang et al. 2016).

6.5 NONMETAL OXIDE AS PHOTOELECTRODE IN PEC WATER SPLITTING

6.5.1 NITRIDE IN PEC

Metal nitrides have received considerable attention in PEC cell for the production of H_2 through water splitting. Metal nitrides compared with metal oxides have many advantages such as increasing STH conversion efficiency, resistance to corrosion in acid/alkaline aqueous solution, and a bandgap that covers the full solar spectrum region (Wu 2009; Moses and Van de Walle 2010; Huygens, Strubbe, and Gomes 2000). Gallium nitride (GaN) and Si-doped GaN were used as a photoelectrode for generating H_2 through the photoelectrochemical process (Alotaibi et al. 2013a). The doping on GaN was achieved by using radio-frequency plasma assisted molecular beam epitaxy. The doped GaN has reduced surface depletion region with increased current conduction. Hydrogen bromide and potassium bromide were used separately as electrolytes. H_2 can be produced without any applied bias. If the bias was applied, the photocurrent varied depending on the dopant in GaN. The charge properties and surface band structure of GaN can also be influenced by the incorporation of dopant. IPCE of Si-GaN NW in HBr under ~350 nm light irradiation was ~18%. An Xe lamp was used as the light source. The H_2 production rate and IPCE value can be increased by altering the density and length of nanowire or by introducing a co-catalyst like NiOx (Yotsuhashi et al. 2012), or by combining with other nanowire structures such as InGaN (Alotaibi et al. 2013b).

Figure 6.2a shows the variation of photocurrent density with an applied voltage under dark illumination, light irradiation at wavelength of 375 nm, and AM 1.5G simulated sunlight. The photocurrent density was increased up to ~23% under illumination from dark to simulated sunlight. The higher value of core–shell nanostructures was attributed to excellent charge carrier transport properties, upward surface band bending that enhances charge separation, and defect free structure. Figure 6.2b shows the IPCE value at different wavelengths. The IPCE value was increased with a decrease in the wavelength, which was related to the spectra absorbance of InGaN. The maximum IPCE value of 27.6% was observed at 350 nm due to the activated GaN and InGaN. There is the formation of strong ionic bonds between different III-nitride semiconductors that increased the stability of n-type InGaN/GaN nanowire for a period of time. GaN also protected InGaN from corrosion and oxidation, additional reasons for the stability of the core–shell structure. A NiO nanoparticle-deposited InGaN/GaN nanorod array was also used for solar H_2 production through PEC (Benton, Bai, and Wang 2013a). The PEC measurements were performed using NaOH electrolyte medium under the illumination of sunlight using LOT-Oriel solar simulator. The photocurrent was increased as compared with that of InGaN/GaN nanorod without NiO NPs that was attributed to the overlap of NiO NPs bandgap with nitride nanorod array that leads to an increase in the separation of electron–hole pairs. The use of NiO NPs also inhibited photoelectrochemical etching.

Graphitic carbon nitride (g-C_3N_4), (E_g 2.7 eV), a metal-free semiconductor, has attracted considerable attention in fabricating photoelectronic devices such as light-emitting diodes, solar cells, and PEC cells. Su et al. (2015) studied the PEC performance of H_2 production using phosphorus-doped g-C_3N_4 modified TiO_2 nanotube array. g-C_3N_4 makes efficient charge transfer and separation due to the higher conduction band potential. This array was synthesized using electrochemical anodization, wet dipping, and thermal polymerization. The photocurrent density of array was 3 times higher than that of TiO_2 NT in 0.1 M NaOH solution as electrolyte. The rate of H_2 evolution for P-g-C_3N_4/TiO_2 NT under simulated sunlight was found to be 36.6 µmol h^{-1} cm^{-2} at 1.0 V vs. RHE, which was higher than that for TiO_2 NT. To enhance the performance of P-g-C_3N_4/TiO_2 NT for H_2 generation was attributed to the increase in light absorption in the visible region and increased the photogenerated electron and hole separation.

A multilayer tantalum nitride (Ta3N5) hollow sphere nanofilm with a bandgap of 2.1 eV was used as photoelectrodes for PEC H_2 production (Gao et al. 2014). The performance of PEC depends on the number of layers formed in Ta_3N_5. This electrode was fabricated using oil–water interfacial

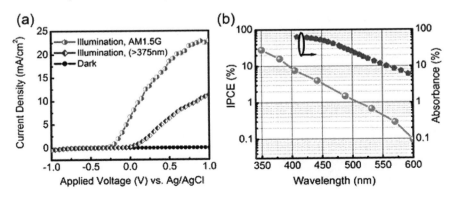

FIGURE 6.2 (a) The plot of current density vs. applied voltage (vs. Ag/AgCl) (scan rate 20 mV/S). (b) Measurement of IPCE for InGaN/GaN nanowire photoelectrodes (electrolyte, 1 mol/L HBr; applied bias, 1 V). (Adopted from Alotaibi 2013 with permission.)

self-assembly and sol–gel methods. The PEC properties were investigated by modifying the electrodes with IrO$_2$ acting as the co-catalyst that reduces the overpotential related to the redox reaction and increases the charge separation. The highest photocurrent for monolayer and Ta$_3$N$_5$ with four layers was found to be 6.1 mA·cm^{-2} and 17.9 mA cm^{-2} at 0.65 V vs. Ag/AgCl. This indicated that the increase in the number of layers will increase the photocurrent density due to the increases in light absorption. If the layer was further increased, the decrease in the photocurrent was observed due to the decrease in electric field and increase in film resistance that results in electron–hole recombination. IPCE for Ta$_3$N$_5$ containing four layers at 1.23 V was estimated to be 43.5% under the illumination of light at wavelength of 470 nm.

6.5.2 Phosphide in PEC

A solar fuel based on molecular catalyst (cobaloxime) immobilized semiconductor was synthesized (Beiler et al. 2016). Here, III-V semiconductor, gallium phosphide (GaP) with the indirect bandgap of 2.26 eV was used. The two faces of GaP (100) and (111) were easily exposed to oxidation and corrosion. This can be avoided by stabilizing GaP with polyvinylimidazole (PVI) polymer. This material was used as the photocathode in PEC H$_2$ generation. The light source used for PEC studies was AM 1.5 filters equipped 100 W Oriel Solar Simulator. Phosphate buffer (PBS) with pH 7 and 100 mV/s was used as the supporting electrolyte and scan rate, respectively. Figure 6.3 shows the linear sweep voltammogram of photoelectrodes with and without cobaloxime-modified GaP. As compared with unmodified electrodes, the photocurrent density of the modified electrode was greater at 0 V vs. RHE. The I_{ph} of PVI-GaP electrodes was 0.32 ± 0.04 and 0.30 ± 0.08 mA cm^{-2}. The I_{ph} of Co-PVI-GaP electrodes was 0.89 ± 0.02 and 0.89 ± 0.03 mA cm^{-2} for the A- and B-faces, respectively. Thus, it shows that the electrodes with both A- and B-faces have identical current and voltage response. This photocurrent was also stable for 1 h.

Nann et al. (2010) discovered robust photoelectrode comprised of an inorganic light absorber and earth abundant element containing electrocatalyst for the production of H$_2$. Both materials are nontoxic and inexpensive. First, a layer-by-layer assembly method was used for constructing a cross-linked indium phosphide (InP) nanocrystal array followed by the incorporation of an iron–sulfur electrocatalyst, namely [Fe$_2$S$_2$(CO)$_6$]. The InP was bound with electrocatalyst through the sulfide bridges present in the electrocatalyst. This InP-[Fe$_2$S$_2$(CO)$_6$] system on gold substrate was used as the photocathode for the production of H$_2$ using 0.1m aqueous NaBF$_4$ as electrolyte in neutral pH. More than 60% efficiency was achieved under the illumination of light at wavelength of 395 nm. The indium-phosphide-based core shell nanorod structure (InP-CdS and InP-ZnTe) was also investigated for PEC H$_2$ production (Yu et al. 2005).

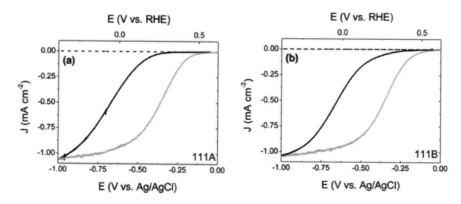

FIGURE 6.3 LSV of (a) GaP(111)A and (b) GaP(111)B; polyvinylimidazole grafting (blue) and cobaloxime attachment (red). The data was observed in the dark (dashed-dotted lines) or under 100 mW cm^{-2} illumination (solid lines) in neutral pH. (Adopted from Beiler 2016 with permission.)

6.5.3 Oxynitride in PEC

Tantalum oxynitride (TaON) is an n-type semiconductor with a bandgap of 2.5 eV. Even though it is nontoxic, the poor stability and low electron mobility offer the synthesis of nanostructure oxynitrides. p-Cu_2O/n-TaON heterojunction nanorod array as photoanodes was discovered for photoelectrochemical water splitting after passivating with ultrathin carbon sheath (Hou et al. 2014). The passivation was achieved by a solution based process for protecting the surface of oxynitrides. Due to the shape anisotropy and p-n heterojunction structure, the photocurrent density of carbon–Cu_2O/TaON heterojunction nanorod arrays reached 3.06 mA cm^{-2} under AM 1.5G simulated sunlight at 1.0 V vs. RHE with a maximum IPCE of 59% at 400 nm. The photocurrent remained at about 87.3% of the initial activity after 60 min irradiation, exhibiting that not only the onset potential is negatively shifted but also the photocurrent and photostability are significantly improved in comparison to that of TaON and Cu_2O/TaON that was due to the high built-in potential in the protective p-n heterojunction device encapsulated in an ultrathin graphitic carbon sheath from the electrolyte.

6.5.4 Sulfide in PEC

Molybdenum disulfide (MoS_2) has attracted considerable attention due to its low cost and high electrocatalytic activity. Jang et al. (2016) fabricated carbon and molybdenum sulfide (MoS_2) modified zinc telluride (ZnTe) as photocathode for the production of H_2 from solar energy and water by the photoelectrochemical process. This electrode was fabricated via carbonization and annealing. The PEC measurements were performed in 0.5 M Na_2SO_4 as the electrolyte under simulated sunlight. The photocurrent density and ABPE of C-MoS_2/ZnTe were found to be −1.48 mA cm^{-2} and 0.12%, respectively. Rabbani and coworkers (2016) developed a new photoelectrocatalytic reactor using zinc sulfide as a photocatalyst for the production of hydrogen along with chlorine and sodium hydroxide from saltwater. They studied the influence of different parameters, namely applied voltage, amount of catalyst, and light intensity, on solar-to-hydrogen conversion efficiency. They reported that the increase in light intensity will increase the production of hydrogen but decrease the efficiency of the process.

Carbide-like tungsten carbides were also used as photoelectrode for the generation of H_2 through the PEC process (Gong et al. 2016).

6.6 QUANTUM-DOT-BASED PHOTOELECTRODE

Quantum-dot-based photoelectrodes have garnered much attention for PEC hydrogen generation due to their extraordinary physical properties such as tunable light absorption, excellent photostability, easy surface binding, and high extinction coefficient. Hydrogen can be produced through photoelectrochemical water splitting using giant core–shell quantum-dot-sensitized mesoporous TiO_2 nanoparticles as the photoanode (Adhikari et al. 2016). Herein, a thin film of Cu_2S nanoflakes was used as counter electrodes instead of standard Pt film without affecting the PEC performance. This quantum dot (QD) was synthesized via successive ionic layer adsorption reaction and sensitization with TiO_2 films was achieved by electrophoretic deposition. Two different QDs were synthesized with different core sizes and the photoelectrochemical properties were studied. Lv et al. (2017) also employed CdSe QD-sensitized NiO-coated indium tin oxide photoelectrodes for hydrogen production under light irradiation. Herein, QD acted as the photocathode. The electrode was synthesized through the spin coating method. They fabricated a rainbow-like QD electrode by layering coatings of different sized QDs.

6.7 HYDROGEN PRODUCTION FROM BIOMASS AND WATER

PEC H_2 production from renewable biomass such as alcohols, organic acids, and saccharides and water was more efficient as compared with that from water splitting. There is a formation of CO_2 during the H_2 production that was retransformed into carbohydrates through natural photosynthesis

TABLE 6.1
Various Semiconductors Used for PEC H$_2$ Production

Photoelectrode Material	Electrolyte	Light Source	I$_{ph}$ or STH or IPCE or H$_2$ Production Rate or ABPE	Reference
GaN nanorod array	HCl	LOT-Oriel solar simulator	Rate 0.73 m^{-1} h^{-1} cm^2; STH 0.26%	Benton, Bai and, Wang 2013b
Ta$_3$N$_5$ nanorod array/Co(OH)$_x$ NPs	1 M NaOH	Simulated sunlight	I$_{ph}$ 2.8 mA cm^{-2}; IPCE 37.8%; rate 274.6 mmol h^{-1} m^{-2}	Zhen et al. 2013
WO$_3$ nanowire array	0.5 M H$_2$SO$_4$	Simulated solar light	IPCE 85%; ABPE ~0.33%	Chakrapani, Thangala, and Sunkara 2009
WN$_3$	0.5 M H$_2$SO$_4$	Simulated solar light	ABPE ~0.4%	Chakrapani, Thangala, and Sunkara 2009
Branched TiO$_2$ nanorod	1 M KOH	Xenon lamp	I$_{ph}$ 0.83 mA cm^{-2}; IPCE 67%; ABPE −0.49%	Cho et al. 2011
p-Si/NiCoSex core/shell nanopillar array photocathode	0.5 M H$_2$SO$_4$	Simulated sunlight	I$_{ph}$ −37.5 mA cm^{-2}	Zhang et al. 2016
CQD/TiO$_2$ NT	0.25 M Na$_2$S and 0.35 M Na$_2$SO$_3$ aqueous solutions	Simulated sunlight	Rate 14.1 mmol h^{-1}; I$_{ph}$ 1.0 mA cm^{-2}	Zhang et al. 2013
Polymeric CN-CuInS$_2$	0.1 M H$_2$SO$_4$ (pH 1)	Simulated sunlight	I$_{ph}$ 50 mA cm^{-2}	Yang et al. 2013
H:TiO$_2$ NW	1 M NaOH	Simulated sunlight	I$_{ph}$ ~2.5 mA cm^{-2}; STH 1.63%: rate 0.47 mLh^{-1}; IPCE 95%	Wang et al. 2011
CNQDs/TiO$_2$ NTAs	0.1 M Na$_2$SO$_4$ (pH =6.0)	Simulated solar light	Rate 22.0 μmol h^{-1} cm^{-2}; I$_{ph}$ 1.34 mA cm^{-2}; IPCE 6.8%	Su, Zhu and Chen. 2016
Ta$_3$N$_5$:Mg+Zr	0.1 M Na$_2$SO$_4$	Simulated solar light	I$_{ph}$ 2.3 mA cm^{-2}; IPCE 9%–18%; STH 0.59%	Seo et al. 2015
CN-rGO	Triethanolamine in 0.1 M KOH	Simulated solar light	I$_{ph}$ 660 μA cm^{-2}; rate 0.8 mol h^{-1} g^{-1}; IPCE 63.2%	Peng et al. 2018
Co-phosphate/Ba-Ta$_3$N$_5$ nanorods	0.5 M K$_2$HPO$_4$	Simulated solar light	I$_{ph}$ 6.7 mA cm^{-2}; STH 1%; rate 130 mmol cm^{-2}	Li et al. 2013
Heterostructures of chemically exfoliated metallic 1T-MoS$_2$ and planar p-type Si	0.5 M H$_2$SO$_4$	Simulated solar light	I$_{ph}$ 17.6 mA cm^{-2}	Ding et al. 2014

(Continued)

TABLE 6.1 (CONTINUED)
Various Semiconductors Used for PEC H_2 Production

Photoelectrode Material	Electrolyte	Light Source	I_{ph} or STH or IPCE or H_2 Production Rate or ABPE	Reference
$Zn_{0.956}Co_{0.044}O$	0.2 M KNO_3	Xe lamp	I_{ph} 55 µAcm^{-2}; IPCE 0.04%	Jaramillo et al. 2005
Cu_2ZnSnS_4/Mo-mesh thin-film	0.1 M NaOH	300 W Xe lamp	STH 6.7%; IPCE 15.5%	Ma et al. 2011
Porous CdS/Nafion	0.25 M Na_2S (pH 12)	Simulated sunlight	I_{ph} 5.68 mA cm^{-2}; IPCE 75%; rate 98 µmol h^{-1} cm^{-2}	Zheng et al. 2016
Monoclinic WO_3 nanomultilayers	0.1 M Na_2SO_4	Simulated sunlight	I_{ph} 1.62 mA cm^{-2}; IPCE 40%; rate 3 µmol h^{-1} cm^{-2}	Zheng et al. 2015
TNTs-CdS@Sb_2S_3	3 M aq. KCl	100 W Xe-arc lamp	Rate 10.23 mmol h^{-1}	Carrera-Crespo et al. 2017
NiP_2	0.5 M H_2SO_4	Simulated sunlight	I_{ph} 12 mA cm^{-2}; STH-2.6%	Chen et al. 2016
p-Si wire arrays and NiMoZn heterojunction	0.5 M H_2SO_4	Simulated sunlight	I_{ph} ~1.45 mA·cm^{-2}	Choi et al. 2017
SnO_2/TiO_2/CdS multi-heterojunction structure	0.25 M Na_2S and 0.35 M Na_2SO_3 aqueous solution	Simulated sunlight	I_{ph} 8.75 mA cm^{-2}; rate 163.24 mmol h^{-1} cm^{-2}	Gao et al. 2015
nano-CdSe thin film	Na_2SO_3 solution	Simulated sunlight	I_{ph} 4.4 mA cm^{-2}; STH 4.31%; rate 82.7 mmol h^{-1} cm^{-2}	Guo et al. 2016
Bi_2S_3 nanowire @ TiO_2 nanorod	0.35 M Na_2SO_3 and 0.25 M Na_2S (pH 11.5)	150 W xenon lamp	I_{ph} 2.4 mA cm^{-2}; rate 35.97 mol cm^{-2} h^{-1}	Liu et al. 2016
Bi_2S_3/$BiVO_4$ nanostructure	0.1 M Na_2S/0.02 M Na_2SO_3	Simulated sunlight	I_{ph} 3.3 mA cm^{-2}; rate ~417 µmol cm^{-2} h^{-1}	Mahadik et al. 2018

(Ampelli et al. 2011). As compared with other processes such as fast pyrolysis, steam gasification, and supercritical conversion, the PEC process offers the use of sunlight under ambient conditions that leads to fabrication of low-cost PEC devices. Like, PEC water splitting, the production of H_2 from biomass and water uses the same experimental setup. The properties of photoelectrodes, particularly photoanodes, decided the H_2 production rate and the decomposition of biomass derivatives. The ideal photoanode for PEC H_2 production from biomass and water should meet criteria similar for PEC water splitting. The biomass materials can be oxidized easier than water due to its more negative redox potential. At first, Kawai and Sakata (1980) generated H_2 from biomass containing glucose backbone and water using $RuO_2/TiO_2/Pt$ photocatalysts. However, the PEC method was more efficient than photocatalysis.

Various metal oxides such as Fe_2O_3, TiO_2, and WO_3 have been employed for PEC production from biomass and water. The metal oxide sensitized quantum dots were more suitable for the production of hydrogen from biomass and water with improved PEC efficiency rather than the use of metal oxides alone. This was due to the consumption of holes produced after absorption of light irradiation by the biomass acting as the hole scavenger that stabilizes the QDs. If methanol is used as the biomass in solar H_2 production, the oxidation of methanol gives CO_2 and H_2 with the formation of formaldehyde or formic acid. The oxidation of methanol varied for different pH. The oxidation of ethanol produced acetic acid or CO_2 with the formation of H_2. Mesoporous tungsten oxide (WO_3) was employed for the production of H_2 from ethanol and water achieving 100% IPCE (Barczuk et al. 2009). Lu et al. (2014) reviewed the photoelectrochemical H_2 production from biomass and water with their reaction mechanisms. Some recent reports for various semiconductors used for PEC H_2 production are listed in the Table 6.1.

6.8 CONCLUSIONS

Incredible growth in the development of efficient photoelectrochemical methods for production of hydrogen employing metal oxide, nonmetal oxide, nanomaterials, and biomass has been made over the past several years. In this chapter, we mainly discussed the development of a more efficient energy conversion photoelectrochemical cell in terms of reliability and durability for the production of hydrogen. However, many of the oxides, such as TiO_2, ZnO_2, iron-based oxides, sulfides, nitrides, and quantum dots, were used as photoelectrodes in the direct splitting of water using sunlight. Fabrications of semiconductor-materials-based photoelectrochemical cells are more efficient for energy conversion for the production of hydrogen. The attractive design and fabrication of nanomaterials-based photoelectrochemical cells can offer a high efficiency and increased durability for the production of hydrogen. Photoelectrodes produce hydrogen from the direct splitting of water using sunlight, which provides clean fuel for a green environment.

REFERENCES

Adhikari R, Jin L, Pardo FN, Benetti D, Alotaibi B, Vanka S, Zhao H, Mi Z, Vomiero A, and Rosei F. 2016. "High efficiency, Pt-free photoelectrochemical cells for solar hydrogen generation based on "Giant" quantum dots." *Nano Energy* no. 27: 265–274.

Alotaibi B, Harati M, Fan S, Zhao S, Nguyen HPT, Kibria MG, and Mi Z. 2013. "High efficiency photoelectrochemical water splitting and hydrogen generation using GaN nanowire photoelectrode." *Nanotechnology* no. 24: 175401 (5pp).

Alotaibi B, Nguyen HPT, Zhao S, Kibria MG, Fan S, and Mi Z. 2013. "Highly stable photoelectrochemical water splitting and hydrogen generation using a double-band InGaN/GaN core/shell nanowire photoanode." *Nano Letters* no. 13(9): 4356–4361.

Ampelli C, Passalacqua R, Genovese C, Perathoner S, and Centi G. 2011. "A novel photoelectrochemical approach for the chemical recycling of carbon dioxide to fuels." *Chemical Engineering Transactions* no. 25: 683–688.

Barczuk PJ, Lewera A, Miecznikowski K, Kulesza P, and Augustynski J. 2009. "Visible light driven photoelectrochemical conversion of the by-products of the ethanol fuel cell into hydrogen." *Electrochemical and Solid-State Letters* no.12: B165–B166.

Beiler AM, Khusnutdinova D, Jacob SI, and. Moore GF. 2016. "Solar hydrogen production using molecular catalysts immobilized on gallium phosphide (111)A and (111)B polymer-modified photocathodes." *ACS Applied Materials Interfaces* no. 8: 10038–10047.

Benton J, Bai J, and Wang T. 2013a. "Enhancement in solar hydrogen generation efficiency using a GaN-based nanorod structure." *Applied Physics Letters* no. 102: 173905.

Benton J, Bai J, and Wang T. 2013b. "Significantly enhanced performance of an InGaN/GaN nanostructure based photo-electrode for solar power hydrogen generation." *Applied Physics Letters* no. 103: 133904: 1–4.

Bicer Y, Chehade G, and Dincer I. 2017. "Experimental investigation of various copper oxide electrodeposition conditions on photoelectrochemical hydrogen production." *International Journal of Hydrogen Energy* no. 42(10): 6490–6501.

Bicer Y, and Dincer I. 2017. "Performance assessment of electrochemical ammoniasynthesis using photoelectrochemically produced hydrogen." *International Journal of Energy Research* no. 41(14): 1987–2000.

Carrera-Crespo JE, Ramos-Sanchez G, De la Luz V, Gonzalez F, Barrera E, and Gonzalez I. 2017. "Photoelectrochemical hydrogen generation on TiO_2 nanotube arrays sensitized with $CdS@Sb_2S_3$ core shell particles." *International Journal of Hydrogen Energy* no. 42(51): 30249–30256.

Chakrapani V, Thangala J, and Sunkara MK. 2009. "WO_3 and W_2N nanowire arrays for photoelectrochemical hydrogen production." *International Journal of Hydrogen Energy* no. 34: 9050–9059.

Chen F, Zhu Q, Wang Y, Cui W, Su X, and Li Y. 2016. "Efficient photoelectrochemical hydrogen evolution on silicon photocathodes interfaced with nanostructured NiP_2 cocatalyst films." *Applied Materials & Interfaces* no. 8(45): 31025–31031.

Cho IS, Chen Z, Forman AJ, Kim DR, Rao PM, Jaramillo TF, and Zheng X. 2011. "Branched TiO_2 nanorods for photoelectrochemical hydrogen production." *Nano Letters* no. 11: 4978–4984.

Choi SK, Piao G, Choi W, and Park H. 2017. "Highly efficient hydrogen production using p-Si wire arrays and NiMoZn heterojunction photocathodes." *Applied Catalysis B: Environmental* no. 217: 615–621.

Dincer I, and Acar C. 2015. "Review and evaluation of hydrogen productionmethods for better sustainability." *International Journal of Hydrogen Energy* no. 40(34): 11094–11111.

Ding Q, Meng F, English CR, Cabán-Acevedo M, Shearer MJ, Liang D, Daniel AS, Hamers RJ, and Jin S. 2014. "Efficient photoelectrochemical hydrogen generation using heterostructures of Si and chemically exfoliated metallic MoS_2." *Journal of the American Chemical Society* no. 136(24): 8504–8507.

Fan W, Yu X, Lu H-C, Bai H, Zhang C, and Shi W. 2016. "Fabrication of TiO_2/RGO/Cu_2O heterostructure for photoelectrochemical hydrogen production." *Applied Catalysis B: Environmental* no. 181: 7–15.

Fujishima A, and Honda K. 1972. "Electrochemical photolysis of water at a semiconductor electrode." *Nature* no. 238: 37–38.

Fulcheri L, and Schwob Y. 1995. "From methane to hydrogen, carbon black and water." *International Journal of Hydrogen Energy* no. 20: 197–202.

Gao C, Zhang Z, Li X, Chen L, Wang Y, He Y, Teng F, Zhou J, Han W, and Xie E. 2015. "Synergistic effects in three-dimensional SnO_2/TiO_2/CdS multi-heterojunction structure for highly efficient photoelectrochemical hydrogen production." *Solar Energy Materials & Solar Cells* no. 141: 101–107.

Gao R, Hu L, Chen M, and Wu L. 2014. "Controllable fabrication and photoelectrochemical property of multilayer tantalum nitride hollow sphere-nanofilms." *Small* no. 10(15): 3038–3044.

Gong Q, Wang Y, Hu Q, Zhou J, Feng R, Duchesne PN, Zhang P, Chen F, Han N, Li Y, Jin C, Li Y, and Lee S-T. 2016. "Ultrasmall and phase-pure W_2C nanoparticles for efficient electrocatalytic and photoelectrochemical hydrogen evolution." *Nature Communications* no. 7: 13216.

Guo M, Wang L, Xia Y, Huang W, and Li Z. 2016. "Fabrication of nano-CdSe thin films from gas/liquid interface reactions and self-assembly for photoelectrochemical hydrogen production." *International Journal of Hydrogen Energy* no. 41: 2278–2284.

Hou J, Yang C, Cheng H, Jiao S, Takeda O, and Zhu H. 2014. "High-performance p-Cu_2O/n-TaON heterojunction nanorod photoanodes passivated with ultrathin carbon sheath for photoelectrochemical water splitting." *Energy Environ. Sci.* no. 7: 3758–3768.

Huygens IM, Strubbe K, and Gomes WP. 2000. "Electrochemistry and photoetching of n-GaN." *Journal of the Electrochemical Society* no. 147: 1797–802.

Jang YJ, Lee J, Lee J, and Lee JS. 2016. "Solar hydrogen production from zinc telluride photocathode modified with carbon and molybdenum sulfide." *ACS Applied Materials & Interfaces* no. 8(12): 7748–7755.

Jaramillo TF, Baeck S-H, Kleiman-Shwarsctein A, Choi K-S, Stucky GD, and McFarland EW. 2005. "Automated electrochemical synthesis and photoelectrochemical characterization of $Zn_{1-x}Co_xO$ thin films for solar hydrogen production." *Journal of Combinatorial Chemistry* no. 7(2): 264–271.

Kalamaras E, Dracopoulos V, Sygellou L, and Lianos P. 2016. "Electrodeposited Ti-doped hematite photoanodes and their employment for photoelectrocatalytic hydrogen production in the presence of ethanol." *Chemical Engineering Journal* no. 295: 288–294.

Kawai T, and Sakata T. 1980. "Conversion of carbohydrate into hydrogen fuel by a photocatalytic process." *Nature* no. 286: 474–476.

Lewis NS, and Nocera DG. 2006. "Powering the planet: chemical challenges in solar energy utilization." *PNAS* no. 103: 15729–15735.

Li Y, Zhang L, Torres-Pardo A, Gonza'lez-Calbet JM, Ma Y, Oleynikov P, Terasaki O, Asahina S, Shima M, Cha D, Zhao L, Takanabe K, Kubota J, and Domen K. 2013. "Cobalt phosphate-modified barium-doped tantalum nitride nanorod photoanode with 1.5 % solar energy conversion efficiency." *Nature Communications* no. 4: 2566 (7pp).

Liu C, Yang Y, Li W, Li J, Li Y, and Chen O. 2016. "A novel Bi_2S_3 nanowire @ TiO_2 nanorod heterogeneous nanostructure for photoelectrochemical hydrogen generation." *Chemical Engineering Journal* no. 302: 717–724.

Lu X, Xie S, Yang H, Tong Y, and Ji H. 2014. "Photoelectrochemical hydrogen production from biomass derivatives and water." *Chemical Society Reviews* no. 43: 7581–7593.

Lv H, Wang C, Li G, Burke R, Krauss TD, Gao Y, and Eisenberg R. 2017. "Semiconductor quantum dot-sensitized rainbow photocathode for effective photoelectrochemical hydrogen generation." *PNAS* no. 114(43): 11297–11302.

Ma G, Minegishi T, Yokoyama D, Kubota J, and Domen K. 2011. "Photoelectrochemical hydrogen production on Cu_2ZnSnS_4/Mo-mesh thin-film electrodes prepared by electroplating." *Chemical Physics Letters* no. 501: 619–622.

Mahadik MA, Chung H-S, Lee SY, Cho M, and Jang JS. 2018. "In-situ noble fabrication of Bi_2S_3/$BiVO_4$ hybrid nanostructure through photoelectrochemical transformation process for solar hydrogen production." *ACS Sustainable Chemistry & Engineering* no. 6(9): 12489–12501.

McKone JR, Lewis NS, and Gray HB. 2013. "Will solar – driven water-splitting devices see the light of day?" *Chemistry of Materials* no. 26: 407–414.

Morita K, Takijiri K, Sakai K, and Ozawa H. 2017. "A platinum porphyrin modified TiO_2 electrode for photoelectrochemical hydrogen production from neutral water driven by the conduction band edge potential of TiO_2." *Dalton Transactions* no.46: 15181–15185.

Moses PG, and Van de Walle CG. 2010. "Band bowing and band alignment in InGaN alloys." *Applied Physics Letters* no. 96: 021908.

Nann T, Ibrahim SK, Woi P-M, Xu S, Ziegler J, and Pickett CJ. 2010. "Water splitting by visible light: a nano-photocathode for hydrogen production." *Angewandte Chemie International Edition* no. 49: 1574–1577.

Peng G, Volokh M, Tzadikov J, Sun J, and Shalom M. 2018. "Carbon nitride/reduced graphene oxide film with enhanced electron diffusion length: an efficient photoelectrochemical cell for hydrogen generation." *Advanced Energy Materials* no. 8(23): 1800566 (7pp).

Rabbani M, Dincer I, and Naterer GF. 2016. "Experimental investigation of hydrogen production in a photoelectrochemical chloralkali processes reactor." *International Journal of Hydrogen Energy* no. 41(19): 7766–7781.

Sánchez-Tovar R, Fernández-Domene RM, Montañés MT, Sanz-Marco A, and García-Antón J. 2016. "ZnO/ZnS heterostructures for hydrogen production by photoelectrochemical water splitting." *RSC Advances* 6: 30425–30435.

Saraswat SK, Rodene DD, and Gupta RB. 2018. "Recent advancements in semiconductor materials for photoelectrochemical water splitting for hydrogen production using visible light." *Renewable and Sustainable Energy Reviews* no. 89: 228–248.

Seo J, Takata T, Nakabayashi M, Hisatomi T, Shibata N, Minegishi T, and Domen K. 2015. "Mg-Zr co-substituted Ta_3N_5 photoanode for lower-onset potential solar-driven photoelectrochemical water splitting." *Journal of the American Chemical Society* no. 137: 12780–12783.

Shaner MR, Atwater HA, Lewis NS, and McFarland EW. 2016. "A comparative technoeconomic analysis of renewable hydrogen production using solar energy." *Energy & Environmental Science* no. 9: 2354–2371.

Su J, Geng P, Li X, Zhao O, Quan X, and Chen G. 2015. "Novel phosphorus doped carbon nitride modified TiO_2 nanotube arrays with improved photoelectrochemical performance." *Nanoscale* no. 7: 16282–16289.

Su J, Zhu L, and Chen G. 2016. "Ultrasmall graphitic carbon nitride quantum dots decorated self-organized TiO_2 nanotube arrays with highly efficient photoelectrochemical activity." *Applied Catalysis B: Environmental* no.186: 127–135.

Turner JA. 2004. "Sustainable hydrogen production." *Science* no. 305: 972–974.

Veras TS, Mozer TS, Santos DCRM, Cesar AS. 2016. "Hydrogen: trends, production and characterization of the main process worldwide." *International Journal of Hydrogen Energy* no. 42(4): 2018–2033.

Walter MG, Warren EL, McKone JR, Boettcher SW, Mi Q, Santori EA, and Lewis NS. 2010. "Solar water splitting cells." *Chemical Reviews* no.110: 6446–6473.

Wang G, Wang H, Ling Y, Tang Y, Yang X, Fitzmorris RC, Wang C, Zhang JZ, and Li Y. 2011. "Hydrogen-treated TiO_2 nanowire arrays for photoelectrochemical water splitting." *Nano Letters* no. 11: 3026–3033.

Wu JQ. 2009. "Group-III nitrides go infrared: new properties and perspectives." *Journal of Applied Physics* no. 106: 011101.

Yang F, Kuznietsov V, Lublow M, Merschjann C, Steigert A, Klaer J, Thomase A, and Schedel-Niedrig T. 2013. "Solar hydrogen evolution using metal-free photocatalytic polymeric carbon nitride/$CuInS_2$ composites as photocathodes." *Journal of Materials Chemistry A* no. 1: 6407.

Yang X, Wolcott A, Wang G, Sobo A, Fitzmorris RC, Qian F,. Zhang JZ, and Li Y. 2009. "Nitrogen-doped ZnO nanowire arrays for photoelectrochemical water splitting." *Nano Letters* no. 9(6): 2331–2336.

Yang Y, Xu D, Wu Q, and Diao P. 2016. "Cu_2O/CuO bilayered composite as a high-efficiency photocathode for photoelectrochemical hydrogen evolution reaction." *Scientific Reports* no.6: 35158 (13pp).

Yotsuhashi S, Deguchi M, Zenitani Y, Hinogami R, Hashiba H, Yamada Y, and Ohkawa K. 2012. "CO_2 conversion with light and water by GaN photoelectrode." *Japanese Journal of Applied Physics* no. 51(2S): 02BP07.

Yu ZG, Pryor CE, Lau WH, Berding MA, and MacQueen DB. 2005. "Core-shell nanorods for efficient photoelectrochemical hydrogen production." *The Journal of Physical Chemistry B* no. 109: 22913–22919.

Zhang H, Ding Q, 'He D, Liu H, Liu W, Li Z, Yang B, Zhang X, Lei L, and Jin S. 2016. "A Si/NiCoSex core/shell nanopillar array photocathode for enhanced photoelectrochemical hydrogen production." *Energy and Environmental Science* no. 9(10): 3113–3119.

Zhang J, Zhang P, Wang, and Gong J. 2015. "Monoclinic WO_3 nanomultilayers with preferentially exposed (002) facets for photoelectrochemical water splitting." *Nano Energy* no. 11: 189–195.

Zhang X, Wang F, Huang H, Li H, Han X, Yang Liu Y, and Kang Z. 2013. "Carbon quantum dot sensitized TiO_2 nanotube arraysfor photoelectrochemical hydrogen generation undervisible light." *Nanoscale* no. 5(6): 2274–2278.

Zhang Z, Hossain MF, and Takahashi T. 2010. "Photoelectrochemical water splitting on highly smooth and ordered TiO_2 nanotube arrays for hydrogen generation." *International Journal of Hydrogen Energy* no. 35: 8528–8535.

Zhen C, Wang L, Liu G, Qing G, and Cheng H-M. 2013. "Template free synthesis of Ta_3N_5 nanorod arrays for efficient photoelectrochemical water splitting." *Chemical Communications* no. 49(29): 3019–3021.

Zheng X, Shen S, Ren F, Cai G, Xing Z, Liu Y, liu D, Zhang G, Xiao X, Wu W, and Jiang C. 2015. "Irradiation-induced TiO_2 nanorods for photoelectrochemical hydrogen production." *International Journal of Hydrogen Energy* no. 40: 5034–5041.

Zheng X-L, Song J-P, Ling T, Hu ZP, Yin P-F, Davey K, Du X-W, and Qiao S-Z. 2016. "Strongly coupled nafion molecules and ordered porous CdS networks for enhanced visible-light photoelectrochemical hydrogen evolution." *Advanced Materials* no. 28: 4935–4942.

7 Recent Trends in Chemiresistive Gas Sensing Materials

Baskaran Ganesh Kumar, J. Nimita Jebaranjitham, and Saravanan Rajendran

CONTENTS

7.1 Introduction and Motivation 121
7.2 Chemiresistive Sensors 122
7.3 Device Design and Mechanism 122
7.4 Analytes 124
7.5 Sensing Material 124
    7.5.1 Metal Oxide Sensors 126
    7.5.2 Polymers 126
    7.5.3 Composites 126
    7.5.4 Complexes 127
    7.5.5 Nanoparticles 127
    7.5.6 MOFs 128
7.6 Sensing Platform 129
7.7 Important Parameters 129
7.8 Summary 130
7.9 Chemiresistive Gas Sensing Materials: Future Directions 130
References 131

7.1 INTRODUCTION AND MOTIVATION

The emergence and demise of economically powerful nations depend on technological advances. Electronics are the central heart of any novel technological invention. We are now living in the age of electronic components and they influence every aspect of life from cooking to aircraft. In our increasingly electronics-reliant society, sensors play a major role in our daily lives. By definition, a sensor is a functional device that responds to any type of stimulus and generates the functionally related output (Joo and Brown 2008). Any digital action starts with sensing, and sensors are expected to be involved in every scientific and technological discipline in the near future (Patolsky et al. 2006). Another important contribution of sensors is sensing the harmful chemicals in the air and environment (Zhang et al. 2017b). Our atmosphere contains various toxic substances from cars, explosives, industrial toxic vapors, and biowar chemicals. These toxic substances are harmful to people and affect masses, animals, and plants. These chemical species should be identified and removed. Sensors are ideal for gas sensing, and toxic substances can be effectively controlled to a desired toxic limit. Hence, sensors are exciting and very relevant to all aspects of life including safety, monitoring, health devices, microorganism detection, the perfume industry, security, and surveillance. So many improvements and advancements in technology have moved the world, which depends completely on sensors. The sensing can originate from mechanical, electrical, thermal, electrical, magnetic, or radiant, as well as chemical change. The scientific community uses many terms to describe sensors, namely, "sensors," "electronic noses," and "transducer," throughout the literature (Amjadi et al. 2016; Fratoddi et al. 2015; Joshi et al. 2018; Pandey 2016; Timmer et al. 2005). Here, we are going to discuss chemiresistive sensors exclusively (Baruah and Dutta 2009; Di Natale et al. 2014; Fratoddi et al. 2015; Joo and Brown 2008; Joshi et al. 2018; Kumar et al. 2015; Lieberzeit and Dickert 2007; Llobet 2013; Mahadeva et al. 2015; Pandey 2016; Pandey et al. 2013; Timmer et al. 2005; Wang et al. 2013). Overall, chemiresistive sensing is important technology and an indispensable tool for next-generation electronic devices.

Throughout the chapter, we discuss the device design, mechanism, innovative materials, and factors influencing sensing activity. Along with many research efforts, we describe prospective materials and conventional analytes for chemiresistive sensing. At the end, we discuss next directions, key challenges, prospects for commercialization, and opportunities for next-generation chemiresistive sensors.

7.2 CHEMIRESISTIVE SENSORS

In 1962, Seiyama et al. reported the first ZnO-based gas sensor using chemiresistive sensing technology. Since then, chemiresistive sensing has been widely adopted and its performance is continuously being increased by researchers. The chemiresistive sensor produces the modified signal to a specific measurand (or analyte) (Fennell Jr. et al. 2016). Simply, the chemiresistive sensor uses electrical detection for identifying analyte molecules. The modified signal in our case is change in resistance. The chemiresistive sensors are generally used to identify the gases and volatile compounds by monitoring the voltage drop due to resistance changes. Chemiresistive sensors offer simple and low-cost technologies for detecting and identifying gases and volatile organic compounds. The ideal chemiresistive sensor is expected to exhibit accuracy, physical stability, chemical stability, high selectivity, high sensitivity, high resolution, linearity, dynamic range, and better repeatability. The simplest sensing mechanism facilitates low energy consumption that would lead to simple possible technology for sensing application, which is ideal for commercialization. Due to its simplicity, the scientific community exploited it to various real-world applications (Table 7.1). The real-world applications include diabetics monitoring, water quality analyses, explosive sensing, and homeland security. This chapter discusses the operating principles, common materials, and nexus of opportunities with various disciplines.

7.3 DEVICE DESIGN AND MECHANISM

Chemiresistive sensors are based on the simple change in resistance in response to the binding of analytes. The ideal simple sensor design contains active materials as the thin film, substrate, and an electrical setup (Figure 7.1) (Fennell Jr. et al. 2016; Liu et al. 2015b; Wang et al. 2008). If the device exhibits change in resistance, this indicates the presence of an analyte. The conductance is proportional to amount of analyte present. The active materials of chemiresistive can be any material type such as metal oxides, polymers, composites, and nanoparticles. The active materials are generally prepared as thin film on the substrate. The substrate of the sensors can be glass, polymer, textile, paper, or any flexible substrate. Simply, the sensor senses gases producing meaningful resistance-change as the signal. The sensed gas molecules are called analytes and they can be small gas molecules like oxygen or large molecules such as DNA. Various researchers have exploited the device design and have sensed many types of organic matter in various medium including air, water, and land (Di Natale et al. 2014; Llobet 2013). The mechanism is that the binding analyte creates interference and creates a new conducting path, which then induces conductivity change. The performance of the device is based on the partition coefficients between the substrate-active substance and analyte vapor (Ancona et al. 2006). The sensitivity is completely

TABLE 7.1
Real-World Applications of Chemiresistive Sensors

Real-World Sensing Applications	Sensing Materials	Reference
Spoilage of meat	Single-walled carbon nanotubes with cobalt meso-aryl porphyrin complexes	Liu et al. 2015b
Public safety	Enzyme alcohol oxidase	Mitsubayashi et al. 1994
Fuel diagnosis	Poly[(2-bromo-5-hexyloxy-1,4-phenylenevinylene)-Co-(1,4-phenylenevinylene)] (BHPPV-co-PPV) copolymer	Benvenho et al. 2009
Water quality	Phenyl-capped aniline tetramer to dope single-walled carbon nanotubes	Hsu et al. 2015
Explosive sensing	Single-walled carbon nanotubes coated with a carbazolylethynylene	Zhang et al. 2015
Medical diagnosis	Polypyrrole nanowire immune sensor	Shirale et al. 2010
Diabetics monitoring	Guar gum and gold nano composite	Pandey and Nanda 2015
Homeland security	Polythiphene-carbon nanotube based sensors	Wang et al. 2008
Environmental protection	Poly(ethyleneimine) functionalized polyaniline nanothin film	Srinives et al. 2015
Detection of nerve agent simulants drugs	Carbon black/poly(ethylene oxide)	Hopkins and Lewis 2001
Fish freshness detection	Tin-oxide film	Hammond et al. 2002
Plant pathogen detection	Polypyrrole nanoribbon	Chartuprayoon et al. 2013
Breath sensors	Hierarchical SnO_2 fibers	Shin et al. 2013
DNA fingerprint sensing	ZnS nanocrystals decorated single-walled carbon nanotube	Rajesh et al. 2011
Human rheumatic heart disease detection	Single-walled carbon nanotubes with *Streptococcus pyogenes* gene	Singh et al. 2014
Plant fungus infection	Poly(9,9-dioctyl-2,7-fluorenyleneethylene)	Gruber et al. 2013
Ovarian cancer	Gold nanoparticles	Kahn et al. 2015

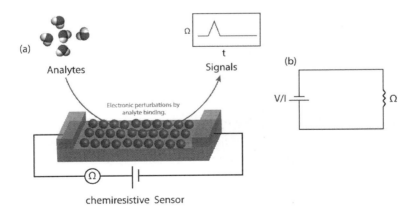

FIGURE 7.1 (a) Device design and sensing of chemiresistive sensors. (b) The circuit diagram of an ideal chemiresistive sensor.

based on resistivity or on conductivity impedance. The interaction zone (between the analyte and active substance) over the cross-sectional area of the device controls the electrical properties of the device (Zhang et al. 2008). The electrical resistance changes of the sensor array are measured by exposing them to various concentrations of different analytes. It is to be mentioned that the analytes also increase the conductivity instead of decreasing it by facilitating charge transfer, lowering barriers, or charge-injection by accidental doping.

Three influencing factors, namely, the receptor, transducer, and utility, control the device performance (Yamazoe 2005). First, the receptor is present in the sensor, which is responsible for the interaction between the analyte and active layer. The receptor generally binds functional groups to anchor the target analyte. Second, the transducer is responsible for interaction-led conductivity change. The utility generally means effective use of the surface to the observed signal. The nature of the material surface decides the degree of diffusion/interaction of the analyte on the surface of the sensor and it leads to better kinetics of the sensing. Generally, sensor response (S) is explained by Equation 7.1. It can be defined as the ratio of resistance change ($R_g - R_a$) due to exposure of analyte to the resistance (R_a is baseline resistance) (Pandey 2016). Since the device design is simple it has several advantages including easy operation, quick troubleshooting, reproducibility, fast response, portability, and low cost. Overall, the purpose the design is to sense the analyte, quantify the analytes, and display the resistive values of analytes.

$$S = \frac{R_g - R_a}{R_g} \times 100\% \qquad (7.1)$$

7.4 ANALYTES

The analyte is a chemical that is to be sensed by chemiresistive sensors. Analytes are various types of chemicals and carry information about various environments, the body, and many other organic systems (Figure 7.2). The analytes in the air (explosives, biowarfares, toxic substances) provide broad information about environments. Similarly, analytes in the body or biological system provide warning information possible dangers and disease detection. The analytes can be either solid, liquid, or gas. Based on the structure, the analytes can be classified broadly to small and large molecules. The small molecules are normally gases and small size structured molecules (e.g., ammonia). The second category is large molecules, which are generally biological molecules (e.g., DNA). A comprehensive but not exhaustive list of analytes is shown in Figure 7.2. The analytes' structure, physical state, functional groups, vapor pressure, concentration, binding kinetics, and binding/interacting nature with substrate greatly influence the sensor performance. To get a social perspective on real-time application, an alcohol test to assess for drunk driving involves sensing of ethanol and aldehyde from the breath (Di Natale et al. 2014).

7.5 SENSING MATERIAL

Sensing materials are the heart of chemiresistive sensors and they interact and sense the analyte and produce the response. In 1985, the best chemiresistive application was from copper phthalocyanine complex (Wohltjen et al. 1985). Later similar smarter materials such as metal oxides, nanoparticles, metal organic frameworks (MOFs), and composites were developed (Table 7.2). The chemical purity

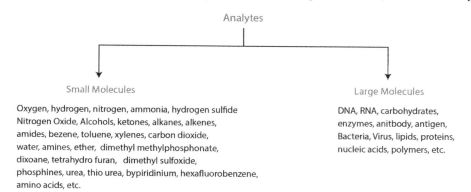

FIGURE 7.2 Classification of analytes based on the physical structure.

of the sensitive material is critical and affects every performance parameter. The choice of active material is based on various parameters such as ease of preparation, coordinating nature with the analyte, ease of film formation, thermodynamic stability, adsorption kinetics of analytes, tunable conductivity, and material compatibility with the system. Generally, active materials are solid surfaces or amorphous polymers. Herein, we classify the sensing material based on the composition as well as properties. In the following are representative examples for each type of sensing material with a focus on their merits and demerits (Mutkule et al. 2016).

TABLE 7.2
Types of Chemiresistive Sensing Materials

Sensing Material	Target Gas	Detection Limit	Reference
Metal Oxides			
(i) WO_3	NO_x	75 ppm	Marquis and Vetelino 2001
(ii) SnO_2	H_2	0.1–10 ppm	Katoch et al. 2016
(iii) CoO_3	Ethanol	1 ppm	Balouria et al. 2013
(iv) ZnO	TNT	8 ppb	Chen et al. 2010
(v) Ga_2O_3	NH_3	0.5 ppm	Pandeeswari and Jeyaprakash 2014
Polymers			
(i) Polyaniline	HCl and NH_3	100 ppm	Huang et al. 2003
(ii) PEDOT:PSS	Ethanol	325 ppm	Kriván et al. 2000
(iii) Polythiophene	Benzoic acid	10 ppm	Chang et al. 2006
(iv) PANI fibers	Ammonia	0.5 ppm	Sutar et al. 2007
(v) PEDOT	NO_2	200 ppb	Le Ouay et al. 2016
Composites			
(i) Gold nanoparticle/dendrimer	1-Propanol	100 ppm	Krasteva et al. 2002
(ii) rGO/polymer	NO_2	0.28 ppm	Yun et al. 2014
(iii) SnO_2/grapheme	LPG	50 ppm	Nemade and Waghuley 2014
(iv) Pd/PEDOT-TAA	Toluene	2 ppm	Vaddiraju and Gleason 2010
(v) CNT/DNA	Propionic acid	150 ppm	Staii et al. 2005
Complexes			
(i) Pd-porphyrine complex	NH_3	100 ppm	Wang et al. 2011
(ii) Iron (II) phthalocyanine	NO	100 ppm	Shu et al. 2010
(iii) Cobalt phthalocyanine	O_2	15 ppb	Bohrer et al. 2007
(iv) Cu_3(hexaiminotriphenylene)$_2$	NH_3	1 ppm	Campbell et al. 2015
(v) Re-thiyl radical	Ethylene	1 ppm	Chauhan et al. 2014
Nanoparticles			
(i) PbS	H_2S	50 ppm	Liu et al. 2015a
(ii) Si nanowire	TNT	1 ppb	Engel et al. 2010
(iii) ZnS	H_2S	50 ppm	Zhang et al. 2017a
(iv) Cu_2O	NO_2	100 ppb	Zhou et al. 2017
(v) CdSe	$CHCl_3$	100 ppm	Mondal et al. 2012
MOFs			
(i) $Cu_3(HHTP)_2$	NH_3	100 ppm	Yao et al. 2017
(ii) ZnO@ZIF-CoZn	Acetone	10 ppm	Yao et al. 2016
(iii) Cobalt imidazolate	Trimethylamine	2 ppm	Chen et al. 2014
(iv) $[Zn_4O]^{6+}$ terephthalic acid	Formaldehyde	5 ppm	Iswarya et al. 2012
(v) Co zeolitic imidazolate	Toluene	5 ppm	Jo et al. 2018

7.5.1 METAL OXIDE SENSORS

Metal-oxide-based sensors are the first and foremost choice for chemiresistive sensing due to their exceptional stability over atmosphere (Afzal et al. 2012). Other metal oxides include SnO_2, ZnO, TiO_2, In_2O_3, CuO, and V_2O_5 (Afzal et al. 2012; DMello et al. 2018; Hammond et al. 2002; Joshi et al. 2018; Li et al. 2012; Marquis and Vetelino 2001). The sensing mechanism depends upon the redox reaction involved during the sensing, and the concentration of gases is monitored by varying resistance. The performance of the device can be enhanced by the crystal structure and morphology. Even though metal oxides have various advantages, they suffer from selectivity. Metal-oxide-based sensors normally sense similar types of analytes present in a gas atmosphere but lack selectivity. As a result, metal-oxide-based sensors have interfering noise signals (Joshi et al. 2018). Various reports demonstrated that oxides of nitrogen (NO_x) are ideal analytes for metal-oxide-based sensors. Additionally, the metal oxide sensor exhibits peak performance at high temperatures instead of room temperature. For example, we will discuss the sensing of TNT using ZnO-based chemiresistive sensors (Chen et al. 2010). The sensors exhibited 8 ppm detection limit of TNT. Chen et al. explained that NO_2 molecules are strongly electron-withdrawing and thus reduced conductance for ZnO substrates. Hence NO_2 increased the hole concentration in the devices. The ZnO-based sensor enabled the electronic nose for explosive materials. Li et al. (2012) reported on a CuO-based H_2S sensor with the detection limit of 10 ppb. The high sensitivity is due to binding kinetics, temperature dependence, and resistance change. Joshi et al. (2014) demonstrated the polycarbazole films for the detection of H_2S with the detection of 10 ppm. The chemiresistive sensors were made with the Langmuir-Blodgett technique. They improved the performance of the device by surface modification using a gold layer.

7.5.2 POLYMERS

Polymers are the second foremost explored chemiresistive sensor material after metal oxides (Pandey 2016; Song and Choi 2013). The polymers conduct with an alternative single and double bond and hence have better conductivity. The polymers are explored due to their film-forming properties, controllable thickness, chemical stability, and tunable conductance. The film-forming property and mechanical property of polymers attracted many researchers to adapt polymers for chemiresistive sensing. One of the main disadvantages of polymers is their instability toward different pHs and that they work properly only in neutral environments. One other drawback is the slow response time of polymers due to analyte doping. In polymer-based chemiresistive sensors, the analyte HCl gas was doped with polymer films during sensing. Polypyrrole was one of the first conducting polymers utilized for chemiresistive sensing applications (Kriván et al. 2000). As an example, we will describe the PEDOT-based chemiresistive sensor for NO_2 analyte detection (Le Ouay et al. 2016). PEDOT is p-type semiconductor and its Fermi level can be tuned by oxidizing analytes. Hence, a strongly electron-withdrawing group NO_2 is used for analyzing the device performance. Le Ouay et al. observed that the conductivity of the sensor is increased when subjected to three NO_2 gas streams. The response time was 30 s and the detection limit was 200 ppb. Kwon et al. (2012) described the polypyrrole based highly sensitized and selective chemiresistive sensor for detection of ammonia. The sensor was prepared using the polydimethylsiloxane (PDMS) substrate using a dry-transfer method. The sensor had the ability to discriminate ammonia when the detection limit of the sensors is 0.1 ppm. Li et al. (2008) reported on poly(4-hexyloxy-2,5-biphenyleneethylene)-based chemiresistive sensing of 1,2-dichloroethane and other chlorinated compounds. The polymer is hydrophobic and has the advantage of water insensitivity and conduction was not perturbed by the humidity in the atmosphere.

7.5.3 COMPOSITES

Composites-based chemiresistive sensors are evolving in recent times based on the fact that combining two high-performing materials provides peak performance (Pandey 2016). The popular combinations are metal/polymer, metal oxide/polymer, nanoparticle/polymer, polymer/polymer,

CNT/polymer, graphene/quantum dots, graphene/polymer, and metal complex/polymer. Mostly, polymers are used as a second composite material because of their film-forming properties, environmental stability, mechanical bending and conductivity, and ease of device fabrication. The polymer association could help the metal oxide sensors become flexible and conform to any type of surface. The main advantage is room temperature performance of the polymer and hence the device, whereas metal oxides alone exhibit peak performance at high temperatures. Throughout research literature polythiophene, polyaniline, and polypyrrole were used as co-composite materials. These composites are sometimes called hybrid of metal or metal oxides. The mechanism for sensing is similar to the simple metal oxide sensors. But the first adsorption process is a little different. During the sensing process, the vapor molecules of analyte approach and attach to the polymer surface and then react with the metal oxides (Meng et al. 2015). Synergistically, the electronic perturbation occurs in the polymer and then the perturbation transfers to the metal through the polymer. An example of a composite is the SnO_2/graphene-based chemiresistive sensor for LPG sensing (Nemade and Waghuley 2014). Due to high surface-to-volume ratio, the sensor receives and oxidizes the analyte gas, which in turn leads to change in the chemical conductivity. Fu et al. (2013) explored the unusual gold–DNA composite for detecting methanol and other volatile organic compounds. The DNA composite gives the gold–alkanethiol properties of sensitivity and detection limit. Such rare combinations are interesting, but the cost and complexity of DNA compositing mean it has not gained wild popularity among researchers.

7.5.4 COMPLEXES

Complexes were attempted for chemiresistive sensing because of their binding/anchoring properties with gas molecules (Wang et al. 2011). Since the synthesis of complexes is easy and existed vastly in literature, various examples of the chemiresistive sensors were reported. The complexes have the advantage of having a single molecule containing both organic and inorganic parts. The organic part contributes to the solubility and film-forming nature. The inorganic part provides good intrinsic electrical conductivity and adsorption with gas molecules. The metal centers are the coordination site for analytes, and the analytes may create a weak bond with the organic region (ligand) due to weak hydrophobic interaction. The sensing mechanism is similar to the metal oxide semiconductor sensing mechanism. During binding with the oxidizing gases (NO_x), the metal sites gained hole concentration and conductivity of devices. Conversely, with the reducing gases (NH_3), electrons are injected and trapped with the complexes and decreased conductivity was observed in sensing devices. The phthalocyanine-based complexes are the dominant form of sensing and exist throughout literature. Bohrer et al. (2007) studied the cobalt-phthalocyanine complex for oxygen sensing. They reported that analyte could be considered the ligand or base that could bind to the central metal atom by a weak coordination bond. The device had a detection limit of O_2 as low as 15 ppb. Since the analyte is considered as the base, the conductivity can be directly correlated with the Lewis basicity. Similarly, the sensor response and recovery were also directly related with Lewis basicity. At the high concentration of the analyte, chemisorption (ligand binding) dominates. However, at low concentration of analyte, sensitivity responded well to transition from physisorption from chemisorption. Wang et al. (2012) reported on lead phthalocyanine-based sensors for the detection of NH_3 with the detection limit of 0.15 ppm. They enhanced the performance of the device using CNTs by anchoring phthalocyanine. The hybrid complex device structure provided higher response and sensitivity.

7.5.5 NANOPARTICLES

Nanoparticles revived old sensing materials with the perspective of nanoscale. Nanodimensional materials are often considered as next-generation materials for sensing applications (Liu et al. 2014). Nanoparticles with diameters that range from 10 to 100 nm are considered to be nanoscale.

The nanoscale materials have a high surface-to-volume ratio when compared to bulk and provide new properties and new opportunities. Nanoscale materials provide a unique surface, crystal facets, which lead to new electronic and optical properties. Nanoscale materials include quantum dots, nanowires, nanosheets, nanorings, nanohelixes, nanoflowers, and nanoporous materials. Nanoparticles have many advantages such as high surface-to-volume ratio, a large surface area to adsorb, fewer defects, and precise chemical composition. Hence, the performance of existing compounds can be increased to the next level as well as the performance of rejected materials can also be revived to the optimum level. Additionally, quantum dots have various handles to control chemiresistive sensing such as size, shape, morphology, composition, and crystal structure. The mechanism that is similar to bulk oxides is the conductivity change in the device during the analyte binding to the quantum dots (Gurlo 2011). After binding, the analyte is ionized and charge carriers created. The minor disadvantage of nanoscale chemiresistive sensors is irreproducible surfaces and the surfaces generate an irreproducible electronic signal (Gurlo 2011). Nanoparticles exhibit n-type response with the electron donating analytes and exhibiting p-type response with the electron withdrawing analytes. Liu et al. (2015a) described a PbS-based nano chemiresistive sensor for detection of H_2S gas. The sensors had an exemplary detection limit of 50 ppm and with the switching of both p-type and n-type conduction. The chemiresistive sensors showed good sensitivity and recoverable response during sensing. Ibañez et al. (2006) demonstrated the detection of toluene using gold-monolayers-based chemiresistive sensors. The gold nanoparticles had the advantage of the tuning conductivity by cluster-to-cluster distance. Penza et al. (2008) studied Pd–Pt nanoclusters for the detection of CO with the detection limit of 1 ppm. They enhanced performance of the chemiresistive sensor by using carbon nanotubes. The incorporation of Pd–Pt on carbon nanotubes gave better sensitivity, reversibility, and very low limit of ppb detection.

7.5.6 MOFs

Metal organic frameworks (MOFs) are improved counterparts of metal complex materials and often considered as alternative to zeolites. In MOFs, the metal part is called the node and ligands are called linkers. MOFs' unprecedented tunability, porosity, and structural diversity provided a new perspective to metal complexes. MOFs have extensive combination possibilities for making new materials such as organic parts, metal ions, and structural motifs. As a postsynthetic modification, MOFs have the possibility of structural modification and tuning. The porosity of MOFs has the natural ability of concentrating the gas (analyte) inside. Hence, MOFs are natural gas sensors with very good selectivity based on structure of and size of analyte (to fit in pores). One disadvantage is that the conductivity of MOFs suffer due to the large amount of organic contribution in their composition. Additionally, the synthesis demands serious expertise from researchers, preventing scalability and wide adoption by industries and researchers. Hence, MOF-based chemiresistive sensors are moderately developed. The mechanism is slightly different from the conventional mechanism. In MOFs the analyte passes through the MOFs and interact with linkers creating the analyte-induced electrical perturbation. The change in the conductivity arises from formation of surface states by chemisorption of oxidizing/reducing gases at the gas–solid interface. Normally, sensitivity is often high because the analyte spends more time inside the cavity compared to other materials. As an example, there is the cobalt-imidazolate-based metal–organic framework for sensing of trimethylamine (Chen et al. 2014). Trimethylamine evolves during the decomposition of fish and is used to identify its freshness. The concentration level of trimethylamine released from fish gradually increases with a decrease in freshness. Hence, the degradation has a linear relation with the increase in the concentration of trimethylamine. The peak detection of trimethylamine was at low operating temperature (75°C) with the detection limit of 2 ppm. Mello et al. (2018) made ZIF-67 MOF for the detection of CO_2 gas sensing and provided an attractive perspective. When the MOF was used with SnO_2, the performance of the device was multiplied. The device had a response of 16.5% with a detection limit of 5000 ppm.

7.6 SENSING PLATFORM

The scientific community used glass and silicon as a substrate for a chemiresistive sensor without any notion of possible end uses or applications. The sensors are in every part of living and nonliving system. Normally, the human body requires a conformational or curvilinear sensor for better integration. At the same time, disposable or one-time-use sensors are required for disease detection. Conventional platforms suffer from rigidity and nonconformity. Nowadays, electronics are moving in the direction of wearable electronics, smartwatches, smart glasses, wearable cameras, e-textiles, and smart packaging (packed in any space and shape) (Briand et al. 2011). Hence, the pool of substrate should be extended to meet demand. Novel platforms include paper, textile, flexible plastic, transparent plastic, and self-healing plastic. The new platforms offer folding, stretching, conformation, rollable, twistable ergonomics and easy destruction (paper-burning). Kumar et al. (2015) demonstrated the fabrication of a graphene chemiresistive sensor on a paper platform for NO_2 detection with a stunning detection limit of 300 parts per trillion. Surprisingly, the performance of the device is retained even after making with paper. Huynh and Haick (2016) demonstrated a sh-crl-PU self-healing polymer for the detection of hexane with the detection limit of 120 ppb. The self-healing sh-crl-PU platform can conform and regain its original shape by itself. The healing efficiency of the polymer is very high, and the chemiresistive sensor survived even after cutting several times at random positions. There is some real interest among researchers to achieve a flexible chemiresistive sensor for its versatility, and the mechanical properties are increasingly becoming a consideration for device design. The next level challenge will be designing flexible chemiresistive sensors without losing performance.

7.7 IMPORTANT PARAMETERS

It is worth mentioning here the important parameters and key words used in the chemiresistive sensor research (Fennell Jr. et al. 2016).

- Limit of detection (LOD)—It is the minimum amount of analyte that can be identified by chemiresistive sensors. It is expressed in terms of parts per million (ppm). The industrial standard is about parts per million but in a few cases parts per trillion is also observed (Kumar et al. 2015).
- Dynamic range—It is the range in which the minimum and maximum amount of analyte can be detected by chemiresistive sensors. The standard dynamic range frequently reported in literature is 1–100 ppm.
- Sensitivity—It is a ratio of the analyte of interest and interfering gases, e.g., analyte ethyl alcohol can be interfered by moisture present in the atmosphere. A typical sensor should be sensitive to the measurand and insensitive to any other analyte.
- Stability—It is defined as how much time the device can produce the same peak output. Stability is of great importance for commercialization.
- Drift—It is the stimuli-independent change of an observed output over a period of time. The drift in the device can generate uncertain results and false alarm during the device operation.
- Response time—The amount of time required to achieve 90% of intensity after analyte binding.
- Dead time—The amount of time required to reach 10% of intensity after signal observation.
- Reversibility—It is about the ability of a sensor to restore the initial state of the signal after observation.
- Recovery time—The time taken by the device to reach the baseline before the next measurement of analyte.
- Selectivity—The same sensor can sense various types of chemicals. But it should be selective toward a particular gas, which is called selectivity. Sensitivity without selectivity is

described as just noise. A chemiresistive sensor should respond strongly only to desired analytes. The selectivity of chemiresistive sensor to a specific gas is necessary for the success of the sensor.
- Accuracy—It is about how exactly the chemiresistive sensor reflects the amount of analyte present. The accuracy can be affected by preparation standards of chemicals as well as sensors, operation conditions, and concentration of analyte.
- Linear response—This is the concentration of gases and the change in resistance expected to be linear. The linear response helps the sensor to be easily calibrated.

7.8 SUMMARY

Chemiresistive sensors are popular among other sensing technologies due to their simplicity, selectivity, fast response time, and high sensitivity. Chemiresistive sensors have the potential to replace any other type of existing sensor (e.g., capacitive). With the current knowledge of chemiresistive sensing, all the possible directions are explored. Currently, tailoring the sensor to specific industrial applications is being carried out by various research groups. Chemiresistive sensors are mostly designed on a trial-and-error combinatorial basis using physics and chemistry as the interdisciplinary approaches. Chemiresistive sensing includes various processes such as carrier transport, chemical synthesis, thin film formation, material gas–solid interaction, electrical measurements, static, and portable packing. All the processes are intertwined and need expertise from many research fields. Hence, serious interdisciplinary approaches considering the adsorption process, gas chemistry, electronics, synthetic chemistry, device fabrication, mechanics and device physics should be carried out collaboratively. The number of disciplines involved in the research and design of chemiresistive sensors has increased drastically. The interdisciplinary approach moves the current conventional research (chasing sensitivity or new materials detection) into a new, innovative direction. The challenge is hard but the potential of chemiresistive sensors continues to motivate researchers. In summary, to study the chemiresistive sensor, researchers should understand the application and end use for next-generation sensing.

7.9 CHEMIRESISTIVE GAS SENSING MATERIALS: FUTURE DIRECTIONS

Chemiresistive sensors reached the sensitivity required to commercialize as gas sensors and are expected to become real-time sensors. The next logical directions are described as follows:

(1) The scope of sensor applications should be increased. Sensors could be used in practically any feedback system and they are key elements in every electronic product. Hence new avenues like robotics should be identified for better outreach. The application pool can be extended by identifying sensing needs and targeting the required physical properties.
(2) Sensors are expected to solve growing practical problems. For example, for body integration for preventive detection of diseases can be realized in flexible sensors for real-time applications in biological, human, robotics, process control, medical, sports, and rehabilitation fields. The sensor would conform to the skin or any curved-linear surface to integrate in any type of working system.
(3) Chemiresistive sensors should be stand-alone devices free from the energy source along with a wide dynamic range. This is possible through wireless technologies and could lead to very different applications such as one-time-use sensors for health and military applications.
(4) From a technological point of view, sensors are central for many electronic devices and machinery. Recent efforts have been toward taking technologies to third-world countries like Somalia at minimal cost or almost cost free. Hence, cost-effective chemiresistive sensors should be developed.

(5) Sensors are normally designed to collect the data. But the future sensor design should not only consider data collection but also data analysis and decision-making.
(6) Sensitivity of sensors should be increased above the state of the art to take advantage of new nanoscale materials. This would create compactness and allow easy integration with any system. By using nanoscale applications, better values can be achieved.
(7) There are a number of minor issues with existing chemiresistive sensors (e.g., poor sensitivity and peak performance at high temperature). By addressing the minor issues, we should be able to achieve the ideal commercially viable product.
(8) The usable chemicals of sensors should be extended to new types of chemicals, e.g., quaternary compounds, carbon nano types, and MOFs.
(9) An empirical and simple theoretical approach should be developed to identify viable chemical sensors. This unified approach could expand the chemical library, optimizations, and be used to customize applications (such as biofriendly magnesium-based sensors for inside the human body).
(10) For commercial viability, the chemicals should be synthesized at larger quantities. Hence, automated synthesis for chemicals used in sensors should be developed with the required quality and quantity.
(11) The toxicity of the chemicals used in chemiresistive sensors contaminates water, land, animals, and plants. Hence, the materials used in these sensors should be biofriendly and biodegradable.
(12) Challenges in the field of chemiresistive sensors are still many, and the fact that carriers (electrons or holes) have intrinsic charge and can give large interfering signals from a variety of polar and ionic species are general issues this community continues to confront.

REFERENCES

Afzal A, Cioffi N, Sabbatini L, Torsi L. (2012). NOx sensors based on semiconducting metal oxide nanostructures: progress and perspectives. *Sensors and Actuators B: Chemical* 171:25–42.

Amjadi M, Kyung KU, Park I, Sitti M. (2016). Stretchable, skin-mountable, and wearable strain sensors and their potential applications: a review. *Advanced Functional Materials* 26:1678–1698.

Ancona MG, Snow AW, Foos EE, Kruppa W, Bass R. (2006). Scaling properties of gold nanocluster chemiresistor sensors. *IEEE Sensors Journal* 6:1403–1414.

Balouria V et al. (2013). Chemiresistive gas sensing properties of nanocrystalline Co3O4 thin films. *Sensors and Actuators B: Chemical* 176:38–45.

Baruah S, Dutta J. (2009). Nanotechnology applications in pollution sensing and degradation in agriculture: a review. *Environmental Chemistry Letters* 7:191–204.

Benvenho AR, Li RW, Gruber J. (2009). Polymeric electronic gas sensor for determining alcohol content in automotive fuels. *Sensors and Actuators B: Chemical* 136:173–176.

Bohrer FI, Sharoni A, Colesniuc C, Park J, Schuller IK, Kummel AC, Trogler WC. (2007). Gas sensing mechanism in chemiresistive cobalt and metal-free phthalocyanine thin films. *Journal of the American Chemical Society* 129:5640–5646.

Briand D, Molina-Lopez F, Quintero AV, Ataman C, Courbat J, de Rooij NF. (2011). Why going towards plastic and flexible sensors? *Procedia Engineering* 25:8–15.

Campbell MG, Sheberla D, Liu SF, Swager TM, Dincă M. (2015). Cu3 (hexaiminotriphenylene) 2: an electrically conductive 2D metal–organic framework for chemiresistive sensing. *Angewandte Chemie International Edition* 54:4349–4352.

Chang JB et al. (2006) Printable polythiophene gas sensor array for low-cost electronic noses. *Journal of Applied Physics* 100:014506.

Chartuprayoon N, Rheem Y, Ng JC, Nam J, Chen W, Myung NV. (2013). Polypyrrole nanoribbon based chemiresistive immunosensors for viral plant pathogen detection. *Analytical Methods* 5:3497–3502.

Chauhan R, Moreno M, Banda DM, Zamborini FP, Grapperhaus CA. (2014). Chemiresistive metal-stabilized thiyl radical films as highly selective ethylene sensors. *RSC Advances* 4:46787–46790.

Chen E-X, Fu H-R, Lin R, Tan Y-X, Zhang J. (2014). Highly selective and sensitive trimethylamine gas sensor based on cobalt imidazolate framework material. *ACS Applied Materials & Interfaces* 6:22871–22875.

Chen PC, Sukcharoenchoke S, Ryu K, Gomez de Arco L, Badmaev A, Wang C, Zhou C. (2010). 2, 4, 6-Trinitrotoluene (TNT) chemical sensing based on aligned single-walled carbon nanotubes and ZnO nanowires. *Advanced Materials* 22:1900–1904.

Di Natale C, Paolesse R, Martinelli E, Capuano R. (2014). Solid-state gas sensors for breath analysis: a review. *Analytica Chimica Acta* 824:1–17.

DMello ME, Sundaram NG, Kalidindi SB. (2018). Assembly of ZIF-67 metal–organic framework over tin oxide nanoparticles for synergistic chemiresistive CO2 gas sensing. *Chemistry–A European Journal* 24:9220–9223.

Engel Y, Elnathan R, Pevzner A, Davidi G, Flaxer E, Patolsky F. (2010). Supersensitive detection of explosives by silicon nanowire arrays. *Angewandte Chemie International Edition* 49:6830–6835.

Fennell Jr. JF et al. (2016). Nanowire chemical/biological sensors: status and a roadmap for the future. *Angewandte Chemie International Edition* 55:1266–1281.

Fratoddi I, Venditti I, Cametti C, Russo MV. (2015). Chemiresistive polyaniline-based gas sensors: a mini review. *Sensors and Actuators B: Chemical* 220:534–548.

Fu K, Li S, Jiang X, Wang Y, Willis BG. (2013). DNA gold nanoparticle nanocomposite films for chemiresistive vapor sensing. *Langmuir* 29:14335–14343.

Gruber J et al. (2013). A conductive polymer based electronic nose for early detection of Penicillium digitatum in post-harvest oranges. *Materials Science and Engineering: C* 33:2766–2769.

Gurlo A. (2011). Nanosensors: towards morphological control of gas sensing activity. SnO 2, In 2 O 3, ZnO and WO 3 case studies. *Nanoscale* 3:154–165.

Hammond J et al. (2002). A semiconducting metal-oxide array for monitoring fish freshness. *Sensors and Actuators B: Chemical* 84:113–122.

Hopkins AR, Lewis NS. (2001). Detection and classification characteristics of arrays of carbon black/organic polymer composite chemiresistive vapor detectors for the nerve agent simulants dimethylmethylphosphonate and diisopropylmethylphosponate. *Analytical Chemistry* 73:884–892.

Hsu LH, Hoque E, Kruse P, Ravi Selvaganapathy P. (2015). A carbon nanotube based resettable sensor for measuring free chlorine in drinking water. *Applied Physics Letters* 106:063102.

Huang J, Virji S, Weiller BH, Kaner RB. (2003). Polyaniline nanofibers: facile synthesis and chemical sensors. *Journal of the American Chemical Society* 125:314–315.

Huynh TP, Haick H. (2016). Self-healing, fully functional, and multiparametric flexible sensing platform. *Advanced Materials* 28:138–143.

Ibañez FJ, Gowrishetty U, Crain MM, Walsh KM, Zamborini FP. (2006). Chemiresistive vapor sensing with microscale films of gold monolayer protected clusters. *Analytical Chemistry* 78:753–761.

Iswarya N, Kumar MG, Rajan K, Rayappan JBB. (2012). Metal organic framework (MOF-5) for sensing of volatile organic compounds. *Journal of Applied Sciences* 12:1681–1685.

Jo Y-M et al. (2018). Metal–organic framework-derived hollow hierarchical Co3O4 nanocages with tunable size and morphology: ultrasensitive and highly selective detection of methylbenzenes. *ACS Applied Materials & Interfaces* 10:8860–8868.

Joo S, Brown RB. (2008). Chemical sensors with integrated electronics. *Chemical Reviews* 108:638–651.

Joshi N et al. (2014). Flexible H2S sensor based on gold modified polycarbazole films. *Sensors and Actuators B: Chemical* 200:227–234.

Joshi N, Hayasaka T, Liu Y, Liu H, Oliveira ON, Lin L. (2018). A review on chemiresistive room temperature gas sensors based on metal oxide nanostructures, graphene and 2D transition metal dichalcogenides. *Microchimica Acta* 185:213.

Kahn N, Lavie O, Paz M, Segev Y, Haick H. (2015). Dynamic nanoparticle-based flexible sensors: diagnosis of ovarian carcinoma from exhaled breath. *Nano Letters* 15:7023–7028.

Katoch A, Abideen ZU, Kim HW, Kim SS. (2016). Grain-size-tuned highly H2-selective chemiresistive sensors based on ZnO–SnO2 composite nanofibers. *ACS Applied Materials & Interfaces* 8:2486–2494.

Krasteva N, Besnard I, Guse B, Bauer RE, Müllen K, Yasuda A, Vossmeyer T. (2002). Self-assembled gold nanoparticle/dendrimer composite films for vapor sensing applications. *Nano Letters* 2:551–555.

Kriván E, Visy C, Dobay R, Harsányi G, Berkesi O. (2000). Irregular response of the polypyrrole films to H2S. *Electroanalysis: An International Journal Devoted to Fundamental and Practical Aspects of Electroanalysis* 12:1195–1200.

Kumar S, Kaushik S, Pratap R, Raghavan S. (2015). Graphene on paper: a simple, low-cost chemical sensing platform. *ACS Applied Materials & Interfaces* 7:2189–2194.

Kwon OS, Park SJ, Yoon H, Jang J. (2012). Highly sensitive and selective chemiresistive sensors based on multidimensional polypyrrole nanotubes. *Chemical Communications* 48:10526–10528.

Le Ouay B, Boudot M, Kitao T, Yanagida T, Kitagawa S, Uemura T. (2016). Nanostructuration of PEDOT in porous coordination polymers for tunable porosity and conductivity. *Journal of the American Chemical Society* 138:10088–10091.

Li RW, Ventura L, Gruber J, Kawano Y, Carvalho LR. (2008). A selective conductive polymer-based sensor for volatile halogenated organic compounds (VHOC). *Sensors and Actuators B: Chemical* 131:646–651.

Li X, Wang Y, Lei Y, Gu Z. (2012). Highly sensitive H 2 S sensor based on template-synthesized CuO nanowires. *RSC Advances* 2:2302–2307.

Lieberzeit PA, Dickert FL. (2007). Sensor technology and its application in environmental analysis. *Analytical and Bioanalytical Chemistry* 387:237–247.

Liu H et al. (2014). Physically flexible, rapid-response gas sensor based on colloidal quantum dot solids. *Advanced Materials* 26:2718–2724.

Liu H et al. (2015a). Enhancement of hydrogen sulfide gas sensing of PbS colloidal quantum dots by remote doping through ligand exchange. *Sensors and Actuators B: Chemical* 212:434–439.

Liu SF, Petty AR, Sazama GT, Swager TM. (2015b). Single-walled carbon nanotube/metalloporphyrin composites for the chemiresistive detection of amines and meat spoilage. *Angewandte Chemie International Edition* 54:6554–6557.

Llobet E. (2013). Gas sensors using carbon nanomaterials: a review. *Sensors and Actuators B: Chemical* 179:32–45.

Mahadeva SK, Walus K, Stoeber B. (2015). Paper as a platform for sensing applications and other devices: a review. *ACS Applied Materials & Interfaces* 7:8345–8362.

Marquis BT, Vetelino JF. (2001). A semiconducting metal oxide sensor array for the detection of NOx and NH3. *Sensors and Actuators B: Chemical* 77:100–110.

Meng F-L, Guo Z, Huang X-J. (2015). Graphene-based hybrids for chemiresistive gas sensors. *TrAC Trends in Analytical Chemistry* 68:37–47.

Mitsubayashi K, Yokoyama K, Takeuchi T, Karube I. (1994). Gas-phase biosensor for ethanol. *Analytical Chemistry* 66:3297–3302.

Mondal S, Bera S, Narender G, Ray S. (2012). CdSe quantum dots-poly (3-hexylthiophene) nanocomposite sensors for selective chloroform vapor detection at room temperature. *Applied Physics Letters* 101:173108.

Mutkule SU, Navale ST, Jadhav V V, et al. (2016). Solution-processed nickel oxide films and their liquefied petroleum gas sensing activity. *Journal of Alloys Compounds* 695:2008–2015. doi: 10.1016/j.jallcom.2016.11.037.

Nemade K, Waghuley S. (2014). In situ synthesis of graphene/SnO2 quantum dots composites for chemiresistive gas sensing. *Materials Science in Semiconductor Processing* 24:126–131.

Pandeeswari R, Jeyaprakash B. (2014). High sensing response of β-Ga2O3 thin film towards ammonia vapours: influencing factors at room temperature. *Sensors and Actuators B: Chemical* 195:206–214.

Pandey S. (2016). Highly sensitive and selective chemiresistor gas/vapor sensors based on polyaniline nanocomposite: a comprehensive review. *Journal of Science: Advanced Materials and Devices* 1:431–453.

Pandey S, Goswami GK, Nanda KK. (2013). Nanocomposite based flexible ultrasensitive resistive gas sensor for chemical reactions studies. *Scientific Reports* 3:2082.

Pandey S, Nanda KK. (2015). Au nanocomposite based chemiresistive ammonia sensor for health monitoring. *ACS Sensors* 1:55–62.

Patolsky F, Zheng G, Lieber CM. (2006). Nanowire sensors for medicine and the life sciences. *Nanomedicine* 1:51–65.

Penza M, Rossi R, Alvisi M, Cassano G, Signore M, Serra E, Giorgi R. (2008). Pt-and Pd-nanoclusters functionalized carbon nanotubes networked films for sub-ppm gas sensors. *Sensors and Actuators B: Chemical* 135:289–297.

Rajesh, Das BK, Srinives S, Mulchandani A. (2011). ZnS nanocrystals decorated single-walled carbon nanotube based chemiresistive label-free DNA sensor. *Applied Physics Letters* 98:013701.

Seiyama T, Kato A, Fujiishi K, Nagatani M. (1962). A new detector for gaseous components using semiconductive thin films. *Analytical Chemistry* 34:1502–1503.

Shin J et al. (2013). Thin-wall assembled SnO2 fibers functionalized by catalytic Pt nanoparticles and their superior exhaled-breath-sensing properties for the diagnosis of diabetes. *Advanced Functional Materials* 23:2357–2367.

Shirale DJ, Bangar MA, Park M, Yates MV, Chen W, Myung NV, Mulchandani A. (2010). Label-free chemiresistive immunosensors for viruses. *Environmental Science & Technology* 44:9030–9035.

Shu JH, Wikle HC, Chin BA. (2010). Passive chemiresistor sensor based on iron (II) phthalocyanine thin films for monitoring of nitrogen dioxide. *Sensors and Actuators B: Chemical* 148:498–503.

Singh S, Kumar A, Khare S, Mulchandani A, Rajesh. (2014). Single-walled carbon nanotubes based chemiresistive genosensor for label-free detection of human rheumatic heart disease. *Applied Physics Letters* 105:213701.

Song E, Choi J-W. (2013). Conducting polyaniline nanowire and its applications in chemiresistive sensing. *Nanomaterials* 3:498–523.

Srinives S, Sarkar T, Hernandez R, Mulchandani A. (2015). A miniature chemiresistor sensor for carbon dioxide. *Analytica Chimica Acta* 874:54–58.

Staii C, Johnson AT, Chen M, Gelperin A. (2005). DNA-decorated carbon nanotubes for chemical sensing. *Nano Letters* 5:1774–1778.

Sutar D, Padma N, Aswal D, Deshpande S, Gupta S, Yakhmi J. (2007). Preparation of nanofibrous polyaniline films and their application as ammonia gas sensor. *Sensors and Actuators B: Chemical* 128:286–292.

Timmer B, Olthuis W, Van Den Berg A. (2005). Ammonia sensors and their applications—a review. *Sensors and Actuators B: Chemical* 107:666–677.

Vaddiraju S, Gleason KK. (2010). Selective sensing of volatile organic compounds using novel conducting polymer–metal nanoparticle hybrids. *Nanotechnology* 21:125503.

Wang B, Chen Z, Zuo X, Wu Y, He C, Wang X, Li Z. (2011). Comparative NH3-sensing in palladium, nickle and cobalt tetra-(tert-butyl)-5, 10, 15, 20-tetraazaporphyrin spin-coating films. *Sensors and Actuators B: Chemical* 160:1–6.

Wang B, Zhou X, Wu Y, Chen Z, He C. (2012). Lead phthalocyanine modified carbon nanotubes with enhanced NH3 sensing performance. *Sensors and Actuators B: Chemical* 171:398–404.

Wang F, Gu H, Swager TM. (2008). Carbon nanotube/polythiophene chemiresistive sensors for chemical warfare agents. *Journal of the American Chemical Society* 130:5392–5393.

Wang S, Kang Y, Wang L, Zhang H, Wang Y, Wang Y. (2013). Organic/inorganic hybrid sensors: a review. *Sensors and Actuators B: Chemical* 182:467–481.

Wohltjen H, Barger W, Snow A, Jarvis NL. (1985). A vapor-sensitive chemiresistor fabricated with planar microelectrodes and a Langmuir-Blodgett organic semiconductor film. *IEEE Transactions on Electron Devices* 32:1170–1174.

Yamazoe N. (2005). Toward innovations of gas sensor technology. *Sensors and Actuators B: Chemical* 108:2–14.

Yao MS, Lv XJ, Fu ZH, Li WH, Deng WH, Wu GD, Xu G. (2017). Layer-by-layer assembled conductive metal–organic framework nanofilms for room-temperature chemiresistive sensing. *Angewandte Chemie International Edition* 56:16510–16514.

Yao MS, Tang WX, Wang GE, Nath B, Xu G. (2016). MOF thin film-coated metal oxide nanowire array: significantly improved chemiresistor sensor performance. *Advanced Materials* 28:5229–5234.

Yun YJ et al. (2014). A 3D scaffold for ultra-sensitive reduced graphene oxide gas sensors. *Nanoscale* 6:6511–6514.

Zhang B et al. (2017a). Sensitive H2S gas sensors employing colloidal zinc oxide quantum dots. *Sensors and Actuators B: Chemical* 249:558–563.

Zhang X, Gao B, Creamer AE, Cao C, Li Y. (2017b). Adsorption of VOCs onto engineered carbon materials: a review. *Journal of Hazardous Materials* 338:102–123.

Zhang T, Mubeen S, Myung NV, Deshusses MA. (2008). Recent progress in carbon nanotube-based gas sensors. *Nanotechnology* 19:332001.

Zhang Y, Xu M, Bunes BR, Wu N, Gross DE, Moore JS, Zang L. (2015). Oligomer-coated carbon nanotube chemiresistive sensors for selective detection of nitroaromatic explosives. *ACS Applied Materials & Interfaces* 7:7471–7475.

Zhou Y, Liu G, Zhu X, Guo Y. (2017). Cu2O quantum dots modified by RGO nanosheets for ultrasensitive and selective NO2 gas detection. *Ceramics International* 43:8372–8377.

8 Role of Innovative Material in Electrochemical Glucose Sensors

R. Suresh, Claudio Sandoval, Eimmy Ramírez, R. V. Mangalaraja, and Jorge Yáñez

CONTENTS

8.1 Introduction 135
8.2 Enzymatic Glucose Sensor 136
8.3 Innovative Materials as Nonenzymatic Glucose Sensors 137
    8.3.1 Metal Nanoparticles 137
    8.3.2 Metal Oxides 138
    8.3.3 Conducting Polymers 140
    8.3.4 Other Metal Compounds 140
8.4 Improving Strategy in Nonenzymatic Glucose Sensors 142
    8.4.1 Functionalization 142
    8.4.2 Alloying 142
    8.4.3 Morphology 144
    8.4.4 Doping 145
    8.4.5 Nanocomposites 145
8.5 Conclusions 146
Acknowledgments 147
References 147

8.1 INTRODUCTION

D-glucose acts as an energy source in living organisms. Also, it plays a major role in metabolic homeostasis (Wang and Lee 2015). Maintenance of the optimum glucose level in blood is necessary. An abnormal glucose level in blood causes diabetes, a chronic health disorder affecting millions of people worldwide (Ogurtsova et al. 2017). Diabetes is caused by malfunctioning of beta (β) cells of the pancreas, which is responsible for production of insulin, a hormone. Insulin controls the glucose level in the blood. A high blood glucose level can lead to blindness; heart problems; kidney failure; and renal, cerebral, and peripheral vascular diseases in diabetic patients. To avoid these complications of diabetes, the glucose level of diabetes patients must be analyzed. To achieve accurate and rapid glucose blood sugar detection, researchers developed several glucose sensing methods.

The available glucose sensing methods are colorimetry (Kim et al 2017), chemiluminesence (Petersson 1989), microdialysis (Tholance et al. 2011), fluorescence (Chen et al. 2018), field effect transistors (Bhat et al. 2017), photoelectrochemical (Wang et al. 2019a), reverse iontophoresis (Liu et al. 2011), electrochemiluminescence (Wang et al. 2019b), and near-infrared spectroscopy (Mouazen and Walaan 2014). However, these methodologies suffer from complicated principles, matrix interference, tedious fabrication process, high cost, and low reproducibility. On the other hand, an electrochemical method for detection of glucose enjoys high potential due to greater sensitivity, accuracy, specificity, rapidity, simplicity, low cost, easy operation, lower detection limit, real-time detection, stability, biocompatibility, and construction for portable glucose sensing devices (Tian, Prestgard, and Tiwari 2014). The redox property of glucose makes it possible to detect electrochemically (Equation 8.1):

$$\text{D-Glucose} \longrightarrow \text{D-Gluconic acid} + 2H^+ + 2e^- \quad (8.1)$$

Bare working electrodes, carbon paste electrode (CPE), graphite electrode, glassy carbon electrode (GCE), gold (Au) disc, platinum (Pt) disc, silicon (Si) wafer, copper (Cu) foam, stainless steel, nickel (Ni) foam, platinum–iridium (Pt-Ir) wire, indium tin oxide (ITO), F-doped tin oxide (FTO), gold-coated glass, and B-doped diamond electrode were used to detect glucose in solution. However, they have three major drawbacks (Bond 2012; Tian, Prestgard, and Tiwari 2014). They are

1. Poor sensitivity—Kinetics of glucose oxidation at a smooth electrode surface is too slow resulting in poor sensitivity.
2. Poor selectivity—Interfering species are also oxidized in the potential range similar to the oxidation of glucose, resulting in poor selectivity.
3. Electrode fouling—The electroactivity of bare electrodes is strongly affected by adsorption of glucose electrooxidation intermediate compounds and chloride ions.

Hence, modification of electroactive materials on the surface of bare electrodes is mandatory. The suitable electroactive material may be enzymes or engineered nanomaterials. Based on these materials, modified electrodes as electrochemical glucose sensors are classified into two types. They are enzymatic and nonenzymatic sensors. In the first category, the electrode surface is modified with glucose-sensitive enzymes, which reduce oxygen to hydrogen peroxide. The generated hydrogen peroxide is monitored electrochemically. In nonenzymatic sensors, the direct electrooxidation of glucose on electroactive-nanomaterials-modified bare electrode surface is measured. So far, metallic nanoparticles, metal oxides, sulfide, selenide, tungstate, nitrides, phosphides, carbon-based materials, and conducting polymers are widely used as the modifying layer on the bare electrodes. Each and every nanomaterial exhibits its own advantages and disadvantages in glucose detection. To improve the sensing performances of nanomaterials-based glucose sensors, different improving strategies have also been proposed.

In this chapter, a concise review of innovative nanomaterials as electrochemical glucose sensors is outlined.

8.2 ENZYMATIC GLUCOSE SENSOR

In 1962, Clark and Lyons (1962) fabricated the first electrochemical sensor for the detection of glucose concentration. The principle of this method is based on glucose oxidase (GO_x) catalyzed glucose oxidation, which was monitored by using an oxygen anode and platinum cathode (Equations 8.2 and 8.3):

$$\text{D-Glucose} + O_2 \xrightarrow{GO_x} \text{D-Gluconic acid} + H_2O_2 \quad (8.2)$$

Role of Innovative Material in Electrochemical Glucose Sensors

$$O_2 + 4H^+ + 4e^- \rightarrow H_2O \tag{8.3}$$

In 1973 (Guilbault and Lubrano 1973), amperometric determination of glucose concentration based on hydrogen peroxide monitoring was developed (Equation 8.4):

$$H_2O_2 \rightarrow O_2 + 2H + 2e^- \tag{8.4}$$

Apart from GOx, glucose dehydrogenase (GDH) was also used in enzyme-based electrochemical glucose sensors (Iwasa et al. 2018). These enzyme-based sensors show good sensitivity and sufficient selectivity. Nevertheless, several drawbacks of enzyme-modified electrodes (Hwang et al. 2018) were also identified. They are (a) enzymes are much costlier; (b) the activity of enzymes depends on the experimental condition, for example, GOx exhibits its activity in the pH range of only 2–8; (c) enzymes have poor stability, for example, GOx will be denatured at above 40 °C; (d) the sensitivity of enzyme-modified electrodes is severely affected by interfering species, for instance, GDH–pyrroloquinoline quinone is affected by high concentrations of maltose or galactose; (e) an acute operational environmental condition is required to be maintained, for example, GOx-based sensors are seriously influenced by unstable humidity content and hence its level should be constantly maintained; (f) these enzymes are inefficient to transfer electrons to the electrode surface, however, this drawback could be eliminated by immobilizing the enzyme on suitable support materials (Lad, Kale, and Bryaskova 2013). Nanostructured metal oxides, including TiO_2, γFe_2O_3, ZnO, MnO_2, and ZrO_2, are used as support for enzyme immobilization. Conducting polymers like poly(sulfobetaine-3,4-ethylenedioxythiophene), and organic molecules such as benzoquinone and ubiquinone have also been used as support for enzyme-based glucose sensors. For instance, GOx was immobilized on graphene/chitosan composite modified GCE, which exhibits much higher sensitivity. The presence of graphene promotes electron transfer from the redox enzyme to the surface of the electrode.

However, enzyme immobilization is a tedious process and thus creating doubt about the reproducibility of the sensor. Hence, there is a demand for the construction of enzyme-free glucose sensors with high sensibility, stability, and reproducibility.

8.3 INNOVATIVE MATERIALS AS NONENZYMATIC GLUCOSE SENSORS

8.3.1 Metal Nanoparticles

The noble metal nanoparticles such as platinum and gold have larger active surface areas and excellent electronic conductivity. These behaviors make them appropriate to enhance the electron transfer rate between the electrode surface and glucose. The glucose sensing performances of a few nanostructured metal electrodes are given in Table 8.1. The electrochemical detection of glucose

TABLE 8.1
Glucose Sensing Performance of Some Metallic Electrodes

Sensor	Linear Range (mM)	Sensitivity (mA mM^{-1} cm^{-2})	Limit of detection (μM)	Reference
Au	0–8	0.16	0.5	Kurniawan, Tsakova, and Mirsky 2006
Pt	0.2–3.2	0.1377	5	Cao et al. 2007
Ni	0.05–7.35	—	2.2	Lu et al. 2013
Cu	0–4.711	2.432	0.19	Guo et al. 2015

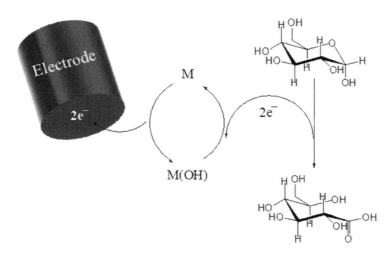

FIGURE 8.1 Electrochemical oxidation of glucose at metallic electrode. M is metal.

by metallic electrodes is explained as follows: (a) The adsorption of glucose onto active sites of the electrode takes place. The factors such as oxidation states and unfilled d-orbitals of metal centers affect the adsorption of glucose. (b) The adsorbed glucose molecule undergoes oxidation and forms gluconic acid through various reaction pathways. During electrocatalysis, the surface-bound reactive hydroxide species (OH$_{ads}$) are generated. They are also involved in the oxidation of glucose molecules (Figure 8.1). (c) When the oxidation state of the redox center is changed, the interaction between the glucose oxidation product and electrode weakens, leading to desorption of the product from the electrode surface.

It is important to mention that the aforementioned model is generally assumed for noble metal electrodes. Therefore, this mechanism is not fully applicable to numerous transition metals or metal-oxide-based electrodes. Instead, the redox behavior of the transition metal centers can explain the glucose oxidation on metallic electrodes.

The advantages of metal electrodes are as follows: (a) metals are free from interfering species like ascorbic acid and uric acid, (b) they are relatively stable under detection conditions, and (c) they have ultrahigh sensitivity. At the same time, there are some disadvantages of metal electrodes as glucose sensors: (a) They are prone to chloride poisoning, and (b) Pt and Au electrodes are expensive. Even though the cost of other transition metals like Cu and Ni electrodes are less expensive, they are unstable in neutral or acidic pH condition, (c) they have insufficient selectivity, and (d) they are active in alkaline media, and thus cannot be used in blood samples directly.

8.3.2 Metal Oxides

The investigation of metal oxides as electrochemical sensors for glucose has gained considerable attention. This is because of their redox behavior, high stability, low cost compared to noble metals, high sensitivity, and rapid response to analytes. Various transition metal oxide nanostructures were studied as electrochemical glucose sensors (Table 8.2). For instance, the glucose sensing behavior of nanostructured NiO electrode has been reported (Heyser, Schrebler, and Grez 2019). At alkaline pH conditions, the electrooxidation of glucose occurs through the Ni(OH)$_2$/NiOOH redox couple. The NiO converts to β-Ni(OH)$_2$, which undergoes oxidation to form β-NiOOH. Then, glucose undergoes oxidation by NiOOH and forms gluconolactone and Ni(OH)$_2$ (Equations 8.5–8.7):

$$NiO + OH^- \rightarrow Ni(OH)_2 \qquad (8.5)$$

TABLE 8.2
Glucose Sensing Performance of Electrochemical Sensors Based on Metal Oxide Nanostructures

Metal Oxide	Linear Rage (mM)	Sensitivity (μA mM^{-1} cm^{-2})	Limit of Detection (μM)	Reference
CuO	0.006–2.5	431.3	0.8	Wang et al. 2009
ZnO	0.01–10	5.6	0.5	Dar et al. 2011
Fe$_2$O$_3$	0.015–8	726	6.0	Cao and Wang 2011
Co$_3$O$_4$	2.04	36.25	0.97	Ding et al. 2010
TiO$_2$	0.01–0.2	201.5	—	He et al. 2018
NiO	0.1–10	206.9	1.16	Heyser, Schrebler, and Grez, 2019
Ag$_2$O	0.2–3.2	—	10	Fang et al. 2009

$$Ni(OH)_2 \rightarrow NiOOH + e^- \tag{8.6}$$

$$NiOOH + C_6H_{12}O_6 \rightarrow C_6H_{10}O_6 + Ni(OH)_2 \tag{8.7}$$

Copper oxide is one of the most extensively studied metal oxides for electrochemical glucose sensing applications. The mechanism of glucose detection (Xu et al. 2015) is based on the following reaction. In a CuO electrode, the Cu(II)/Cu(III) redox couple is responsible for glucose determination (Equations 8.8 and 8.9):

$$Cu^{(II)}O + H_2O + OH \rightarrow Cu^{(III)}(OH)_4^- + e^- \tag{8.8}$$

$$Cu^{(III)}(OH)_4^- + e^- + C_6H_{12}O_6 \rightarrow C_6H_{10}O_6 + Cu^{(II)}O \tag{8.9}$$

The Co$_3$O$_4$-modified GCE also detects glucose concentration through the Co(III)/Co(IV) redox couple (Equation 8.10), i.e., in an alkaline solution, Co$_3$O$_4$ converts as CoO$_2$, which oxidizes glucose into gluconolactone (Ding et al. 2010).

$$2CoO_2 + C_6H_{12}O_6 \rightarrow 2CoOOH + C_6H_{10}O_6 \tag{8.10}$$

Zinc oxide is also an important candidate in the field of sensors owing to its high stability, high isoelectronic point, biocompatibility, and piezoelectric property. Unlike NiO and CuO electrodes, the glucose sensing mechanism of ZnO electrode follows Equations 8.11–8.13 (Dar et al. 2011). This sensor has a detection limit of 0.5 μM.

$$O_2 ads(ZnO) + 2e^-(ZnO) \leftrightarrow 2e^- ads(O^-/O_2^-) \tag{8.11}$$

$$Glucose + O^- \rightarrow Glucanolactone + 2e^- \tag{8.12}$$

$$Glucanolactone \xrightarrow{Hydrolysis} Glucanoic\ acid \tag{8.13}$$

8.3.3 CONDUCTING POLYMERS

Conducting polymers are promising materials in electrochemical sensors due to their excellent electronic conductivity, redox behavior, tunable morphology, and easy deposition on the electrode surface. However, it should be mentioned that reports on pure conducting-polymer-modified electrodes for glucose quantification are very few. In 2013, a sensor based on poly(3-aminophenylboronic acid-co-3-octylthiophene)-modified (Figure 8.2a) GCE was reported (Nia et al. 2015) with a detection limit of 0.5 mM. Moreover, this sensor was not interfered with by dopamine, uric, and ascorbic acids.

Later, a poly(hydroxymethyl-3-4-ethylendioxythiphene) glucose sensor was developed (Figure 8.2b). This sensor showed responses in the range of 1–9 mM with good selectivity, which is comparable to commercial glucose sensors. The reason for the excellent sensing performance of this sensor was attributed to the structure of the polymer, which favors activation of the hydroxyl group and promotes oxidation of glucose molecules (Hocevar et al. 2016).

Recently (Kailasa et al. 2019), a polyaniline nanosheets-based screen-printed electrode was fabricated and applied for glucose monitoring with better selectivity. The glucose sensing mechanism is shown in Equations 8.14 and 8.15:

$$PANI + SO_4^{2-} \rightarrow PANI^{2+}\text{-}SO_4^{2-} + 2e^- \qquad (8.14)$$

$$PANI^{2+}\text{-}SO_4^{2-} + C_6H_{12}O_6 \rightarrow PANI^+ + C_6H_{10}O_6 \qquad (8.15)$$

8.3.4 OTHER METAL COMPOUNDS

Apart from metals, metal oxides, and conducting polymers, various metal compounds such as metal hydroxide, sulfide, phosphate, tungstate, hydroxide, nitride, phosphides, and tellurides have also been used for glucose determination (Figure 8.3). For example, a β-Ni(OH)$_2$ nanowire sensor with a sensitivity of 60.5 A mM^{-1} cm^{-2} was reported. It showed a linear response in the range of 20 μM–0.5 mM glucose solution with a sensitivity of 60.5l A mM^{-1} cm^{-2}. The mechanism of glucose oxidation at β-Ni(OH)$_2$ electrode is similar to Ni or NiO electrodes (Luo et al. 2012), i.e., the Ni(II)/Ni(III) redox couple involved in the glucose oxidation.

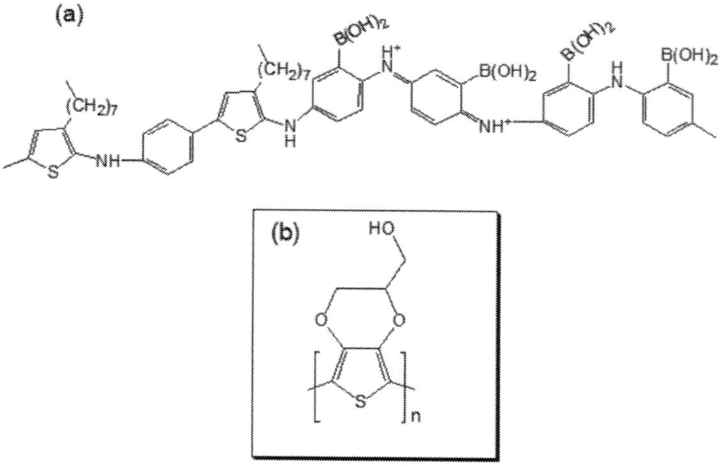

FIGURE 8.2 Structure of (a) poly(3-aminophenylboronic acid-co-3-octylthiophene) and (b) poly(hydroxymethyl-3-4-ethylendioxythiphene).

Role of Innovative Material in Electrochemical Glucose Sensors

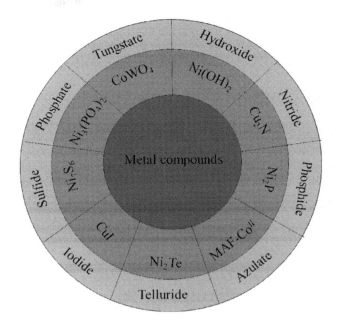

FIGURE 8.3 Different kinds of metal compounds used as electrochemical glucose sensors.

Metal tellurides have garnered significant attention in sensors because of their enhanced metallic behavior, i.e., electronegativity of Te is lower than oxygen (O) atom and thus they possess a greater metallic property. This property offers enhanced electronic transmittability. In this view, porous nickel telluride ($NiTe_2$) nanosheets synthesized by the hydrothermal method with excellent electrocatalytic activity toward glucose oxidation have been reported. This sensor showed good response to glucose in solution as well as blood serum and fruit juices (Li et al. 2018).

The $Ni_3(PO_4)_2$ nanoflakes synthesized by the hydrothermal method were coated on nickel foam electrode and used as a glucose sensor (Padmanathan, Shao, and Razeeb 2018). This sensor shows a very low detection limit of 97 nM with high sensitivity (24.39 mA mM^{-1} cm^{-2}). The proposed mechanism of glucose oxidation at $Ni_3(PO_4)_2$ electrode is given by Equations 8.16 and 8.17:

$$\text{-P-O-Ni-O-P-} + OH^- \rightarrow \text{-P-O-Ni(OH)-O-P-} + e^- \tag{8.16}$$

$$\text{-P-O-Ni(OH)-O-P-} + C_6H_{12}O_6 \rightarrow \text{-P-O-Ni(OH)-O-P-} + C_6H_{10}O_6 \tag{8.17}$$

Binary metal sulfides are remarkable electrocatalysts due to their redox nature, a synergistic effect between two metal centers, and easier electron transportation property. For example, the nickel cobalt sulfide ($NiCo_2S_4$) nanoflowers, synthesized by electrochemical method, are explored as excellent electrochemical glucose sensors in alkaline media with a detection limit of 50 nM. The reason for observed enhanced performance is due to the redox active sites, synergistic effect between Co^{4+} and Ni^{3+}, and rapid electron transportation tuned with the substitution of sulfur (Justice Babu et al. 2018).

Transition metal phosphides having excellent electrical conductivity and thus could perform well as electrocatalysts. For instance, nickel phosphide (Ni_2P) nanoparticles coated on carbon cloth was fabricated and used for glucose determination in alkaline condition (Chen et al. 2016). The Ni_2P sensor showed a sensitivity of 7792 μA mM^{-1} cm^{-2} with satisfactory selectivity and reproducibility.

The CuI nanoplates, covered with teeth-like tips, showed structure-dependent sensitive detection of glucose with a wide linear range (Khan et al. 2018). The glucose oxidation process at a CuI electrode follows Equations 8.18 and 8.19. Further it was found that Cu(III) ions are the main source for electron transfer from glucose to current collector.

$$Cu(I)/Cu(II) \rightarrow Cu(II)/Cu(III) + e^- \qquad (8.18)$$

$$Cu(II)/Cu(III) + C_6H_{12}O_6 \rightarrow Cu(I)/Cu(II) + C_6H_{10}O_6 \qquad (8.19)$$

As an example of metal nitride, copper nitride (Wang et al. 2017) also acts as a superior glucose sensor with a sensitivity of 7600 µA mM^{-1} cm^{-2} and a detection limit of 8.9 nM.

A metal azolate framework is a kind of metal–organic framework. It also acts as a good alternative for conventional metal–organic frameworks owing to its high stability and reproducibility. The existence of hydrophobic liners offers high stability to the metal azolate framework. For instance, (MAF-4-Co) Co(2-methylimidazole), a Co-based azolate-framework-based glucose sensor has been reported (Equation 8.20 and 8.21). The oxidation of glucose to gluconolactone takes place via the MAF-4-Co(II)/MAF-4-Co(III) redox couple (Lopa et al. 2019).

$$MAF\text{-}4\text{-}Co(II) \rightarrow MAF\text{-}4\text{-}Co(III) + e^- \qquad (8.20)$$

$$MAF\text{-}4\text{-}Co(III) + C_6H_{12}O_6 \rightarrow MAF\text{-}4\text{-}Co(II) + C_6H_{10}O_6 \qquad (8.21)$$

Metal tungstates have much attention as electrode materials due to their unique properties. For example, the CoWO$_4$ nanospheres act as excellent glucose sensors with high sensitivity of 1416.2 µA mM^{-1} cm^{-2} (Sivakumar et al. 2016).

8.4 IMPROVING STRATEGY IN NONENZYMATIC GLUCOSE SENSORS

Unlike glucose biosensors, pure electroactive nanomaterials do not satisfactorily performed in the detection of femto- or picomolar concentrations of glucose in solutions. Hence, to improve their sensing performance further the following strategies have been developed.

8.4.1 FUNCTIONALIZATION

The glucose sensing performance of conducting polymers and carbon materials can be significantly enhanced by the functionalization of a suitable moiety. In general, phenylboronic acid is selected for functionalization, because boronic acid can reversibly bind with cis-1,2-diols or cis-1,3-diols to form cyclic boronate ester. At pH > 8, boronic acid converts into boronate ester, which further improves its affinity to diols. Therefore, boronic acid group functionalized material could act as a sensing probe for detection of glucose. For example, pure poly(3,4-ethylenedioxythiophene) functionalized by phenylboronic acid (Figure 8.4a) modified gold chip electrode has been used for detection of glucose. The detection limit is found to be 50 µM (Huang et al. 2018). Similarly, N-phenylboronic acid functionalized polypyrrole (Figure 8.4b) deposited Pt disk electrode was fabricated and was examined for its glucose sensing property. At pH 12.0, this functionalized polypyrrole effectively binds with glucose and thus acts as a good glucose sensor (Aytac et al. 2011).

8.4.2 ALLOYING

The shortcomings of pure metallic electrodes as glucose sensors can be overcome by making alloy nanostructures. The success of alloy-based sensors depends on two important factors. One is the choice of suitable metals. Generally, Pt-based alloys are widely synthesized and used for glucose determination (Figure 8.5). The choice of other metals in Pt alloys is based on biocompatibility, enhancement of active sites, and durability. The 3d-transition metals such as Cu, Ni, Bi, and Cd are

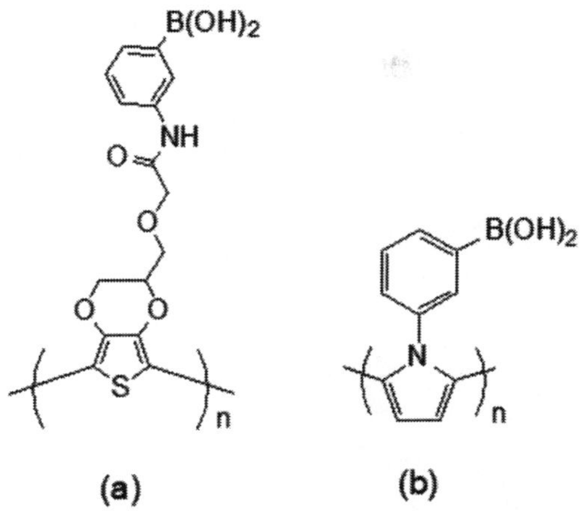

Role of Innovative Material in Electrochemical Glucose Sensors

FIGURE 8.4 Structures of (a) poly(3,4-ethylenedioxythiophene) functionalized by phenylboronic acid and (b) N-phenylboronic acid functionalized polypyrrole.

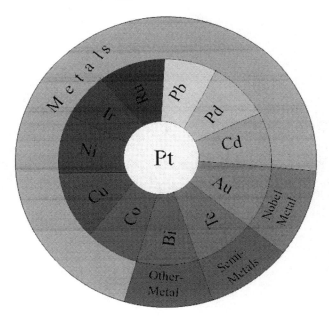

FIGURE 8.5 Pt-based alloys for glucose detection.

used as counter metallic parts in Pt alloys. These Pt alloys generally afford rapid response, good stability, and high electrocatalytic efficiency toward glucose molecules. On the other hand, the optimum composition of metals has also played a vital role in glucose detection. For example, $Pt_{75}Cu_{25}$ alloy is sensitive (135 µA mM^{-1} cm^{-2}) and selective for glucose detection rather than $Pt_{88}Cu_{12}$ and $Pt_{50}Cu_{50}$ alloys (Cao et al. 2013).

In addition to Pt alloys, transition-metal-based alloys are also explored as excellent glucose sensors. For example, Cu/Co alloy (Noh et al. 2012) electrode detects glucose efficiently, mainly due to Co(II) ions while Cu(I) ions provide limited assistance in the glucose oxidation. On the other hand, CuNi dentric nanostructures, synthesized by the electrochemical method also showed enhanced glucose oxidation in alkaline conditions (Qiu et al. 2007). For more clarity, a few metal alloy electrodes with their glucose detection limits are given in Figure 8.6.

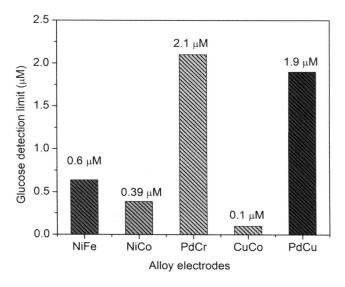

FIGURE 8.6 The NiFe (Bao et al. 2019), NiCo (Ranjani et al. 2015), PdCr (Zhao et al. 2014), CuCo (Noh et al. 2012), and PdCu (Yang et al. 2017) alloy electrodes with their glucose detection limits values.

8.4.3 Morphology

It is known that the bulk material couldn't offer excellent glucose detection due to its lack of active centers. However, nanoparticles have larger surface area and thus will provide larger active sites for glucose oxidation. Furthermore, electrocatalytic activity of nanomaterials also depends on their morphology. This is because the multidimensional nanostructures may have larger surface areas, which offer more channels for electron transfer within the nanoparticles. Also those materials are good for accessibility of glucose molecules. For a better impression, comparisons of electrochemical glucose sensing performances of CuO with various morphologies along with their synthesis methods and working electrodes are given in Table 8.3.

TABLE 8.3
Glucose Sensing Performances of CuO with Different Morphologies

Morphology	Synthesis Method	Working Electrode	Sensitivity ($\mu A\ mM^{-1}\ cm^{-2}$)	Limit of Detection	Reference
Microfibers	Electrospinning	FTO	2321	2.2 nm	Cao, Gong 2012
Nanoflowers	Template-assisted	—	2217	0.96 μM	Kong et al. 2018
Nanoplatelets	Template-free process	Cu foil	3490.7	0.50 μM	Wang and Zhang 2011
Nanorods	Electrochemical	GCE	1523.5	1 μM	Kim et al. 2019
Nanosheets	Electrochemical	Cu foam	33.95 mA cm^{-1} cm^{-2}	330 nM	Zhang et al. 2019
Nanowires	Wet-chemical	Cu foam	32330	20 nM	Liu et al. 2017
Nanopetals	Wet-chemical	GCE	7546.37	0.259 μM	Wang et al. 2018
Nanorose	Hydrothermal	Silver electrode	4.640	0.39 mM	Kim, Umar, and Hwang 2015
Nano needle	Hydrothermal	GCE	637	1.7 μM	Ma et al. 2017
Nanodisk	Microwave	Carbon	627.3	0.2 μM	Jagadeesan et al. 2019
Nano urchins	Templated growth	GCE	1634	1.97 μM	Sun et al. 2013
Dandelion	Electrochemical	Cu foil	5368	1.2 μM	Li et al. 2014

8.4.4 Doping

In general, metal ion doping in semiconductor metal oxides can induce structural defects that increase electronic conductivity. The materials with greater conductivity could exhibit enhanced electrochemical sensing performances. Hence, metal ion doping in transition metal oxides is also an important strategy to improve glucose sensors. For example, it was found that doping of Cu^{2+} ions in Co_3O_4 significantly improves its glucose sensing property (Harry et al. 2019). Similarly Zn^{2+} ion doping in $Ni(OH)_2$ also enhances the sensitivity, selectivity, and long-term stability toward glucose detection (Nguyen, Bach, and Bui 2019). Also, Li^+-ion doped NiO nanofibers also showed greater glucose sensing features than pure NiO nanofibers (Luo et al. 2015). Mahmoud et al. (2019) have reported that when compared to pure ZnO, copper-doped ZnO exhibits superior glucose sensing properties. The reasons for obtaining improved performances of Cu-doped ZnO are due to high surface area, large number of available electroactive sites for glucose oxidation and greater conductivity.

8.4.5 Nanocomposites

Nanocomposites have wide interest in many fields, including glucose sensors. This is due to integration of desired properties by combination of various kinds of materials. The materials generally used for making composites are metals, metal compounds, carbon-based materials, and organic molecules with selective binding to glucose molecules. According to the number of constituent materials, nanocomposites are classified as binary and ternary nanocomposites. In binary composites, two kinds of materials are coupled. Generally, binary composites will be metal–metal, metal–metal compound, metal compound–metal compound, metal–conducting polymer, metal–carbonaceous material, metal compound–conducting polymer, or metal compound–carbonaceous material. It has been reported that binary compounds exhibit better sensing performances than their constituent materials. Binary nanocomposite based electrodes with their glucose sensing performances are given in Table 8.4.

On the other hand, ternary composites are designed by coupling three different nanomaterials. Many researchers have found that ternary-composite-based glucose sensors are pronounced due to their relatively high conductivity, large active surface area, and fast electron transfer rate. Hence,

TABLE 8.4
Performance of Binary-Composite-Based Electrochemical Glucose Sensors

Composite Sensor[a]	Linear Range (mM)	Sensitivity ($\mu A\ mM^{-1}\ cm^{-2}$)	Limit of Detection (μM)	Reference
Pt/Polyaniline	0.01 to 8	96.1	0.7	Zhai et al. 2013
Pd–SWNTs	0.5–17	160	0.2 ± 0.05	Meng et al. 2009
Ag–CuO	100 to 1000 µM	2528.6	1.5	Felix, Kollu, and Grace 2019
PSAC/Co_3O_4	—	34.2 $mA\ mM^{-1}\ cm^{-2}$	21 nM	Madhu et al. 2015
CuO-ZnO	8.00×10^{-7} to 3.88×10^{-3} M	463.7	0.126	Wu and Yin 2013
Cu–Cu_2S	2 µM–8.1 mM	361.58	0.1	Zhang et al 2012
SDCNs/ $Ni(OH)_2$	0.0001–10.22	—	28 nM	Karikalan et al. 2016
Ni_2P/Graphene	5 µM to 1.4 mM	—	0.44	Zhang et al. 2018
Co–CoOOH	Up to 0.5	967	10.9	Lee et al. 2012

[a] SWNTs, single-walled nanotubes; PSAC, pongam-seed-shells-derived activated carbon; SDCNs, sulfur-doped carbon nanoparticles.

they show good response in wide linear range, very low detection limit, and high sensitivity and selectivity. The electrochemical activity of a few ternary-nanocomposite-based glucose sensors with their results is given in Table 8.5.

8.5 CONCLUSIONS

Based on the necessity of glucose determination, researchers have developed many quantification methods. Among those methods, the electrochemical method has attention worldwide due to its simplicity, sensitivity, portability, and cost effectiveness. The advantages and disadvantages of enzymatic and nonenzymatic electrochemical methods were discussed in this chapter. Major drawbacks of enzymatic sensors are poor stability, high cost, and stringent experimental conditions. Hence, focus on nonenzymatic sensors has been given by researchers worldwide. Early, nonenzymatic glucose sensors were fabricated by using nanostructured noble metals like Pt, Au, and Ag. Then, other transition metals like Ni and Cu electrode were also utilized as sensors. A mechanism of glucose oxidation at metallic electrodes was highlighted. Later, much effort has been given to NiO, Cu_2O/CuO, Co_3O_4, Fe_2O_3/Fe_3O_4, Ag_2O, and RuO_2. Reports on pure conducting polymer electrodes for glucose determination are also available. Various metal compounds like metal hydroxide, telluride, tungstate, nitride, phosphate, and phosphide were also examined for their electrocatalytic activity toward glucose determination. Nanomaterial-based glucose sensors offer high conductivity, active surface area, selective binding ability, and easy synthesis. Hence, their glucose sensing performances are quite appreciable. However, these sensors also have some drawbacks such as unsatisfactory performance in pico- or femtomolar concentration of glucose solution and the effect of interfering species. In order to further improve performances of nanomaterials, strategies such as functionalization, alloys, morphology tuning, and doping of metal ions have been adopted. Furthermore, the fabrication of nanocomposites by using various metals and/or metal compounds, conducting polymers, and carbon materials has been explored for their promising electrochemical glucose sensing properties.

TABLE 8.5
Performance of Ternary-Composite-Based Electrochemical Glucose Sensors

Composite Sensor[a]	Linear Range (mM)	Sensitivity ($\mu A\ mM^{-1}\ cm^{-2}$)	Detection Limit (μM)	Reference
TiO_2/PAPBA–Au	0.5 and 11	66.8 mA cm^{-2} mM^{-1}	9.3	Muthuchamy et al. 2018
Cu/graphene-resorcinol benzaldehyde	0.30–30.0 µM	—	—	Khalid, Meng, and Cao (2015)
CuO/NiO-Carbon	100 nM-4.5 mM	586.7	37 nM	Archana et al. 2019
NiO/Fe_3O_4-SH/para-amino hippuric acid	0.1–10.0 µM and 10.0–300.0 µM	—	0.13	Baghayeri et al. 2018
Ag-CuO/rGO	0.01–28	214.37	0.76	Xu et al. 2018
NiO/CuO/PANI	20–2500 µM	—	2.0	Ghanbari and Babaei 2016
Cu/Cu_2O/CuO	—	8726	0.39	Lin et al. 2018
Cu@Cu_2O/Pd	0.00–10	1.157 mA cm^{-2} mM^{-1}	742 nM	Ji et al. 2017
Ni/Co/Fe_3O_4	1 µM–11 mM	2171	0.19	Vennila et al. 2017
Pt_xCo_{1-x}/C	0.10 to 14.20	73.60	—	Sheng et al. 2015

[a] PAPBA, poly(3-aminophenyl boronic acid); rGO, reduced graphene oxide; PANI, polyaniline

ACKNOWLEDGMENTS

The authors acknowledge the National Commission for Scientific and Technological Research (CONICYT), Santiago, Chile, for financial assistance in the form of FONDECYT Post-Doctoral projects Nos. 3160499 and 3190534 and CONICYT Doctoral Fellowship 21180934.

REFERENCES

Archana V, Xia Y, Fang R, Gnana Kumar G (2019) Hierarchical CuO/NiO-carbon nanocomposite derived from metal organic framework on cello tape for the flexible and high performance nonenzymatic electrochemical glucose sensors. *ACS Sustain. Chem. Eng.* 7: 6707–6719. doi: 10.1021/acssuschemeng.8b05980

Aytac S, Kuralay F, Boyacı IH, Unaleroglu C (2011) A novel polypyrrole–phenylboronic acid based electrochemical saccharide sensor. *Sens. Actuators B Chem.* 160: 405–411. doi: 10.1016/j.snb.2011.07.069

Baghayeri M, Amiri A, Alizadeh Z, Veisi H, Hasheminejad E (2018) Non-enzymatic voltammetric glucose sensor made of ternary NiO/Fe$_3$O$_4$-SH/para-amino hippuric acid nanocomposite. *J. Electroanal. Chem.* 810: 69–77. doi: 10.1016/j.jelechem.2018.01.007

Bao C, Niu Q, Cao X, Liu C, Wang H, Lu W (2019) Ni–Fe hybrid nanocubes: an efficient electrocatalyst for nonenzymatic glucose sensing with a wide detection range. *New J. Chem.* 43:11135–11140. doi: 10.1039/C9NJ01792E

Bhat KS, Ahmad R, Yoo JY, Hahn YB (2017) Nozzle-jet printed flexible field-effect transistor biosensor for high performance glucose detection. *J. Colloid Interface Sci.* 506: 188–196. doi: 10.1016/j.jcis.2017.07.037

Bond GC (2012) Chemisorption and reactions of small molecules on small gold particles. *Molecules* 17: 1716–1743. doi: 10.3390/molecules17021716

Cao F, Gong J (2012) Nonenzymatic glucose sensor based on CuO microfibers composed of CuO nanoparticles. *Anal. Chim. Acta* 723: 39–44. doi: 10.1016/j.aca.2012.02.036

Cao X, Wang N (2011) A novel non-enzymatic glucose sensor modified with Fe$_2$O$_3$ nanowire arrays. *Analyst* 136: 4241–4246. doi: 10.1039/C1AN15367F

Cao X, Wang N, Jia S, Shao Y (2013) Detection of glucose based on bimetallic PtCu nanochains modified electrodes. *Anal. Chem.* 85: 5040–5046. doi: 10.1021/ac400292n

Cao Z, Zou Y, Xiang C, Sun LX, Xu F (2007) Amperometric glucose biosensor based on ultrafine platinum nanoparticles. *Anal. Lett.* 40: 2116–2127. doi: 10.1080/00032710701566909

Chen C, Zhang Y, Zhang Z, He R, Chen Y (2018) Fluorescent determination of glucose using silicon nanodots. *Anal. Lett.* 51: 2895–2905. doi: 10.1080/00032719.2018.1456547

Chen T, Liu D, Lu W, Wang K, Du G, Asiri AM, Sun X (2016) Three-dimensional Ni$_2$P nanoarray: an efficient catalyst electrode for sensitive and selective nonenzymatic glucose sensing with high specificity. *Anal. Chem.* 88: 7885–7889. doi: 10.1021/acs.analchem.6b02216

Clark JrLC, Lyons C (1962) Electrode systems for continuous monitoring in cardiovascular surgery. *Ann. N. Y. Acad. Sci.* 102: 29–45. doi: 10.1111/j.1749-6632.1962.tb13623.x

Dar GN, Umar A, Zaidi SA, Baskoutas S, Kim SH, Abaker M, Al-Hajry A, Al-Sayari SA (2011) Fabrication of highly sensitive non-enzymatic glucose biosensor based on ZnO nanorods. *Sci. Adv. Mater.* 3: 901–906. doi: 10.1166/sam.2011.1242

Ding Y, Wang Y, Su L, Bellagamba M, Zhang H, Lei Y (2010) Electrospun Co$_3$O$_4$ nanofibers for sensitive and selective glucose detection. *Biosens. Bioelectron.* 26: 542–548. doi: 10.1016/j.bios.2010.07.050

Fang B, Gu AX, Wang GF, Wang W, Feng YH, Zhang CH, Zhang XJ (2009) Silver oxide nanowalls grown on Cu substrate as an enzymeless glucose sensor. *ACS Appl. Mater. Interfaces* 1: 2829–2834. doi: 10.1021/am900576z

Felix S, Kollu P, Grace AN (2019) Electrochemical performance of Ag–CuO nanocomposites towards glucose sensing. *Mater. Res. Innov.* 23: 27–32. doi: 10.1080/14328917.2017.1358507

Ghanbari K, Babaei Z (2016) Fabrication and characterization of non-enzymatic glucose sensor based on ternary NiO/CuO/polyaniline nanocomposite. *Anal. Biochem.* 498: 37–46. doi: 10.1016/j.ab.2016.01.006

Guilbault GG, Lubrano GJ (1973) An enzymatic electrode for the amperometric deteremination of glucose. *Anal. Chim. Acta* 64: 439–455. doi: 10.1016/S0003-2670(01)82476-4

Guo MM, Xia Y, Huang W, Li ZL (2015) Electrochemical fabrication of stalactite-like copper micropillar arrays via surface rebuilding for ultrasensitive nonenzymatic sensing of glucose. *Electrochim. Acta* 151: 340–346. doi: 10.1016/j.electacta.2014.11.041

Harry M, Chowdhury M, Cummings F, Arendse CJ (2019) Elemental Cu doped Co$_3$O$_4$ thin film for highly sensitive non-enzymatic glucose detection. *Sens. Biosensing Res.* 23: 100262. doi: 10.1016/j.sbsr.2019.100262

He L, Liu Q, Zhang S, Zhang X, Gong C, Shu H, Wang G, Liu H, Wen S, Zhang B (2018) High sensitivity of TiO nanorod array electrode for photoelectrochemical glucose sensor and its photo fuel cell application. *Electrochem. Commun.* 94: 18–22. doi: 10.1016/j.elecom.2018.07.021

Heyser C, Schrebler R, Grez P (2019) New route for the synthesis of nickel (II) oxide nanostructures and its application as non-enzymatic glucose sensor. *J. Electroanal. Chem.* 832: 189–195. doi: 10.1016/j.jelechem.2018.10.054

Hocevar MA, Fabregat G, Armelin E, Ferreira CA, Alemán C (2016) Nanometric polythiophene films with electrocatalytic activity for non-enzymatic detection of glucose. *Eur. Polym. J.* 79: 132–139. doi: 10.1016/j.eurpolymj.2016.04.032

Huang PC, Shen MY, Yu HH, Wei SC, Luo SC (2018) Surface engineering of phenylboronic acid-functionalized poly(3,4-ethylenedioxythiophene) for fast responsive and sensitive glucose monitoring. *ACS Appl. Bio. Mater.* 1: 160–167. doi: 10.1021/acsabm.8b00060

Iwasa H, Ozawa K, Sasaki N, Kinoshita N, Yokoyama K, Hiratsuka (2018) A Fungal FAD-dependent glucose dehydrogenases concerning high activity, affinity, and thermostability for maltose-insensitive blood glucose sensor. *Biochem. Eng. J.* 140: 115–122. doi: 10.1016/j.bej.2018.09.014

Jagadeesan MS, Movlaee K, Krishnakumar T, Leonardi SG, Neri G (2019) One-step microwave-assisted synthesis and characterization of novel CuO nanodisks for non-enzymatic glucose sensing. *J. Electroanal. Chem.* 835: 161–168. doi: 10.1016/j.jelechem.2019.01.024

Ji Y, Liu J, Liu X, Yuen MMF, Fu XZ, Yang Y, Sun R, Wong CP (2017) 3D porous Cu@CuO films supported Pd nanoparticles for glucose electrocatalytic oxidation. *Electrochim. Acta* 248: 299–306. doi: 10.1016/j.electacta.2017.07.100

Justice Babu K, Raj Kumar T, Yoo DJ, Phang SM, Gnana Kumar G (2018) Electrodeposited nickel cobalt sulfide flowerlike architectures on disposable cellulose filter paper for enzyme-free glucose sensor applications. *ACS Sustainable Chem. Eng.* 6: 16982–16989. doi: 10.1021/acssuschemeng.8b04340

Kailas S, Geeta B, Jayarambabu N, Kumar Reddy RK, Sharma S, Venkateswara Rao K (2019) Conductive polyaniline nanosheets (CPANINS) for a non-enzymatic glucose sensor. *Mater. Lett.* 245: 118–121. doi: 10.1016/j.matlet.2019.02.103

Karikalan N, Velmurugan M, Chen SM, Karuppiah C (2016) Modern approach to the synthesis of Ni(OH)$_2$ decorated sulfur doped carbon nanoparticles for the nonenzymatic glucose sensor. *ACS Appl. Mater. Interfaces* 8: 22545–22553. doi: 10.1021/acsami.6b07260

Khalid B, Meng QH, Cao B (2015) A non-enzymatic thermally reduced Cu nanoparticle based graphene-resorcinol benzaldehyde glucose sensor. *Mater. Res. Innov.* 19: 91–96. doi: 10.1179/1433075X14Y.0000000219

Khan MU, You H, Zhang D, Zhang L, Fang J (2018) One-step synthesis of non-symmetric CuI nanoplates for a highly sensitive non-enzymatic glucose biosensor. *Cryst. Eng. Comm.* 20: 7582–7589. doi: 10.1039/c8ce01387j

Kim JS, Oh HB, Kim AH, Kim JS, Lee ES, Baek JY, Lee KS, Chung SC, Jun JH (2017) A study on detection of glucose concentration using changes in color coordinates. *Bioengineered* 8: 99–104. doi: 10.1080/21655979.2016.1227629

Kim K, Kim S, Lee HN, Park YM, Bae YS, Kim HJ (2019) Electrochemically derived CuO nanorod from copper-based metal organic framework for non-enzymatic detection of glucose. *Appl. Surf. Sci.* 479: 720–726. doi: 10.1016/j.apsusc.2019.02.130

Kim SH, Umar A, Hwang SW (2015) Rose-like CuO nanostructures for highly sensitive glucose chemical sensor application. *Ceram. Int.* 41: 9468–9475. doi: 10.1016/j.ceramint.2015.04.003

Kong C, Lv J, Hu X, Zhao N, Liu K, Zhang X, Meng G, Yang Z, Yang S (2018) Template-synthesis of hierarchical CuO nanoflowers constructed by ultrathin nanosheets and their application for non-enzymatic glucose detection. *Mater. Lett.* 219: 134–137. doi: 10.1016/j.matlet.2018.02.067

Kurniawan F, Tsakova V, Mirsky VM (2006) Gold nanoparticles in nonenzymatic electrochemical detection of sugars. *Electroanalysis* 18: 1937–1942. doi: 10.1002/elan.200603607

Lad U, Kale GM, Bryaskova R (2013) Glucose oxidase encapsulated polyvinyl alcohol–silica hybrid films for an electrochemical glucose sensing electrode. *Anal. Chem.* 85: 6349–6355. doi: 10.1021/ac400719h

Lee KK, Loh PY, Sow CH, Chin WS (2012) CoOOH nanosheets on cobalt substrate as a non-enzymatic glucose sensor. *Electrochem. Commun.* 20: 128–132. doi: 10.1016/j.elecom.2012.04.012

Li K, Fan G, Yang L, Li F (2014) Novel ultrasensitive non-enzymatic glucose sensors based on controlled flower-like CuO hierarchical films. *Sens. Actuators B Chem.* 199: 175–182. doi: 10.1016/j.snb.2014.03.095

Li Y, He X, Guo M, Lin D, Xu C, Xie F, Sun X (2018) Porous NiTe nanosheet array: an effective electrochemical sensor for glucose detection. *Sens. Actuators B Chem.* 274: 427–432. doi: 10.1016/j.snb.2018.07.172

Lin LY, Karakocak BB, Kavadiy S, Soundappan T, Biswas P (2018) A highly sensitive non-enzymatic glucose sensor based on Cu/Cu$_2$O/CuO ternary composite hollow spheres prepared in a furnace aerosol reactor. *Sens. Actuators B Chem.* 259: 745–752. doi: 10.1016/j.snb.2017.12.035

Liu J, Jiang L, Liu H, Cai X (2011) A bifunctional biosensor for subcutaneous glucose monitoring by reverse iontophoresis. *J. Electroanal. Chem.* 660: 8–13. doi: 10.1016/j.jelechem.2011.05.012

Liu X, Yang W, Chen L, Jia J (2017) Three-dimensional copper foam supported CuO nanowire arrays: an efficient non-enzymatic glucose sensor. *Electrochim. Acta* 235: 519–526. doi: 10.1016/j.electacta.2017.03.150

Lopa NS, Rahman MM, Ahmed F, Ryu T, Lei J, Choi I, Kim DH, Lee YH, Kim W (2019) A chemically and electrochemically stable, redox-active and highly sensitive metal azolate framework for non-enzymatic electrochemical detection of glucose. *J. Electroanal. Chem.* 840: 263–271 doi: 10.1016/j.jelechem.2019.03.081

Lu WB, Qin XY, Asiri AM, Al-Youbi AO, Sun XP (2013) Ni foam: a novel three-dimensional porous sensing platform for sensitive and selective nonenzymatic glucose detection. *Analyst* 138: 417–420. doi: 10.1039/C2AN36138H

Luo X, Zhang Z, Wan Q, Wu K, Yang N (2015) Lithium-doped NiO nanofibers for non-enzymatic glucose sensing. *Electrochem. Commun.* 61: 89–92. doi: 10.1016/j.elecom.2015.10.005

Luo Z, Yin S, Wang K, Li H, Wang L, Xu H, Xia J, (2012) Synthesis of one-dimensional β-Ni(OH)$_2$ nanostructure and their application as nonenzymatic glucose sensors. *Mater. Chem. Phys.* 132: 387–394. doi: 10.1016/j.matchemphys.2011.11.041

Ma X, Zhao Q, Wang H, Ji S (2017) Controlled synthesis of CuO from needle to flower-like particle morphologies for highly sensitive glucose detection. *Int. J. Electrochem. Sci.* 12: 8217–8226. doi: 10.20964/2017.09.37

Madhu R, Veeramani V, Chen SM, Manikandan A, Lo AY, Chueh YL (2015) Honeycomb-like porous carbon–cobalt oxide nanocomposite for high-performance enzymeless glucose sensor and supercapacitor applications. *ACS Appl. Mater. Interfaces* 7: 15812–15820. doi: 10.1021/acsami.5b04132

Mahmoud A, Echabaane M, Omri K, El Mir L, Ben Chaabane R (2019) Development of an impedimetric non enzymatic sensor based on ZnO and Cu doped ZnO nanoparticles for the detection of glucose. *J. Alloys Compd.* 786: 960–968. doi: 10.1016/j.jallcom.2019.02.060

Meng L, Jin J, Yang G, Lu T, Zhang H, Cai C (2009) Nonenzymatic electrochemical detection of glucose based on palladium–single-walled carbon nanotube hybrid nanostructures. *Anal. Chem.* 81: 7271–7280. doi: 10.1021/ac901005p

Mouazen AM, Walaan NA (2014) Glucose adulteration in Saudi honey with visible and near infrared spectroscopy. *Int. J. Food Prop.* 17: 2263–2274. doi: 10.1080/10942912.2013.791837

Muthuchamy N, Gopalan A, Lee KP (2018) Highly selective non-enzymatic electrochemical sensor based on a titanium dioxide nanowire–poly(3-aminophenyl boronic acid)–gold nanoparticle ternary nanocomposite. *RSC Adv.* 8: 2138–2147. doi: 10.1039/c7ra09097h

Nguyen DM, Bach LG, Bui QB (2019) Hierarchical nanosheets based on zinc-doped nickel hydroxide attached 3D framework as free-standing nonenzymatic sensor for sensitive glucose detection. *J. Electroanal. Chem.* 837: 86–94. doi: 10.1016/j.jelechem.2019.02.019

Nia PM, Lorestani F, Woi PM, Alias Y (2015) A novel non-enzymatic H$_2$O$_2$ sensor based on polypyrrole nanofibers–silver nanoparticles decorated reduced graphene oxide nano composites. *Appl. Surf. Sci.* 332: 648–656. doi: 10.1016/j.apsusc.2015.01.189

Noh HB, Lee KS, Chandra P, Won MS, Shim YB (2012) Application of a Cu–Co alloy dendrite on glucose and hydrogen peroxide sensors. *Electrochim. Acta* 61: 36–43. doi: 10.1016/j.electacta.2011.11.066

Ogurtsova K, Fernandes JDDR, Huang Y, Linnenkamp U, Guariguata L, Cho NH, Cavan D, Shaw JE, Makaroff LE (2017) IDF diabetes atlas: global estimates for the prevalence of diabetes for 2015 and 2040. *Diabetes Res. Clin. Pract.* 128: 40–50. doi: 10.1016/j.diabres.2017.03.024

Padmanathan N, Shao H, Razeeb KM (2018) Multifunctional nickel phosphate nano/microflakes 3d electrode for electrochemical energy storage, nonenzymatic glucose, and sweat pH sensors. *ACS Appl. Mater. Interfaces* 10: 8599–8610. doi: 10.1021/acsami.7b17187

Petersson BA (1989) Evaluation of an enzymatic method for determination of glucose in whole blood using flow injection analysis with detection by chemiluminescence. *Anal. Lett.* 22: 83–100. doi: 10.1080/00032718908051187

Qiu R, Zhang XL, Qiao R, Li Y, Kim YI, Kang YS (2007) CuNi dendritic material: synthesis, mechanism discussion, and application as glucose sensor. *Chem. Mater.* 19: 4174–4180. doi: 10.1021/cm070638a

Ranjani M, Sathishkumar Y, Lee YS, Yoo DJ, Kim AR, Gnana kumar G (2015) Ni–Co alloy nanostructures anchored on mesoporous silica nanoparticles for non-enzymatic glucose sensor applications. *RSC Adv.* 5: 57804–57814. doi: 10.1039/c5ra08471g

Sheng Q, Mei H, Wu H, Zhang X, Wang S (2015) A highly sensitive non-enzymatic glucose sensor based on Pt$_x$Co$_{1-x}$/C nanostructured composites. *Sens. Actuators B Chem.* 207: 51–58. doi: 10.1016/j.snb.2014.09.079

Sivakumar M, Madhu R, Chen SM, Veeramani V, Manikandan A, Hung WH, Miyamoto N, Chueh YL (2016) Low-temperature chemical synthesis of CoWO$_4$ nanospheres for sensitive nonenzymatic glucose sensor. *J. Phys. Chem. C* 120: 17024–17028. doi: 10.1021/acs.jpcc.6b04116

Sun S, Zhang X, Sun Y, Zhang J, Yang S, Song X, Yang Z (2013) A facile strategy for the synthesis of hierarchical CuO nanourchins and their application as non-enzymatic glucose sensors. *RSC Adv.* 3: 13712–13719. doi: 10.1039/C3RA41098F

Tholance Y, Barcelos G, Quadrio I, Renaud B, Dailler F, Liaudet AP (2011) Analytical validation of microdialysis analyzer for monitoring glucose, lactate and pyruvate in cerebral microdialysates. *Clin. Chim. Acta* 412: 647–654. doi: 10.1016/j.cca.2010.12.025

Tian K, Prestgard M, Tiwari A (2014) A review of recent advances in nonenzymatic glucose sensors. *Mater. Sci. Eng. C* 41: 100–118. doi: 10.1016/j.msec.2014.04.013

Vennila P, Yoo DJ, Kim AR, Gnana kumar G (2017) Ni-Co/Fe$_3$O$_4$ flower-like nanocomposite for the highly sensitive and selective enzyme free glucose sensor applications. *J. Alloys Compd.* 703: 633–642. doi: 10.1016/j.jallcom.2017.01.044

Wang D, Liang Y, Su Y, Shang Q, Zhang C (2019a) Sensitivity enhancement of cloth-based closed bipolar electrochemiluminescence glucose sensor via electrode decoration with chitosan/multi-walled carbon nanotubes/graphene quantum dots-gold nanoparticles. *Biosens. Bioelectron.* 130: 55–64. doi: 10.1016/j.bios.2019.01.027

Wang HC, Lee AR (2015) Recent developments in blood glucose sensors. *J. Food Drug Anal.* 23: 191–200. doi: 10.1016/j.jfda.2014.12.001

Wang J, Zhang WD (2011) Fabrication of CuO nanoplatelets for highly sensitive enzyme-free determination of glucose. *Electrochim. Acta* 56: 7510–7516. doi: 10.1016/j.electacta.2011.06.102

Wang S, Li S, Wang W, Zhao M, Liu J, Feng H, Chen Y, Gu Q, Du Y, Hao W (2019b) A non-enzymatic photoelectrochemical glucose sensor based on BiVO$_4$ electrode under visible light. *Sens. Actuators B Chem.* 291: 34–41. doi: 10.1016/j.snb.2019.04.057

Wang W, Zhang LL, Tong SF, Li X, Song WB (2009) Three-dimensional network films of electrospun copper oxide nanofibers for glucose determination. *Biosens. Bioelectron.* 25: 708–714. doi: 10.1016/j.bios.2009.08.013

Wang X, Ge CY, Chen K, Zhang YX (2018) An ultrasensitive non-enzymatic glucose sensors based on controlled petal-like CuO nanostructure. *Electrochim. Acta* 259: 225–232. doi: 10.1016/j.electacta.2017.10.182

Wang Z, Cao X, Liu D, Hao S, Kong R, Du G, Asiri AM, Sun X (2017) Copper-nitride nanowires array: an efficient dual-functional catalyst electrode for sensitive and selective non-enzymatic glucose and hydrogen peroxide sensing. *Chem. Eur. J.* 23: 4986–4989. doi: 10.1002/chem.201700366

Wu J, Yin F (2013) Easy fabrication of a sensitive non-enzymatic glucose sensor based on electrospinning CuO-ZnO nanocomposites. *Integr. Ferroelectr.* 147: 47–58, doi: 10.1080/10584587.2013.790695

Xu D, Zhu C, Meng X, Chen Z, Li Y, Zhang D, Zhu S (2018) Design and fabrication of Ag-CuO nanoparticles on reduced graphene oxide for nonenzymatic detection of glucose. *Sens. Actuators B Chem.* 265: 435–442. doi: 10.1016/j.snb.2018.03.086

Xu WN, Dai S, Wang X, He XM, Wang MJ, Xi Y, Hu CG, (2015) Nanorod-aggregated flower-like CuO grown on a carbon fiber fabric for a super high sensitive non-enzymatic glucose sensor. *J. Mater. Chem. B* 3: 5777–5785. doi: 10.1039/C5TB00592B

Yang H, Wang Z, Li C, Xu C (2017) Nanoporous PdCu alloy as an excellent electrochemical sensor for H$_2$O$_2$ and glucose detection. *J. Colloid Interface Sci.* 491: 321–32. doi: 10.1016/j.jcis.2016.12.041

Zhai D, Liu B, Shi Y, Pan L, Wang Y, Li W, Zhang R, Yu G (2013) Highly sensitive glucose sensor based on Pt nanoparticle/polyaniline hydrogel heterostructures. *ACS Nano* 7: 3540–3546. doi: 10.1021/nn400482d

Zhang L, Liang H, Ma X, Ye C, Zhao G (2019) A vertically aligned CuO nanosheet film prepared by electrochemical conversion on Cu-based metal-organic framework for non-enzymatic glucose sensors. *Microchem. J.* 146: 479–485. doi: 10.1016/j.microc.2019.01.042

Zhang X, Wang L, Ji R, Yu L, Wang G (2012) Nonenzymatic glucose sensor based on Cu–Cu$_2$S nanocomposite electrode. *Electrochem. Commun.* 24: 53–56. doi: 10.1016/j.elecom.2012.08.014

Zhang Y, Xu J, Xia J, Zhang F, Wang Z (2018) MOF-derived porous Ni$_2$P/graphene composites with enhanced electrochemical properties for sensitive nonenzymatic glucose sensing. *ACS Appl. Mater. Interfaces* 10: 39151–39160. doi: 10.1021/acsami.8b11867

Zhao D, Wang Z, Wang J, Xu C (2014) The nanoporous PdCr alloy as a nonenzymatic electrochemical sensor for hydrogen peroxide and glucose. *J. Mater. Chem. B* 2: 5195–5201. doi: 10.1039/c3tb21778g

9 Electrocatalysts for Wastewater Treatment

Prasenjit Bhunia, Kingshuk Dutta, and M. Abdul Kader

CONTENTS

9.1 Introduction 153
9.2 Fundamentals of Electrocatalytic Oxidation 154
    9.2.1 Configuration of Electrocatalytic Reactor 155
    9.2.2 Mechanism of Electrocatalytic Oxidation 156
        9.2.2.1 Direct Oxidation 158
        9.2.2.2 Indirect Oxidation 159
    9.2.3 Categorization of Electrocatalytic Anode Material 160
9.3 Recent Advances in Electrocatalysts for the Decontamination of Wastewater 161
9.4 Superiorities and Inferiorities of Electrocatalytic Approach 163
9.5 Conclusion 164
References 164

9.1 INTRODUCTION

In recent years, pollution of the aquatic environment is a major emerging problem caused by the discharged organic and inorganic chemicals from municipal and industrial sources (Albadarin et al. 2017; Daneshvar et al. 2017). Especially, industries are continuously engaged with the production of enormous volumes of wastewater containing organic and inorganic contaminants, which are extremely detrimental for the aquatic environment. The poor biodegradability of the discharged chemicals from the industries has been demonstrated bearing a high degree of pollutants, such as high dissolved solids, chemical oxygen demand (COD), color, and chloride content (Bisschops and Spanjers 2003, Kim et al. 2003, Ölmez, Kabdaşlı, and Tünay 2007). Therefore, untreated industrial effluent causes severe damage to the aquatic environment. These issues have led to a number of scientific studies and development of industrial techniques with increasing awareness of the potential risk of contaminants present in industrial effluents. Generally, the decontamination techniques are selected depending on the nature and concentration of the contaminants. Due to the extreme diverse features of the industrial wastewater, which usually carries a complicated mixture of organic as well as inorganic contaminants, decontamination techniques are urgently required (Woisetschläger et al. 2013; Sirés et al. 2014). After a thorough investigation, several decontamination techniques, including biological processes, adsorption, membrane processes, and chemical coagulation, have been developed for the treatment of industrial wastewater (Dutta and De 2017a,b; Dutta and Rana 2019; Naushad et al. 2016). However, the critical drawbacks of the aforementioned processes are expensive chemical coagulants and adsorbents, membrane fouling, and generation of secondary pollutants, which limit their scale-up applications (Kaur, Sangal, and Kushwaha 2015; Georgiou, Melidis, and Aivasidis 2002; Cañizares et al. 2006; Kaur, Kushwaha, and Sangal 2018). Although the biological processes are more favorable in terms of cost-effective, long-term treatment and environment friendliness, the limited range of operational pH of the microorganisms with extremely toxic contaminants, nonbiodegradability of the synthetic dyes, or reappearance of the refractory by-products are the major disadvantages. Similarly, a high concentration of contaminants can be eliminated through incineration and chemical oxidation, but serious emission problems and generation of toxic by-products set them back (Sirés et al. 2014). In this regard, electrocatalytic oxidation (EO), which involves degradation through the interaction of electrocatalysts and contaminants, is attracting enormous attention for the decontamination of industrial effluents nowadays. The essence of EO as an encouraging alternative technique, due to its wide application for the decontamination of effluents, was elucidated long back in 1970s (Martínez-Huitle et al. 2015). This technique is recognized as being environmentally benign, because of its deep involvement of very clean reagent—"the electron." Of course, there are other advantages related to its versatility, viz. high energy efficiency, amenability to automation, and safety owing to its mild operation conditions (Rajeshwar, Ibanez, and Swain 1994). In practice, there are several hurdles to be faced to attain high conversion efficiencies. Consequently, several strategies, such as the use of suitable electrocatalysts, the promotion of turbulent regimes, preconcentration, and use of multistage systems, need to be adopted (Rajeshwar and Ibanez 1997).

As the wastewater can either be reused for other purposes or recycled through proper management, "wastewater" can be reconsidered as "a used resource" rather than "a waste." It should be mentioned here that the nutrients can be recovered from wastewater electrochemically through sustainable decontamination of detrimental chemicals or recovery of some essential chemicals (Yoshino, Cong, and Sakakibara 2013; Lahav et al. 2013). The versatility of the EO technique has been investigated with effluents containing several contaminants/ingredients, including dyes and dye effluents, pesticides and herbicides, phenolic compounds, pharmaceuticals, and personal care products (Wu, Huang, and Lim 2014). In addition, conventional water quality parameters, such as COD, biochemical oxygen demand (BOD), suspended solids (SS), and nutrients (nitrogen, phosphorus, and phosphate), are also evaluated through application of the EO technique. A large number of electrocatalytic anodes, including metal oxides (MO_x) such as ruthenium dioxide (RuO_2) (Kim et al. 2010; Arikawa, Murakami, and Takasu 1998; Feng and Li 2003; Ardizzone, Fregonara, and Trasatti 1990), iridium dioxide (IrO_2) (Li et al. 2009; Chen, Chen, and Yue 2001), tantalum-doped iridium dioxide (Ta-IrO_2) (Ren et al. 2015); antimony-doped tin oxide (Sb-SnO_2) (Shao et al. 2014; Yang et al. 2012; Borras et al. 2007; Yang, Kim, and Park 2014; Yang, Choi, and Park 2015), lead dioxide (PbO_2) (Panizza and Cerisola 2007; Iniesta et al. 2002) and bismuth-doped titanium oxide (Bi-TiO_2) (Park, Vecitis, and Hoffmann 2008, 2009; Park et al. 2008, 2012; Ji et al. 2009; Kim et al. 2013; Cho and Hoffmann 2014), and nonoxides such as platinum (Pt) (Li et al. 2009; Iniesta et al. 2002), carbon (Kuramitz et al. 2002; Haque, Cho, and Kwon 2014) and boron-doped diamond (BDD) (Cañizares et al. 2002), have been examined for the electrocatalytic decontamination of wastewater (Cong et al. 2016). Mostly the metal oxides are coated on Ti foil. Depending on the onset potential as well as the catalytic efficiency for single or multiple electron transfer kinetics, the catalysts are segregated as either low or high oxygen evaluation potential anodes (Panizza and Cerisola 2009). It should be noted here that the strong oxidizing intermediates, including hydroxyl radical (•OH) (H_2O/•OH, +2.80 V vs. SHE), ozone (O_3) (O_2/O_3, +2.07 V vs. SHE), peroxodisulfate ($S_2O_8^{2-}$) (SO_4^{2-}/$S_2O_8^{2-}$, +2.01 V vs. SHE), and chloride radical (•Cl) (Cl^-/ClO_2^-, +1.57 V vs. SHE), are generally corresponded to decompose organic contaminants in electrocatalytic advance oxidation processes (EAOPs) (Chen 2004; Martínez-Huitle and Andrade 2011). The most attractive features of EAOPs for the treatment of wastewater are high treatment performances, accumulation of less toxic by-products, and finally its environmentally benign technology. This chapter does not intend to cover all the literature, but to focus on the fundamentals of EO and to demonstrate an overview regarding the electrocatalysts employed in the technique for decontamination of wastewater, so as to offer to the readers an easy grasp and integral view of the field.

9.2 FUNDAMENTALS OF ELECTROCATALYTIC OXIDATION

Electrochemistry has a significant role in the EO technique, and the efficiency of the electrolytic cell depends on a number of parameters. The efficiency of the EO technology has focused on the configuration of the electrolytic cell through the proper modification of electrodes and electrocatalysts.

Electrocatalysts for Wastewater Treatment

9.2.1 Configuration of Electrocatalytic Reactor

A conceptual configuration of an electrocatalytic reactor (Figure 9.1) for the decontamination of wastewater through electrooxidation includes a current supply, a cathode, an anode, and the electrolyte (the wastewater under investigation). The representative reactions involved at the respective electrodes are also displayed in Figure 9.1 (Cong et al. 2016; Martínez-Huitle and Andrade 2011). More precisely, the anode/cathode couple is dipped in the supporting electrolyte, which usually is the effluent under investigation. A direct current (DC) power supply is used to maintain a constant cell voltage or current in between the anode/cathode couple. Therefore, the treatment of effluents as well as the development of new techniques or the less harmful by-products through EO is often mentioned as the process-integrated environmental protection. Generally, the EO technique delivers two functions during decontamination with the aim of oxidizing the contaminants, not only to CO_2 and water (also called electrochemical combustion or mineralization) but also to biodegradable products (Martínez-Huitle and Ferro 2006).

As the electrocatalytic reaction takes place on the electrode surfaces, the most important challenge in the cell configuration is to maintain high mass transfer rates. Thus, common techniques, such as high fluid velocity, gas purging, introduction of several types of turbulence promoters, and use of baffles, are applied to improve the mass transfer to the surface of the electrode. Sonoelectrochemical processes may also be employed to assist the mass transfer (Chen 2004). The cell should be constructed in such a way that all the components are easily accessible and exchangeable to the electrolyte/effluent. In this regard, two types of electrodes, mainly two-dimensional and three-dimensional, are known to exist. Obviously, the latter electrode ensures a high value of electrode surface-to-volume ratio. Again, the electrodes can be divided into two types, namely static and moving electrodes. Due to turbulence promotion, the latter type of electrode leads to improved values of the mass transport coefficient. In recent trends in reactors for EO, the static parallel and cylindrical electrodes are designed and employed in two-dimensional electrodes for the decontamination of wastewater. For scale-up cell designs, the parallel plate geometry in a filter press arrangement is widely used to obtain a larger electrode size through the use of either more electrode pairs or an increased number of stacked cells (Rajeshwar and Ibanez 1997). It should be mentioned here that during the design of the reactor, the configuration of the

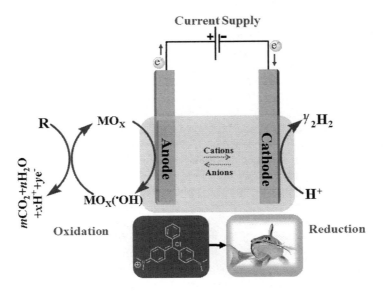

FIGURE 9.1 Speculative fundamental structure of an electrocatalytic reactor with representative reactions on the respective electrode surfaces.

cell, i.e., either divided or undivided, must be considered. Especially, in case of divided cells, a porous diaphragm or an ion-conducting membrane must be deployed to separate the anolyte and the catholyte. As the separators are expensive and reduce the electrode distance, the use of divided cells should be avoided as much as possible. Besides, they encounter a host of mechanical and corrosion problems (Wendt and Kreysa 1999). Finally, the limiting hydrodynamic behavior, including plug flow and perfect mixing, should be considered to maintain the flow characteristics in a reactor.

9.2.2 Mechanism of Electrocatalytic Oxidation

In 1990, the mechanism of oxidation of organic contaminants was first proposed by Feng and Johnson (1990). According to their investigation, a reactive oxygenated intermediate was generated at high potential during oxygen evaluation reaction (OER). During water discharge, the anodic oxygen from H_2O was transferred to the organic contaminants through the adsorbed hydroxyl radicals (•OH) according to the following reactions:

$$MO_x[\,] + H_2O \rightarrow MO_x[\cdot OH] + H^+ + e^- \quad (9.1)$$

$$MO_x[\cdot OH] + R \rightarrow MO_x[\,] + R_{Ox} + H^+ + e^- \quad (9.2)$$

where, $MO_x[\,]$ represents the adsorption sites on the surface of MO_x electrocatalysts for the sorption of the •OH intermediates and R represents the organic contaminants present in wastewater. In addition to these reactions, due to the oxidation of water through OER, an undesirable and inevitable reaction may also occur:

$$MO_x[\cdot OH] + H_2O \rightarrow MO_x[\,] + O_2 + 3H^+ + 3e^- \quad (9.3)$$

However, the oxidation of organic contaminants may also take place directly on the surface of the anode, according to the tentative indication by Feng and Johnson (1990).

In principle, the generated reactive intermediates, for instance •OH, adsorbed on the MO_x electrode surfaces, are responsible for the initiation of the electrocatalytic anodic reactions. The intermediates are consumed through the surface reactions with the aqueous contaminants (Panizza and Cerisola 2009; Chen 2004; Kesselman et al. 1997; Lee 2015). Also, some mobile reactive intermediates, such as Cl_2^- and $HOCl/OCl^-$, directly get produced from the supporting electrolytes, which are responsible for the reaction in the bulk (Park, Vecitis, and Hoffmann 2008; Park et al. 2008). Usually, high energy is consumed for the generation of these reactive species due to oxidation of supporting ions and water via a single electron. Based on the catalytic activities of the MO_xs for single-electron and multielectron transfer kinetics as well as their onset potential, they are differentiated into two categories, namely low and high oxygen evolution potential (Panizza and Cerisola 2009). For instance, due to higher oxygen evolution potential of $Sb\text{-}SnO_2$ anode compared to dimensionally stable anodes RuO_2 and IrO_2, it manifests high electrocatalytic activities toward phenol degradation in sulfate electrolyte. This may be due to the efficient generation of surface-adsorbed •OH according to Reactions 9.1 and 9.2 (Kim et al. 2010). However, indistinguishable electrocatalytic activities have been observed for RuO_2 and $Sb\text{-}SnO_2$ anodes in chloride electrolyte, due to superficial generation of •Cl according to Reaction 9.4a (Kim et al. 2010). Finally, various species, such as hypochlorous acid (HOCl), chlorine (Cl_2), dichloride anions (Cl_2^-), and high-valent chlorine species (ClO_4^-, ClO_3^-, and ClO_2), are converted from the fugitive •Cl according to Reactions 9.4b to 9.4d. Concomitantly, the molecular oxygen (O_2) is usually evolved in both electrolytes, following Reaction 9.5.

$$MO_x[\cdot OH] + Cl^- \rightarrow MO_x[\cdot Cl] + OH^- \quad (9.4a)$$

$$2MO_x[\cdot Cl] \rightarrow 2MO_x[] + Cl_2 \quad (9.4b)$$

$$MO_x[\cdot Cl] + Cl- \rightarrow MO_x[] + Cl_2^- \quad (9.4c)$$

$$Cl_2 + H_2O \rightarrow HOCl + Cl^- + H^+ \quad (9.4d)$$

$$MO_x[\cdot OH] \rightarrow MO_x[] + \tfrac{1}{2}O_2 + H^+ + e^- \quad (9.5)$$

Meanwhile, with the application of various anode materials (e.g., Ti/SnO$_2$, Ti/IrO$_2$, and Pt), the electrocatalytic combustion and conversion of organic contaminants along with parallel oxygen evolution were studied. Based on the generation of a considerable quantity of •OH at the anode surfaces and, consequently, their reaction with the pollutants, Comninellis (1994) suggested the corresponding inceptive mechanism of electrocatalytic oxidation. The proposed pathway was evident through the recognition of •OH created by water discharge at the anodes manipulated by electron spin resonance through spin trapping. The investigation transparently demonstrated that the SnO$_2$ anode surface accumulates a much higher concentration of •OH, whereas the surfaces of the Pt and IrO$_2$ anodes manifest almost zero concentration. However, the surface concentration of •OH intermediates is governed through the intermediate step of oxidation reaction as well as the interaction between metal ions and oxygen. In 1996, Comninellis and Battisti reported new evidence of electrocatalytic oxidation pathways depending on the aforementioned observation. Hence, relying on the suggested pathways by Feng and Johnson (1990), Comninellis (1994), and Comninellis and De Battisti (1996), the EO for decomposing organic contaminants approaches through two different pathways. Firstly, the contaminants are degraded at the surface of the anode, known as direct anodic oxidation (Figure 9.2), and, secondly, an electrochemically formed intermediary (e.g. H$_2$S$_2$O$_8$, HClO, H$_2$O$_2$, etc.) conducts the oxidation, called indirect oxidation (Figure 9.3). They are elaborated in the forthcoming sections. It should be noted here that the two pathways may synchronize during electrooxidation of aqueous contaminants (Chiang, Chang, and Wen 1995).

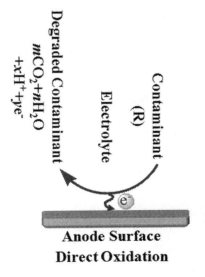

FIGURE 9.2 A conceptual scheme of direct electrolytic decontamination.

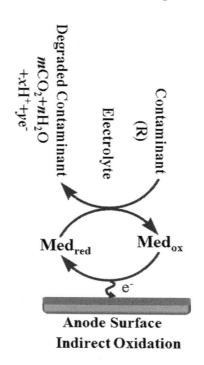

FIGURE 9.3 A conceptual scheme of indirect electrolytic decontamination.

9.2.2.1 Direct Oxidation

In this pathway, either the generated chemisorbed "active oxygen" (incorporated oxygen atom in the metal oxide crystal lattice, i.e., MO_{x+1}) or the physisorbed "active oxygen" (•OH) are highly responsible for the EO of contaminants (Martínez-Huitle and Ferro 2006; Kuhn 1971). The method is also named "anodic oxidation" or "direct oxidation," which propagates in two steps: (1) diffusion of the contaminants from the bulk to the surface of the anode and (2) oxidation thereof. Practically, the highly active electrodes, for instance, graphite, boron-doped diamond (BDD), or Ebonex, have been effectively employed for the investigation. It is highly significant to mention here that the fundamental model suggested by Comninellis is the clearest mechanistic pathway to express and comprehend the electrolytic reactions through the EO approach. The simplified mechanism conceives the characteristics of the electrode material as well as the sensibility of the •OH and is represented in Scheme 9.1. The •OH gets generated from the discharge of H_2O (or OH^-) adsorbed on the MO_x anode in the first step of the mechanism, according to Reaction 9.6a:

$$MO_x + H_2O \rightarrow MO_x(\cdot OH) + H^+ + e^- \quad (9.6a)$$

$$MO_x(\cdot OH) \rightarrow MO_{x+1} + H^+ + e^- \quad (9.6b)$$

The surface-adsorbed •OH interfere with the oxygen atom of the MO_x in the second step, resulting in a pretended higher oxide MO_{x+1} as displayed in Reaction 9.6b. This may be possible when the MO_x anode contained a higher oxidation state whose standard potential is higher compared to that of oxygen evolution. Nevertheless, both states of "active oxygen," namely chemisorbed "active oxygen" (MO_{x+1}) and physisorbed "active oxygen" (•OH), coexist on the anode surface. Henceforth, two scenarios may arise, namely (1) absence of oxidizable organics, which are transformed into molecular oxygen by the physisorbed and chemisorbed "active oxygen" according to Reactions 9.6c

Electrocatalysts for Wastewater Treatment

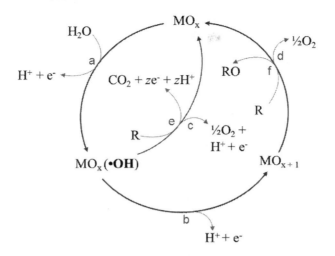

SCHEME 9.1 Schematic representation of the mechanism of electrocatalytic oxidation of contaminants with concomitant evaluation of oxygen. Gradual changes are as follows: (a) generation of •OH on the catalyst surface, (b) reorganization of oxygen atom from •OH to the metal oxide lattice, (c and d) evaluation of molecular oxygen due to lack of oxidizable organic contaminants in the electrolyte, (e) electrocatalytic degradation of organic contaminants via generated •OH, and (f) electrocatalytic transformation of organic pollutants (R). (Reproduced with permission from Comninellis 1994, © Elsevier).

and 9.6d; and (2) presence of oxidizable organics (R), in which the organic contaminants are completely mineralized according to Reactions 9.6e and 9.6f.

$$MO_x(\bullet OH) \rightarrow MO_x + \tfrac{1}{2}O_2 + H^+ + e^- \qquad (9.6c)$$

$$MO_{x+1} \rightarrow MO_x + \tfrac{1}{2}O_2 \qquad (9.6d)$$

$$R + MO_x(\bullet OH)_z \rightarrow MO_x + CO_2 + zH^+ + ze^- \qquad (9.6e)$$

$$R + MO_{x+1} \rightarrow MO_x + RO \qquad (9.6f)$$

It is clear from the reactions that most of the organic pollutants are degraded to CO_2 by the •OH through the transfer of generated electrons to the surface of the anode (Reaction 9.6e and Figure 9.2). Contrarily, the RO can be formed from the organic contaminants (R) with simultaneous conversion of the electrode into its original form MO_x. Thus, the physisorbed •OH effectively mineralizes the organic contaminants to CO_2, H_2O, and inorganic ions through electrochemical incineration (Comninellis 1994; Comninellis and Battisti 1996), whereas, the chemisorbed "active oxygen" selectively converts the refractory organics into biodegradable by-products (Brillas and Martínez-Huitle 2015). However, this oxidation pathway results in very poor decontamination of pollutants.

9.2.2.2 Indirect Oxidation

In this pathway, the contaminants are decomposed in the bulk solution through the electrochemical formation of a robust oxidizing agent at the surface of the anode (Anglada, Urtiaga, and Ortiz 2009). The electro-generated oxidant plays the role of a mediator, which conducts the total or partial decontamination process (Figure 9.3). It is presumed that due to enough availability of chloride ions in wastewater, the chlorine is the most familiar electrogenerated oxidant at the anode surface. However, the function of active chlorine in the indirect oxidation process is not clear and is still under investigation. Peroxodisulfuric acid ($H_2S_2O_8$), hydrogen peroxide (H_2O_2), and ozone (O_3) are the other familiar electrogenerated oxidants. For instance, the H_2O_2 mediator is formed through the indirect consumption (i.e., dimerization) of surface-adsorbed •OH:

$$2MO_x(•OH) \rightarrow 2MO_x + H_2O_2 \tag{9.7}$$

Alternatively, for the formation of •OH, metal ion catalysts, such as Ag^{2+}, Co^{3+}, and Fe^{3+}, may also be used as the external mediators. Of course, the remediation techniques are required to recover the metallic intermediates, as they result in more toxic effluent compared to the initial one (Martínez-Huitle and Ferro 2006).

9.2.3 Categorization of Electrocatalytic Anode Material

The efficiency and selectivity of the electrocatalytic oxidation method are strictly directed by the nature of the electrode material. In the meantime, the BDD electrode was found to be highly reactive toward electrooxidation of organic contaminants. The organic incineration with high current efficiency indicated that the electrogenerated •OH at the surface of the BDD anode were found to be unusually reactive compared to the other conventional electrode materials. Based on the aforementioned observation on BDD electrode, Comninellis (1994) suggested the conclusive mechanism. The predictions in the model were nicely fitted with the obtained data. Depending upon the oxidation capability, the electrocatalysts as anode material are classified in two different categories: active and nonactive anodes. However, in the suggested model, it is presumed that the oxidation of H_2O molecules resulting in the generation of physisorbed •OH, i.e., $MO_x(•OH)$, is considered to be the instigative reaction in both categories of anodes (Reaction 9.6a).

In the active anodes, the oxygen evolution overpotential of the electrocatalysts is low, which is suitable for oxygen evolution reaction (OER); and as a result, they oxidize the organic contaminants selectively. These anodes assemble with the electrogenerated •OH in a comparatively stronger way to form a higher oxide or super oxide (MO_{x+1}) (Reaction 9.6b). Carbon, graphite, IrO_2, RuO_2, and platinum-based anodes are common examples of this category. It should be mentioned here that the redox couple MO_{x+1}/MO_x functions as the mediator during the oxidation of organic contaminants (Reaction 9.6f) along with the competitive OER as a subsidiary reaction (Reaction 9.6d).

In contrast, due to the high oxygen evolution overpotential of the nonactive anodes, they are relatively poor electrocatalysts for OER and are expected to perform direct electrochemical oxidation. The nonactive anode surfaces interact with •OH so feebly that $MO_x(•OH)$ favors the direct reaction with the organic contaminants to produce CO_2 (Reaction 9.6e), leading to complete combustion reactions. Only SnO_2 and PbO_2 are considered to be the nonactive anodes, and obviously are the most suitable for complete electrochemical incineration reactions. Similarly, the generated surface redox couple MO_{x+1}/MO_x (Reaction 9.6f) by oxidative reaction is more exciting compared to the mineralization reaction (Reaction 9.6e) via the physisorbed •OH. The surface-adsorbed •OH, i.e., $MO_x(•OH)$, also competes with either the direct oxidation to O_2 (Reaction 9.6d) or through dimerization to hydrogen peroxide (Reaction 9.7) via indirect consumption along with the mineralization reaction. Although a nonactive electrode neither supplies any active catalytic site nor participates in the direct anodic reaction of organic contaminants, it functions as an electron sink and an inert substrate. The distinctive overpotential values applying to the most extensively studied anode materials, along with the generated typical oxidants for OER in acidic medium, are listed elsewhere in the literature in order to gain better understanding of the performance (Martínez-Huitle and Andrade 2011). As the MO_x–•OH interaction is the governing factor for the electrochemical activity and the chemical reactivity; the BDD anode is found to be the best nonactive electrode for the degradation of organic contaminants (Martínez-Huitle and Andrade 2011). Thus, the oxidation pathways can be tuned through proper selection of the anode materials.

Electrocatalysts for Wastewater Treatment

9.3 RECENT ADVANCES IN ELECTROCATALYSTS FOR THE DECONTAMINATION OF WASTEWATER

According to the aforementioned information, it is clear that the selectivity and the efficiency of an electrocatalytic process for the oxidation of organic contaminants entirely depend upon the nature of the electrocatalysts. So, in order to find a suitable electrocatalyst for the decontamination of wastewater, a thorough literature survey will be needed. In this regard, a number of electrocatalytic anode materials applied to various toxic contaminants are enlisted in Table 9.1. The table summarizes the important parameters, namely the applied electrocatalysts, the initial concentration of the contaminants (C_0), and removal efficiency (RE).

Some of the MO_x anodes, for instance, PbO_2, IrO_2, SnO_2, and RuO_2, have been extensively employed for the decontamination of a wide range of pollutants. It has been observed that for the contaminants, such as tannin, methyl red, and 4-chloropheneol, RE of almost 100% was obtained applying PbO_2 coating on titanium substrate anode (Ti/PbO_2) (Panizza and Cerisola 2004; Panizza and Cerisola 2008; Duan, Ma, and Chang 2012). A significant enhancement in conductivity and stability of the MO_x electrode has been noticed after coating on titanium substrate. Concomitantly, the generation of •OH on the electrode surface is responsible for the remarkable degradation of tannin:

$$PbO_2 + H_2O \rightarrow PbO_2(\cdot OH) + H^+ + e^- \quad (9.8)$$

$$C_{27}H_{22}O_9 + PbO_2(\cdot OH) \rightarrow PbO_2 + \text{degraded products} \quad (9.9)$$

As discussed in the preceding section, the large oxygen evolution overpotential of PbO_2 electrode enhanced the complete degradation of tannin through formation of electroactive chloride ions. Selective electrocatalytic degradation of tetracycline in raw cow milk and oxytetracycline in Na_2SO_4 and NaCl electrolyte, respectively, through the application of Ti/IrO_2 anode were reported by Kitazono et al. (2012). Also, packed-bed granular platinum anode has been demonstrated for the phenolic compound degradation with nominal energy consumption at a high RE (Sakakibara, Kounoike, and Kashimura 2010). Although titanium-based electrode, namely Ebonex (Ti_4O_7), has been reported by Zaky and Chaplin (2014) for the removal of phenol derivatives, the RE was very poor (~30.2%). Not only have metal or metal oxide electrodes, but also some nonmetal electrodes have been demonstrated for the decontamination process. A thin layer of carbon nanotube (CNT)–Prussian blue (PB) nanocomposite was employed by Nossol et al. (2013) for decontamination of methyl orange with high RE of 96% to 98%. It is to be mentioned here that the BDD electrodes exhibited excellent conversion efficiency of 99% and significantly higher removal performance in chloride-free solution. This is because of the weak sorption affinity of the •OH toward the BDD electrode surface, resulting in outstanding chemical reactivity for the decomposition of organic pollutants (Costa et al. 2010). It also exhibited complete removal of β-estradiol, oxalic/oxamic acid, and bisphenol A in 0.05 M to 0.1 M Na_2SO_4 electrolyte. For the selective decontamination of toxic contaminants, Cong, Iwaya, and Sakakibara (2014) demonstrated a granular glassy carbon electrode. In this decontamination technique, the synthetic and natural estrogen in wastewater was eliminated through electropolymerization. Especially, the Sb-SnO_2 anode is uncommon due to its lower toxicity, whereas PbO_2 has large oxygen evolution potential compared to RuO_2 and IrO_2. The BDD and Pt electrodes are highly expensive to apply in the field. Therefore, because of the efficient generation of •OH on Sb-SnO_2 anodes, it manifests remarkably high electrocatalytic activities for the decontamination of phenol derivatives in sulfate electrolyte (Niu et al. 2013). Although the Bi-TiO_2 electrode is not included in the table, it is known to be an efficient electrocatalyst in a chloride electrolyte (Ahn et al. 2017). Based on the survey, it has been found that remarkable improvement has been achieved through the application of mixed-metal oxides on recalcitrant pollutants in wastewater, which are extremely hard to remove through the conventional techniques (Wu, Huang, and Lim 2014).

TABLE 9.1
Electrocatalysts for the Treatment of Wastewater

Electrode	Pollutant	Initial concentration (C_0) (mg L^{-1})	RE (%)	Reference
Modified Metal Oxide Electrode, Dimensionally Stable Electrode, Ti Substrate					
PbO$_2$	Textile wastewater	COD: 470, PhOH: 0.11	59, 70	Mukimin, Vistanty, and
	Real textile wastewater	COD: 550	95	Zen 2015
	4-chlorophenol	10–90	100	Aquino et al. 2014
	Methyl red	COD: 220	≈100	Duan, Ma, and Chang 2012
	Tannin	Tannin: 1500, COD: 6000	100	Panizza and Cerisola 2008
				Panizza and Cerisola 2004
IrO$_2$	Oxytetracycline	100	83	Kitazono et al. 2012
	Tetracycline	100	99.4	Miyata et al. 2011
SnO$_2$	17-alpha-ethinylestradiol	0.5–10	96.5	Feng et al. 2010
RuO$_2$	Tannin	1500	100	Panizza and Cerisola 2004
	Real textile wastewater	COD: 1156	81	Kaur, Kushwaha, and
	Highly acidic wastewater	COD: 6,43,000	41.83	Sangal 2017
				Goyal and Srivastava 2017
SnO$_2$-Sb	Pentachlorophenol	100	99.8	Niu et al. 2013
Ru$_{0.3}$Ti$_{0.7}$O$_2$	Atrazine	20	100	Malpass et al. 2006
Sn-Sb-Ni-O	4-chlorophenol	1028	100	Wang et al. 2006
RuIrO$_2$	Real batik printing wastewater	COD: 253	66.4	Mukimin et al. 2017
Other Electrode Materials					
Graphite carbon	Bisphenol-A	500	78.3	Govindaraj et al. 2013
	Bovine serum albumin	2500	91.02	Sundarapandiyan et al. 2010
Boron-doped diamond (BDD)	Oxalic/oxamic acid	2.08 mM	≈100	Garcia-Segura and Brillas 2011
	Tannery wastewater	COD: 2370	70–90	
	Chlorobenzene	50	94.3	Costa et al. 2010
	17β-Estradiol	0.5	100	Liu et al. 2009
	Bisphenol A	20	100	Yoshihara and Murugananthan 2009
				Murugananthan et al. 2007
CNT/Prussian blue	Methyl orange	2.5×10^{-7} M	96–98	Nossol et al. 2013
Ebonex Ti$_4$O$_7$	P-substituted phenol	2.5 mM p-nitrophenol	30.2	Zaky and Chaplin 2014
Pt	Chlorobenzene	50	85.2	Liu et al. 2009
	17β-Estradiol, bisphenol A	0.08–0.14	96	Sakakibara et al. 2006
	Reactive orange 91/red 184/ Blue 182/black 5	5×10^{-4} M (each dye)	94.4	Sakalis et al. 2005
Glassy carbon	Estrone, allylamine	0.01 mmol L^{-1}, 10 mmol L^{-1}	93	Cong, Iwaya, and Sakakibara 2014; Bardini et al. 2013

In principle, the property of the polymers has to be remarkably different from the conventional polymers for the decontamination through electrocatalytic reduction. Accordingly, the polymers have to be conductive enough for the passage of electrons and have a large surface area as well as exceptional stability (Yang et al. 2014). Although the application of polymer in electrocatalytic water treatment is limited, some conductive polymers such as polyaniline (PAni) and polypyrrole (PPy), either functionalized or unfunctionalized, have been employed for the decontamination of

Electrocatalysts for Wastewater Treatment

hexavalent chromium (Cr^{6+}), bromate (BrO_3^-), azo dyes, etc. Because of the reversible conductivity, by changing the oxidation state PAni can be transformed to be the appropriate catalyst for the electroreduction of contaminants from water (Ruotolo, Santos-Júnior, and Gubulin 2006; Ding et al. 2010). In this regard, Ding et al. (2010) reported a PAni-film-modified electrode for the electrocatalytic reduction of bromate ion. Accordingly, the decontamination of bromate ion was investigated through electrochemical treatment and the BrO_3^- removal efficiency in 0.10 mol L^{-1} Na_2SO_4 supporting electrolyte was 99% at pH 7 in 25 min. In addition, a graphene-functionalized PAni was also reported in order to improve the catalytic efficiency (Ji et al. 2015; Yang et al. 2014). Accordingly, the Cr^{6+} reduction efficiency was found to be more than 99%, and graphene sheets enhanced the effective electron transfer rate. Furthermore, Haque, Smith, and Wong (2015) explored PPy films for the decontamination of azo dyes from textile industry effluent through electrochemical treatment. Thus, in green technology, conducting PPy films can potentially be applied for industrial effluent treatment.

9.4 SUPERIORITIES AND INFERIORITIES OF ELECTROCATALYTIC APPROACH

Of course, the EO technique has advantages as well as disadvantages, and both of them are elaborated herewith. In order to discuss the superiority of the EO approach for the treatment and avoidance of pollution complications, it should be mentioned here that since this technique employs a clean reagent—"the electron"—it is environmentally compatible. Simultaneously, the use of detrimental chemicals can be avoided in the process, leading to mild operation and straightforward equipment. Likewise, the technique is also recognized through its versatility, cost-effectiveness, ease of automation, and remarkable efficiency. In addition, the process can be operated through renewable energy from solar sources and wind. A wide variety of effluents has been examined for the decontamination of wastewater through this technique by several groups of researchers (Anglada, Urtiaga, and Ortiz 2009). For instance, (a) urban and domestic wastewater; (b) effluents from chemical factories, such as paper and pulp, petrochemical, textile, fine chemical industry, pharmaceutical, food industry, and tannery industry; and (c) the agro-industry, namely dairy, manure, and olive oil, have also been investigated (Anglada, Urtiaga, and Ortiz 2009). Nevertheless, the scale-up application of the technique is found to be challenging because of the operational restrictions for the competent applicability of the EO approach exceedingly regulating the involved pathways. More precisely, the advantages of the process can be shown by the following points:

- Controllable automation—The electrical parameters applied in the methods (particularly, current [I] and voltage [V]) convenient for boosting data process automation, acquisition, and control can be regulated.
- Robustness—In very short time the degradation reactions can be either restarted or terminated through turning off the power supply.
- Mild operation condition—As this process is operated in lower pressure and temperature compared to nonelectrochemical techniques (e.g., super critical oxidation, incineration), discharge of unreacted wastes and volatilization of toxic components can be overridden.
- Adaptability—A wide variety of industrial and domestic wastewater can be decontaminated from microliters to millions of liters through this technique.

Even after having such superiorities, the EO approach also has several inferiorities. The foremost demerits are summarized in the following:

- Extrinsic limitations—Under certain circumstances the current efficiency is low and some electrodes are expensive and/or possess a short life span. In addition, the potential or current distribution, exhibition of unwanted reactions, influence of pH, type and concentration of contaminants, design of the electrocatalytic reactors, and arrangement of electrodes on

the efficiency of the technique are still not completely resolved. Moreover, because of the adherence of the by-products on the surface, electrode fouling may take place.
- Intrinsic limitations—Until now, low surface-to-volume ratio, low space–time yield, gradual temperature enhancement, and mass transport limitations have not been adequately elucidated. Also, the electrical conductivity of the effluent is an unavoidable condition; however, all waste streams are not conductive enough for the applicability of the technology. Thus, the addition of external electrolyte may complicate the process.

9.5 CONCLUSION

In this chapter, fundamental principles, recent progress, and superiorities of EO technology for wastewater decontamination have been described for better understanding of the readers. A wide variety of effluents can be treated efficiently through this process, manifesting its versatility. With the remarkable advancement of material science, a wide variety of electrocatalytic electrode material, including metal, metal oxides, mixed-metal oxides, carbon, and BDD, have been developed ensuring future suitability for the decontamination technique. Among them, superior contribution of diamond electrodes for their capability of treating high volumes of effluent ensured the scale-up application of the EO process. Also, this process indicated its capability for decontamination of specific pollutants in a complicated matrix. Before scale-up implementation of the process, the principal barriers, such as evolution-stable and efficient electrocatalysts as well as operating cost reduction, must be addressed and should be other major areas for future research. It is needless to mention that evaluation of efficient electrocatalysts, innovative electrocatalytic reactors, and unconventional design of the pilot plant will be more supportive of fruitful industrial implementation.

REFERENCES

Ahn, Yong Yoon, So Young Yanga, Chimyung Choi, Wonyong Choi, Soonhyun Kim, and Hyunwoong Park. 2017. "Electrocatalytic activities of Sb-SnO$_2$ and Bi-TiO$_2$ anodes for water treatment: effects of electrocatalyst composition and electrolyte." *Catalysis Today* no. 282 (1):57–64.

Albadarin, AB, MN Collins, M Naushad, et al. 2017. "Activated lignin-chitosan extruded blends for efficient adsorption of methylene blue." *Chemical Engineering Journal* no. 307:264–272. doi: 10.1016/j.cej.2016.08.089

Anglada, Ángela, Ane Urtiaga, and Inmaculada Ortiz. 2009. "Contributions of electrochemical oxidation to wastewater treatment: fundamentals and review of applications." *Journal of Chemical Technology & Biotechnology* no. 84 (12):1747–1755.

Aquino, M José, Romeu CRocha-Filho, Luís AM Ruotolo, Nerilso Bocchi, and Sonia R Biaggio. 2014. "Electrochemical degradation of a real textile wastewater using β-PbO$_2$ and DSA anodes." *Chemical Engineering Journal* no. 251 (17):138–145.

Ardizzone, S, G Fregonara, and S Trasatti. 1990. "Inner and outer active surface of RuO$_2$ electrodes." *Electrochimica Acta* no. 35 (1):263–267.

Arikawa, T, Y Murakami, and Y Takasu. 1998. "Simultaneous determination of chlorineand oxygen evolving at RuO$_2$/Ti and RuO$_2$-TiO$_2$/Ti anodes by differential electrochemical mass spectroscopy." *Journal of Applied Electrochemistry* no. 28 (5):511–516.

Bardini, Luca, Marcel Ceccato, Mogens Hinge, Steen U Pedersen, Kim Daasbjerg, Massimo Marcaccio, and Francesco Paolucci. 2013. "Electrochemical polymerization of allylamine copolymers." *Langmuir* no. 29 (11):3791–3796.

Bisschops, IAE, and H Spanjers. 2003. "Literature review on textile wastewater characterization." *Environmental Technology* no. 24 (11):1399–1411.

Borrás, Carlos, C Berzoy, Jorge Mostany, Julio Herrera, and Benjamin Ruben Scharifker. 2007. "A comparison ofthe electrooxidation kinetics of p-methoxyphenol and p-nitrophenol on Sb-doped SnO$_2$ surfaces: concentration and temperature effects." *Applied Catalysis B: Environmental* no. 72 (1–2):98–104.

Brillas, Enric, and Carlos A Martínez-Huitle. 2015. "Decontamination of wastewaters containing synthetic organic dyes by electrochemical methods: an updated review." *Applied Catalysis B: Environmental* 166–167 (5):603–643.

Cañizares, Pablo, Fabiola Martínez, Carlos Jiménez, Justo Lobato, and Manuel A Rodrigo. 2006. "Coagulation and electro coagulation of wastes polluted with dyes." *Environmental Science and Technology* no. 40 (20):6418–6424.

Cañizares, P, F Martinez, M Diaz, J Garcia-Gomez, and MA Rodrigo. 2002. "Electrochemical oxidation of aqueous phenol wastes using active and nonactive electrodes." *Journal of Electrochemical Society* no. 149 (8):D118–D124.

Chen, Guohua. 2004. "Electrochemical technologies in wastewater treatment." *Separation and Purification Technology* no. 38 (1):11–41.

Chen, Xueming, Guohua Chen, and Po Lock Yue. 2001. "Stable Ti/IrO$_x$-Sb$_2$O$_5$-SnO$_2$ anode for O$_2$ evolution with low Ir content." *Journal of Physical Chemistry B* no. 105 (20):4623–4628.

Chiang, Li-Choung, Juu-En Chang, and Ten-ChinWen. 1995. "Indirect oxidation effect in electrochemical oxidation treatment of landfill leachate." *Water Research* no. 29 (2):671–678.

Cho, Kangwoo, and Michael R. Hoffmann. 2014. "Urea degradation by electrochemically generated reactive chlorine species: products and reaction pathways." *Environmental Science amd Technology* no. 48 (19):11504–11511.

Comninellis, Christos. 1994. "Electrocatalysis in the electrochemical conversion/combustion of organic pollutants for waste water treatment." *Electrochimica Acta* no. 39 (11–12):1857–1862.

Comninellis, Christos, and Achille P De Battisti. 1996. "Electrocatalysis in anodic oxidation of organics with simultaneous oxygen evolution." *Journal of Physical Chemistry* no. 93 (4):673–679.

Cong, Vo Huu, Sota Iwaya, and Yutaka Sakakibara. 2014. "Removal of estrogens by electrochemical oxidation process." *Journal of Environmental Sciences (China)* no. 26 (6):1355–1360.

Cong, Vo Huu, Yutaka Sakakibara, Masahito Komori, Naoyuki Kishimoto, Tomohide Watanabe, Iori Mishima, Ikko Ihara, Tsuneo Tanaka, Yukihito Yoshida, and Hiroaki Ozaki. 2016. "Recent developments in electrochemical technology for water and wastewater treatments." *Journal of Water and Environment Technology* no. 14 (2):25–36.

Costa, Carla Regina, Francisco Montilla, Emilia Morallón, and Paulo Olivi. 2010. "Electrochemical oxidation of synthetic tannery wastewater in chloride-free aqueous media." *Journal of Hazardous Materials* no. 180 (1–3):429–435.

Daneshvar, E, A Vazirzadeh, A Niazi, et al. 2017. "Desorption of methylene blue dye from brown macroalga: effects of operating parameters, isotherm study and kinetic modeling." *Journal of Cleaner Production* no. 152:443–453. doi: 10.1016/j.jclepro.2017.03.119

Ding, Liang, Qin Li, Hao Cui, Rong Tang, Hui Xu, Xianchuan Xie, and Jianping Zhai. 2010. "Electrocatalytic reduction of bromate ion using a polyaniline-modified electrode: an efficient and green technology for the removal of BrO$_3^-$ in aqueous solutions." *Electrochimica Acta* no. 55 (28):8471–8475.

Duan, Xiaoyue, F Ma, and LM Chang. 2012. "Electrochemical degradation of 4-chlorophenol in aqueous solution using modified PbO$_2$ anode." *Water Science and Technology* no. 66 (11):2468–2474.

Dutta, Kingshuk, and Dipak Rana. 2019. "Polythiophenes: an emerging class of promising water purifying materials." *European Polymer Journal* no. 116:370–385.

Dutta, Kingshuk, and Sirshendu De. 2017a. "Aromatic conjugated polymers for removal of heavy metal ions from wastewater: a short review." *Environmental Science: Water Research and Technology* no. 3 (5):793–805.

Dutta, Kingshuk, and Sirshendu De. 2017b. "Smart responsive materials for water purification: an overview." *Journal of Materials Chemistry A* no. 5 (42):22095–22112.

Feng, Jianren, and Dennis C Johnson. 1990. "Electrocatalysis of anodic oxygen-transfer reactions Fe-doped beta-lead dioxide electrodeposited on noble metals." *Journal of Electrochemical Society* no. 137 (2):507510.

Feng, Yujie, Ce Wang, Junfeng Liu, and Zhaohan Zhang. 2010. "Electrochemical degradation of 17-alpha-ethinylestradiol (EE2) and estrogenic activity changes." *Journal of Environmental Monitoring* no. 12 (2):404–408.

Feng, Yujie, and Xing Yi Li. 2003. "Electro-catalytic oxidation of phenol on several metal-oxideelectrodes in aqueous solution." *Water Research* no. 37 (10):2399–2407.

Garcia-Segura, Sergi, and Enric Brillas. 2011. "Mineralization of the recalcitrant oxalic and oxamic acids by electrochemical advanced oxidation processes using a boron-doped diamond anode." *Water Research* no. 45 (9):2975–2984.

Georgiou, Dimitri, Paraschos Melidis, and A Aivasidis. 2002. "Use of a microbial sensor: inhibition effect of azo-reactive dyes on activated sludge." *Bioprocess and Biosystems Engineering* no. 25 (2):79–83.

Govindaraj, M, R Rathinam, C Sukumar, M Uthayasankar, and S Pattabhi. 2013. "Electrochemical oxidation of bisphenol-A from aqueous solution using graphite electrodes." *Environmental Technology* no. 34 (4):503–511.

Goyal, Akash, and Vimal Chandra Srivastava. 2017. "Treatment of highly acidic wastewater containing high energetic compounds using dimensionally stable anode." *Chemical Engineering Journal* no. 325 (20):289–299.

Haque, Md Niamul, Daechul Cho, and Sunghyun Kwon. 2014. "Performances of metallic (sole, composite) and non-metallic anodes to harness power in sediment microbial fuel cells." *Environmental Engineering Research* no. 19 (4):363–367.

Haque, MM, WT Smith, and DKY Wong. 2015. "Conducting polypyrrole films as a potential tool for electrochemical treatment of azo dyes in textile wastewaters." *Journal of Hazardous Materials* no. 283:164–170.

Hartmut, Wendt, and Kreysa Gerhard. 1999. *Electrochemical Engineering: Science and Technology in Chemical and other Industries.* Springer: Cham.

Iniesta, Jesús, Eduardo Exposito, J González-García, V Montiel, and A Aldaz. 2002. "Electrochemical treatment of industrial wastewater containing phenols." *Journal of Electrochemical Society* no. 149 (5):D57–D62.

Ji, Qinghua, Dawei Yu, Gong Zhang, Huachun Lan, Huijuan Liu, and Jiuhui Qu. 2015. "Microfluidic flow through polyaniline supported by lamellar-structured graphene for mass-transfer-enhanced electrocatalytic reduction of hexavalent chromium." *Environmental Science and Technology* no. 49 (22):13534–13541.

Ji, Tianhao, Fang Yang, Yuanyuan Lv, Jiaoyan Zhou, and Jiayue Sun. 2009. "Synthesis and visible-light photocatalytic activity of Bi-doped TiO_2 nanobelts." *Materials Letters* no. 63 (23):2044–2046.

Kaur, Parminder, Jai Prakash Kushwaha, and Vikas Kumar Sangal. 2017. "Evaluation and disposability study of actual textile wastewater treatment by electro-oxidation method using Ti/RuO_2 anode." *Process Safety and Environmental Protection* no. 111 (7):13–22.

Kaur, Parminder, Jai Prakash Kushwaha, and Vikas Kumar Sangal. 2018. "Electrocatalytic oxidative treatment of real textile wastewater in continuous reactor: degradation pathway and disposability study." *Journal of Hazardous Materials* no. 346 (15):242–252.

Kaur, Parminder, Vikas Kumar Sangal, and Jai Prakash Kushwaha. 2015. "Modeling and evaluation of electrooxidation of dye wastewater using artificial neural networks." *RSC Advances* no. 44 (5):34663–34671.

Kesselman, M Janet, Oleh Weres, Nathan S Lewis, and Michael R Hoffmann. 1997. "Electrochemical production of hydroxyl radical at polycrystalline Nb-doped TiO_2 electrodesand estimation of the partitioning between hydroxyl radical and direct holeoxidation pathways." *Journal of Physical Chemistry B* no. 101 (14):2637–2643.

Kim, Jungwon, Won Joon, K Choi, Jina Choi, Michael R Hoffmann, and Hyunwoong Park. 2013. "Electrolysis of urea andurine for solar hydrogen." *Catalysis Today* no. 199 (1):2–7.

Kim, Sangyong, Chulhwan Park, Tak-Hyun Kim, Jinwon Lee, and Seung-Wook Kim. 2003. "COD reduction and decolorization of textile effluent using a combined process." *Journal of Bioscience and Bioengineering* no. 95 (1):102–105.

Kim, Soonhyun, Sung Kyu Choi, Bok YoungYoon, Sang Kyoo Lim, and Hyunwoong Park. 2010. "Effects of electrolyte on the electrocatalytic activities of RuO_2/Ti and $Sb-SnO_2$/Ti anodes for watertreatment." *Applied Catalysis B: Environmental* no. 97 (1–2):135–141.

Kitazono, Yumika, Ikko Ihara, Gen Yoshida, Kiyohiko Toyoda, and Kazutaka Umetsu. 2012. "Selective degradation of tetracycline antibiotics present in raw milk by electrochemical method." *Journal of Hazardous Materials* no. 243 (23):112–116.

Kuhn, AT. 1971. "Electrolytic decomposition of cyanides, phenols and thiocyanates in effluent streams–a literature review." *Journal of Applied Chemistry and Biotechnology* no. 21 (2):29–34.

Kuramitz, Hideki, Jun Saitoh, Toshiaki Hattori, and Shunitz Tanaka. 2002. "Electrochemical removal of p-nonyl phenol from dilute solutions using a carbon fiber anode." *Water Research* no. 36 (13):3323–3329.

Lahav, Ori, Yuval Schwartz, Paz Nativ, and Youri Gendel. 2013. "Sustainable removal of ammonia from anaerobic-lagoon swine waste effluents using an electrochemically-regenerated ion exchange process." *Chemical Engineering Journal* no. 218 (4):214–222.

Lee, Changha. 2015. "Oxidation of organic contaminants in water by iron-induced oxygenactivation: a short review." *Environmental Engineering Research* no. 20 (3):205–211.

Li, Miao, Chuanping Feng, Weiwu Hu, Zhenya Zhang, and Norio Sugiura. 2009. "Electrochemical degradation of phenol using electrodes of Ti/RuO_2-Pt and Ti/IrO_2-Pt." *Journal of Hazardous Materials* no. 162 (1):455–462.

Liu, L, G Zhao, M Wu, Y Lei, and R Geng. 2009. "Electrochemical degradation of chlorobenzene on borondoped diamond and platinum electrodes." *Journal of Hazardous Materials* no. 168 (1):179–186.

Malpass, Geoffroy Roger Pointer, Douglas Miwa, SAS Machado, Paulo Olivi, and Artur De Jesus Motheo. 2006. "Oxidation of the pesticide atrazine at DSA electrodes." *Journal of Hazardous Materials* no. 137 (1):565–572.

Martínez-Huitle, A Carlos, and Leonardo S Andrade. 2011. "Electrolysis in wastewater treatment: recent mechanism advances." *Química Nova* no. 34 (5):850–858.

Martínez-Huitle, A Carlos, Manuel A Rodrigo, Ignasi Sirés, and Onofrio Scialdone. 2015. "Single and coupled electrochemical processes and reactors for the abatement of organic water pollutants: a critical review." *Chemcal Reviews* no. 115 (24):13362–13407.

Martínez-Huitle, A Carlos, and Sergio Ferro. 2006. "Electrochemical oxidation of organic pollutants for the wastewater treatment: direct and indirect processes." *Chemical Society Review* no. 35 (12):1324–1340.

Miyata, Marcelo, Ikko Ihara, G Yoshid, K Toyod, and Kazutaka Umetsu. 2011. "Electrochemical oxidation of tetracycline antibiotics using a Ti/IrO$_2$ anode for wastewater treatment of animal husbandry." *Water Science and Technology* no. 63 (3):456–461.

Mukimin, Aris, Hanny Vistanty, and Nur Zen. 2015. "Oxidation of textile wastewater using cylinder Ti/β-PbO$_2$ electrode in electrocatalytic tube reactor." *Chemical Engineering Journal* no. 259 (24):430–437.

Mukimin, Aris, Nur Zen, Agus Purwanto, Kukuh Aryo Wicaksono, Hanny Vistanty, and Abdul Syukur Alfauzi. 2017. "Application of a full-scale electrocatalytic reactor as real batik printing wastewater treatment by indirect oxidation process." *Journal of Environmental Chemical Engineering* no. 5 (5):5222–5232.

Muthu, Murugananthan, S Yoshihara, T Rakuma, N Uehara, and T Shirakashi. 2007. "Electrochemical degradation of 17β-estradiol (E2) at boron-doped diamond (Si/BDD) thin film electrode." *Electrochimica Acta* no. 52 (9):3242–3249.

Naushad, M, Z Abdullah ALOthman, M Rabiul Awual, et al. 2016. "Adsorption of rose Bengal dye from aqueous solution by amberlite Ira-938 resin: kinetics, isotherms, and thermodynamic studies." *Desalin Water Treat* no. 57:13527–13533. doi: 10.1080/19443994.2015.1060169

Niu, Junfeng, Yueping Bao, Yang Li, and Zhen Chai. 2013. "Electrochemical mineralization of pentachlorophenol (PCP) by Ti/SnO$_2$-Sb electrodes." *Chemosphere* no. 92 (11):1571–1577.

Nossol, Edson, Arlene BS Nossol, Aldo JG Zarbin, and Alan M Bond. 2013. "Carbon nanotube/Prussian blue nanocomposite film as a new electrode material for environmental treatment of water samples." *RSC Advances* no. 3 (16):5393–5400.

Olmez, Tuğba, Işik Kabdaşli, and Olcay Tünay. 2007. "The effect of the textile industry dye bath additive EDTMPA on colour removal characteristics by ozone oxidation." *Water Science and Technology* no. 55 (10):145–153.

Panizza, Marco, and Giacomo Cerisola. 2004. "Electrochemical oxidation as a final treatment of synthetic tannery wastewater." *Environmental Science and Technology* no. 38 (20):5470–5475.

Panizza, Marco, and Giacomo Cerisola. 2007. "Electrocatalytic materials for the electrochemical oxidation of synthetic dyes." *Applied Catalysis B: Environmental* no. 75 (1–2):95–101.

Panizza, Marco, and Giacomo Cerisola. 2008. "Electrochemical degradation of methyl red using BDD and PbO$_2$ anodes." *Industrial and Engineering Chemistry Research* no. 47 (18):6816–6820.

Panizza, Marco, and Giacomo Cerisola. 2009. "Direct and mediated anodic oxidation of organic pollutants." *Chemical Review* no. 109 (12):6541–6569.

Park, Hyunwoong, Ayoung Bak, Yong Yoon Ahn, Jina Choi, and Michael R Hoffmannn. 2012. "Photoelectrochemical performance of multi-layered BiO$_x$-TiO$_2$/Ti electrodes for degradation of phenol and production of molecular hydrogen in water." *Journal of Hazardous Materials* no. 211 (S1):47–54.

Park, Hyunwoong, Chad D Vecitis, and Michael R Hoffmann. 2008. "Solar-powered electrochemical oxidationof organic compounds coupled with the cathodic production of molecular hydrogen." *Journal of Physical Chemistry A* no. 112 (33):7616–7626.

Park, Hyunwoong, Chad D Vecitis, and Michael R Hoffmann. 2009. "Electrochemical water splitting coupled with organic compound oxidation: the role of active chlorine species." *Journal of Physical Chemistry C* no. 113 (18):7935–7945.

Park, Hyunwoong, Chad D Vecitis, Wonyong Choi, Oleh Weres, and Michael R Hoffmann. 2008. "Solar-powered production of molecular hydrogen from water." *Journal of Physical Chemistry C* no. 112 (4):885–889.

Rajeshwar, Krishnan, and Jorge G Ibanez. 1997. *Environmental Electrochemistry: Fundamentals and Applications in Pollution Abatement*. Academic Press: San Diego.

Rajeshwar, Krishnan, Jorge G Ibanez, and Greg M Swain. 1994. "Electrochemistry and the environment." *Journal of Applied Electrochemistry* no. 24 (11):1077–1091.

Ren, Zhandong, Shanshan Quan, Jie Gao, Wenyang Li, Yuchan Zhu, Ye Liu, Bo Chai, and Yourong Wang. 2015. "The electrocatalytic activity of IrO$_2$–Ta$_2$O$_5$ anode materials and electrolyzed oxidizing water preparation and sterilization effect." *RSC Advances* no. 5 (12):8778–8786.

Ruotolo, LAM, DS Santos-Júnior, and JC Gubulin. 2006. "Electrochemical treatment of effluents containing Cr(VI). Influence of pH and current on the kinetic." *Water Research* no. 40 (8):1555–1560.

Sakakibara, Yutaka, T Kounoike, and H Kashimura. 2010. "Enhanced treatment of endocrine disrupting chemicals by a granular bed electrochemical reactor." *Water Science and Technology* no. 62 (10):2218–2224.

Sakakibara, Yutaka, Y Senda, T Obanayama, and R Nagata. 2006. "Enhanced treatment of trace pollutants by a novel electrolytic cell." *Engineering in Life Sciences* no. 6 (6):573–576.

Sakalis, Anastasios, Konstantinos Mpoulmpasakos, Ulrich Nickel, Konstantinos Fytianos, and Anastasios Voulgaropoulos. 2005. "Evaluation of a novel electrochemical pilot plant process for azodyes removal from textile wastewater." *Chemical Engineering Journal* no. 111 (1):63–70.

Shao, Dan, Xiaoliang Li, Hao Xu, and Wei Yan. 2014. "An improved stable Ti/Sb–SnO$_2$ electrode with high performance in electrochemical oxidation processes." *RSC Advances* no. 4 (41):21230–21237.

Sirés, Ignasi, Enric Brillas, Mehmet A Oturan, Manuel A Rodrigo, and Marco Panizza. 2014. "Electrochemical advanced oxidation processes: today and tomorrow: a review." *Environmental Science and Pollution Research* no. 21 (14):8336–8367.

Sundaramoorthy, Sundarapandiyan, R Chandrasekar, B Ramanaiah, S Krishnan, and Palanivel Saravanan. 2010. "Electrochemical oxidation and reuse of tannery saline wastewater." *Journal of Hazardous Materials* no. 180 (1–3):197–203.

Wang, Yun-Hai, Kwong-Yu Chan, XY Li, and SK So. 2006. "Electrochemical degradation of 4-chlorophenol at nickel-antimony doped tin oxide electrode." *Chemosphere* no. 65 (7):1087–1093.

Woisetschläger, D, Bernd Humpl, M Koncar, and Matthaeus Siebenhofer. 2013. "Electrochemical oxidation of wastewater – opportunities and drawbacks." *Water Science and Technology* no. 68 (5):1173–1179.

Wu, Weiyi, Zhao-Hong Huang, and Teik-Thye Lim. 2014. "Recent development of mixed metal oxide anodes for electrochemical oxidation of organic pollutants in water." *Appied Catalysis A: General* no. 480 (13):58–78.

Yang, So Young, Dongseog Kim, and Hyunwoong Park. 2014. "Shift of the reactive species in the Sb–SnO$_2$-electrocatalyzed inactivation of E-coli and degradation of phenol: effects of nickel doping and electrolytes." *Environmental Science and Technology* no. 48 (5):2877–2884.

Yang, So Young, Wonyong Choi, and Hyunwoong Park. 2015. "TiO$_2$ nanotube array photoelectrocatalyst and Ni-Sb-SnO$_2$ electrocatalyst bifacial electrodes: a new type of bi-functional hybrid platform for water treatment." *ACS Applied Material and Interfaces* no. 7 (3):1907–1914.

Yang, So Young, Yeon Sik Choo, Soonhyun Kim, Sang Kyoo Lim, Jaesang Lee, and Hyunwoong Park. 2012. "Boosting the electrocatalytic activities of SnO$_2$ electrodes for remediation of aqueous pollutants by doping with various metals." *Applied Catalysis B: Environmental* no. 111 (1):317–325.

Yang, Yang, Mu-he Diao, Ming-ming Gao, Xue-fei Sun, Xian-wei Liu, Guo-hui Zhang, Zhen Qi, and Shuguang Wang. 2014. "Facile preparation of graphene/polyaniline composite and its application for electrocatalysis hexavalent chromium reduction." *Electrochimica Acta* no. 132:496–503.

Yoshihara, S, and Murugananthan Muthu. 2009. "Decomposition of various endocrine-disrupting chemicals at boron doped diamond electrode." *Electrochimica Acta* no. 54 (7):2031–2038.

Yoshino, Hiroaki, Vo Huu Cong, and Yutaka Sakakibara. 2013. "High efficient phosphate removal and HAP recovery by a multielectrode system." *Journal of Japan Society of Civil Engineers, Ser. G (Environmental Research)* no. 69 (7):III_137–III_143.

Zaky, Amr M, and Brian P Chaplin. 2014. "Mechanism of p-substituted phenol oxidation at a Ti4O7 reactive electrochemical membrane." *Environmental Sciecne and Technology* no. 48(10):5857–5867.

10 Development and Characterization of A-MWCNTs/Tetra Functional Epoxy Coatings for Corrosion and Prevention of Mild Steel

D. Duraibabu, A. Gnanaprakasam, and S. Ananda Kumar

CONTENTS

10.1 Introduction 169
10.2 Experimental Details 172
    10.2.1 Synthesis of N,N,N'N''-Tetrakis(2,3-Epoxypropyl)-4,4'-
        (1,4-Phenylenedioxy) Dianiline 172
    10.2.2 Synthesis of Amine Functionalized Carbon Nanotubes (A-CNTs) 172
    10.2.3 Surface Preparation of the Mild Steel Specimens 172
    10.2.4 Sample Preparation 173
10.3 Results and Discussion 174
    10.3.1 Thermal Properties 175
    10.3.2 Electrochemical Impedance Study 176
    10.3.3 Morphology Studies 176
10.4 Conclusion 177
Acknowledgments 177
Bibliography 178

10.1 INTRODUCTION

Mild steel is low cost and can be widely used in various fields such as chemical processing, mining, construction, production, and marine applications (Nayak and Mohana 2018; Duraibabu et al. 2014). The main drawback is corrosion of mild steel on the metal surface, which leads to structural damage and decreases the chemical properties, mechanical properties, vessels, and pipelines (Kadhum et al. 2014; Zhang et al. 2016). Moreover, mild steel is remarkably sensitive toward corrosion, such as to diffusive chloride ions in saltwater (Manjumeena et al. 2016; Hernandez et al. 2012). In order to overcome the issue, various methods, such as protective coatings like coating the metal substrate, chemical vapor deposition, polymer coating, and electroplating have been developed (Deyab et al. 2016; Raotole et al. 2015). In the last decades, organic coatings are most commonly used for corrosion protection in metallic structures and act as a physical barrier to corrosive agents. However, these coatings do not have long-term corrosion resistance against corrosive ions (Mohammadi et al. 2019; Sørensen et al. 2009). Current coatings research mainly focuses on long-term corrosion resistance performance, among them epoxy resin is one of the outstanding protective coatings in terms of corrosion protection, good quality insulating properties, saltwater resistance, strong affinity and adhesion for heterogeneous materials, great adherence to many substrates, and chemical resistance (Wei et al. 2007; Dhanapal et al. 2015). A general structure of commercially available DGEBA epoxy resin is shown in Figure 10.1. Epoxy coatings are osmotic to corrosive species related to water, oxygen, and ions. The corrosion mechanism is as follows: the hydroxyl ions (OH) generate at the cathode and increase the pH of the coating. This may reduce the adhesion properties and accelerate corrosion of the metal underneath the coating (Ramezanzadeh and Attar 2011; Wu et al. 2014).

Consequently, the adhesion of epoxy coating properties can be improved by the incorporation of inorganic nanoparticles such as Al_2O_3, ZnO, SiO_2, $CaCo_3$, TiO_2, and organoclay (Afzal et al. 2013; Behzadnasab et al. 2011). These inorganic particles display better mechanical, optical, and chemical properties (Bhattacharya and Gupta 2005). Generally, the addition of inorganic nanofillers into the polymer matrix is a challenging task due to the accumulation of inorganic fillers, which leads to reduced coating performance (Toorani et al. 2017; Neisiany et al. 2018). At present, carbon-based materials like carbon nanotubes (CNTs) and graphene oxide (GO) possess greater chemical stability, thermal, electrical, and mechanical properties (Lee et al. 2017; Moitra et al. 2016). The reinforcement of CNTs into the polymer improves the strength, thermal properties, radiation sources, electrical properties, and corrosion protection properties. Over the past years a number of different techniques for coatings have been developed and are summarized in Table 10.1. Moreover, CNTs are difficult to disperse in the polymer; it could be because the high specific surface area CNTs tend to aggregate and enhance the van der Waals π–π interactions (Park and Ruoff 2009; Tasis et al. 2006). The surface functionalization of CNTs can be avoided with the disaggregation of the polymer in a homogeneous coating. Kumar et al. (2017) studied the homogeneous cluster-free dispersion of carbon nanotubes in the polymer matrix. The authors reported a systematic method to achieve a better dispersion in multiwalled carbon nanotubes (MWCNTs) in epoxy by implementing ultrasonic waves and generating shear force in the flow of axial impeller. Field emission scanning electron microscopy (FESEM) results showed that cluster-free uniform dispersion of MWCNTs in the epoxy matrix can lead to the enhancement of toughness by 53%, tensile strength by 35%, and storage modulus by 35% by loading of 0.75 wt% of MWCNTs in epoxy. Likewise, Kumar et al. (2018) analyzed the properties of mechanical and anticorrosion in epoxy MWCNT/TiO_2 hybrid nanofiller toward the high-performance epoxy composite using transmission electron microscopy (TEM). The results of the MWCNT/TiO_2 hybrid epoxy nanocomposite exhibited greater mechanical performance and anticorrosion than the nanocomposites. The tensile storage modulus and tensile strength of epoxy increased up to 43% and 61%, respectively, by implementing MWCNT/TiO_2 hybrid nanofiller. Additionally, from these studies, the corrosion rate decreased to 2.5×10^{-3} MPY and 0.87×10^{-3} MPY from 16.81 MPY (Kumar et al. 2018) and the protection efficiency increased to 99.99% by a process of coated MWCNT/TiO_2 hybrid epoxy nanocomposite on mild steel. Enhancement of the epoxy adhesion observed between the steel plates by treatment method and CNT/short-fiber reinforcement was studied by Wang et al. (2016). The shear strength of the epoxy adhesive improved 50% using the treatment method. The results further observed enhanced shear strength by incorporating MWCNTs and short Kevlar fibers into the adhesive joints, the reinforcing mechanisms along with the interfacial zone, and the toughening mechanisms within the composite adhesive joints. Yeole et al. (2016) analyzed the corrosion protection of prone metals like iron and aluminum. These metals are utilized for anticorrosive coatings for a large number of applications. Nanocontainers have the capability of encapsulating large amounts of guest molecules to aid in the self-healing abilities of the coating to provide active protection. The process involves the deposition of the CNT layer-by-layer process, inhibitor, and polyelectrolytes to enhance the corrosion protection on mild steel (MS) plates. The surface charge, thickness of the

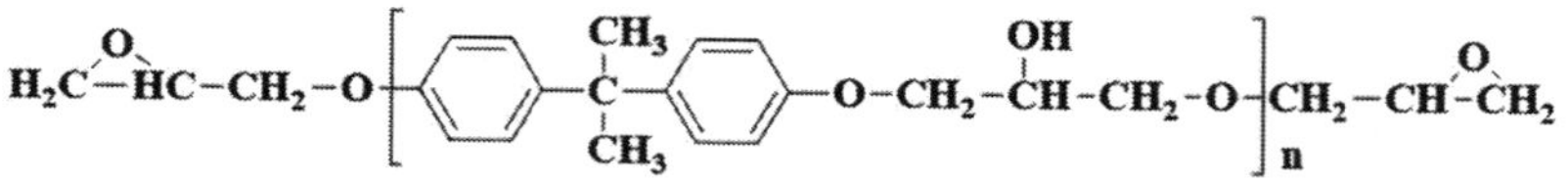

FIGURE 10.1 Chemical structure of epoxy (DGEBA) resin.

TABLE 10.1
Summary of Various Coating Systems Used for Anticorrosive Applications

Resin	Technique Used	Application	Reference
Epoxy resin	Spin coating	Corrosion resistance	Nayak and Mohana 2018
DGEBA epoxy resin (Epon 1001)	Solution process	Corrosion resistance	Kadhum et al. 2014
DGEBA epoxy resin (GY250)	Bar-coater	Anticorrosion and antimicrobial	Manjumeena et al. 2016
Epoxy resin (Epon Shell 1001)	Film applicator coating	Anticorrosion	Mohammadi et al. 2019
Epoxy resin (Cam coat 2071)	Spray coating	Anticorrosion	Kumar et al. 2017
Fusion bonded epoxy (FBE)	Electrostatic spraying coating	Anticorrosion	Ramezanzadeh et al. 2019
NANYA epoxy resin (NPEL-127)	Brushing coating	Corrosion resistance	Hosseini and Aboutalebi 2019
Epoxy resin (Shell)	Film applicator coating	Anticorrosion	Yeganeha et al. 2019
Epoxy resin (E-51)	Spray coating	Anticorrosion	Lou et al. 2019
Epoxy resin (EPON™ 828)	Film applicator	Anticorrosion	Ramezanzadeh et al. 2019
Epoxy resin (1001)	Film applicator	Anticorrosion	Yeganeh et al. 2019
Epoxy ester resin (EE-430CS)	Film applicator	Corrosion resistant	Sanaei et al 2018
DGEBA epoxy resin (BE-188)	Doctor blade film applicator	Anticorrosion	Kabeb et al. 2019
DGEBA epoxy resin (Epiran 6)	Dip coating	Anticorrosion	Pour et al. 2018
Epoxy resin	Spray coating	Anticorrosion	Sambyal et al. 2018
Epoxy ester resin (EE-430CS)	Film applicator	Anticorrosion	Sanaei et al 2017
DGEBA epoxy resin (GY250)	Bar-coater	Anticorrosion and antifouling	Palanivelu et al 2017
Epoxy resin (AR555)	Air spray	Anticorrosion	Wang et al. 2015
Epoxy resin (GZ 7071)	Film applicator	Anticorrosion	Majd et al. 2019
Epoxy resin (E44)	Bar-coater	Anticorrosion	Qiu et al 2017
Epoxy resin	Dip coating	Anticorrosion	Khodair et al. 2018
Epoxy resin (GZ 7071)	Film applicator	Anticorrosion	Vakili et al. 2015
Silicone-epoxy resin (SILIKOPON EF)	Air spray	Anticorrosion	Yuan et al. 2015

layer, and functional groups on each layer were determined using various analytical techniques such as particle size distribution, Fourier transform infrared (FTIR) spectroscopy, zeta potential analysis, and X-ray diffractograms analyses of CNT. These modified CNT techniques were performed to examine the crystallographic properties. From the results, Yeole et al. (2016) concluded that morphology and particle size clearly indicate the development of a nanocontainer. Further, the corrosion rate analysis of nanocontainer epoxy coatings on MS panels was measured by DC polarization and salt spray. In addition to that, the corrosion resistance measured the immersion of the coated samples into the alkali solution.

In this work, a TGBAPB epoxy resin with an ether linkage through the intermediates 1,4'-bis(4-nitrophenoxy) benzene BHPB and 1,4'-bis(4-aminephenoxy)-benzene BAPB was developed. The intermediates and TGBAPB epoxy resin were characterized by means of FTIR and nuclear magnetic resonance (NMR) spectroscopic studies. The TGBAPB epoxy resin was coated in sandblasted mild steel specimens using a room temperature curing agent HY 951 in the presence and in the absence of an amine functionalized carbon nanotube (A-MWCNTs) to a thickness of 100 microns by brush. The anticorrosive behavior of these coated specimens was examined by electrochemical impedance studies (EIS) and the results are discussed in the following.

10.2 EXPERIMENTAL DETAILS

10.2.1 Synthesis of N,N,N'N'''-Tetrakis(2,3-Epoxypropyl)-4,4'-(1,4-Phenylenedioxy) Dianiline

The ether linked tetrafunctional epoxy resin, N,N,N'N''-tetrakis(2,3-epoxypropyl)-4,4'-(1,4-phenylenedioxy) dianiline, was analyzed from 4,4'-(1,4-phenylenedioxy) dianiline according to the method reported in the literature (Duraibabu et al. 2014) (Figure 10.2).

10.2.2 Synthesis of Amine Functionalized Carbon Nanotubes (A-CNTs)

A-MWCNTs were formed via acid functionalized MWCNTs. MWCNTs were treated with a mixture of 3:1 (concentrated sulfuric and nitric acid) by stirring at temperature of 40°C for 4 hours. The amine MWCNTs mixture was filtered and washed with deionized water. Further, the acid-MWCNTs were dried at 100°C for 12 hours in a vacuum oven. Furthermore, the 1 gram of acid-MWCNTs was dispersed in ethyl alcohol solvent using an ultrasonic generator for 10 minutes. Also, 0.2 grams of dodecylamine solution was added into the amine MWCNTs mixture, and the reaction was carried out at 80°C for 24 hours. Then the residue of reaction output was washed with ethanol solvent to remove unreacted organic compounds. The amine functionalized carbon nanotubes were dried at 100°C for 12 hours in a vacuum oven (Jin et al. 2011) (Scheme 10.1).

10.2.3 Surface Preparation of the Mild Steel Specimens

Mild steel specimens (Si-0.01%; C-0.04%; P-0.002%; Mn-0.17%; S-0.005%; Mo-0.03%; Ni-1.31%; Cr-0.04%; Fe-balance) were used for our study. The specimens were degreased with acetone to remove stain from the steel substrate and to condition before use (Saravanan et al. 2018).

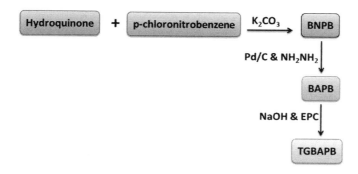

FIGURE 10.2 Preparation of TGBAPB epoxy resin.

A-MWCNT Epoxy Coatings

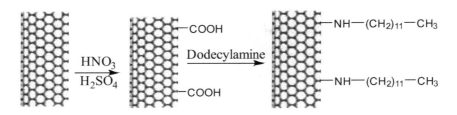

SCHEME 10.1 Surface functionalization of multiwalled carbon nanotubes (A-MWCNTs).

10.2.4 Sample Preparation

The TGBAPB epoxy resin was mixed individually with MWCNTs and A-MWCNTs at a temperature of 80°C for 1 hour and further ultrasonic analysis for 30 minutes. After that, the stoichiometric amount of curing agent was added to HY951 at room temperature (Table 10.2). Then the formulation was applied on the mild steel with the range of thickness of about 100 m using a bar coater technique (Figure 10.3 and Figure 10.4).

TABLE 10.2
Composition of Neat TGBAPB and TGBAPB/MWCNTs/A-MWCNTs Epoxy Coatings

Coating System	TGBAPB/%MWCNTs/%A-MWCNTs	Room Temperature Curing Agent
a	100/0/0	HY951
b	100/0.5/0	HY951
c	100/0/0.5	HY951

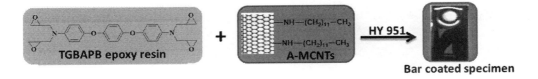

FIGURE 10.3 Schematic diagram of the preparation of TGBAPB/A-MWCNTs epoxy coating.

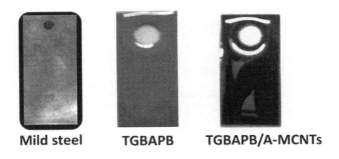

FIGURE 10.4 TGBAPB epoxy and TGBAPB/A-MWCNTs epoxy coating system cured by HY951.

10.3 RESULTS AND DISCUSSION

The FT-IR spectrum of TGBAPB epoxy resin is presented in Figure 10.5a. The band at 913 cm^{-1} appears due to the presence of an oxirane ring. The appearance of bands at 2969 cm^{-1}, 1506 cm^{-1}, and 1612 cm^{-1} corresponds to the aromatic rings of TGBAPB epoxy resin and 1372 cm^{-1} is due to the C-N stretching of TGBAPB epoxy resin. After curing (HY951) with TGBAPB, epoxy resin shows the appearance of the peak at 3449 cm^{-1} and disappearance of oxirane rings at 912 cm^{-1}, as presented in Figure 10.5b. In the ^{1}HNMR spectrum of TGBAPB epoxy resin the chemical shift values obtained are assigned according to the protons of their environment; δ = 6.5–7.0 ppm. This may be due to the presence of aromatic protons (Figure 10.6). The remaining oxirane and methylene

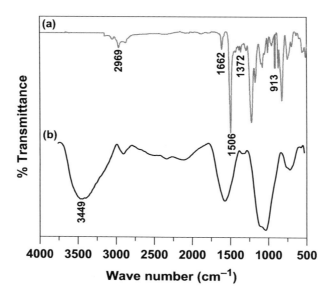

FIGURE 10.5 FT-IR spectra of (a) TGBAPB epoxy resin and (b) TGBAPB/A-MWCNTs epoxy resin cured with HY951.

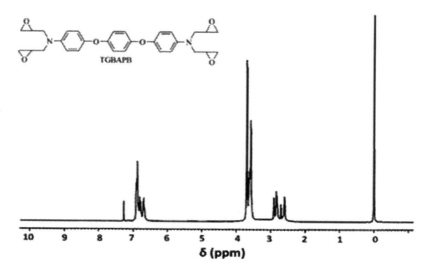

FIGURE 10.6 ^{1}HNMR spectrum of TGBAPB epoxy resin. (From Duraibabu 2014 with permission.)

protons adjacent to the oxirane ring appear at around δ = 2.8–3.8 ppm, as illustrated in Figure 10.6. The ¹³CNMR spectrum of TGBAPB is presented in Figure 10.7. The appearance of signals at 46 ppm and 49.5 ppm was assigned due to the presence of –OCH$_2$ and –CH–, respectively, of TGBAPB epoxy resin (Duraibabu et al. 2014). The signal at 60 ppm may be due to the presence of –N–CH$_2$ carbon. The remaining signals that appear at around 116–153 ppm may be due to the aromatic carbon of TGBAPB (Duraibabu et al. 2014).

10.3.1 Thermal Properties

The samples TGBAPB/MWCNTs, TGBAPB/A-MWCNTs, and neat TGBAPB were thermal analyzed by thermogravimetric analysis (TGA). Figure 10.8 shows that the epoxy system c (TGBAPB/A-MWCNTs) indicates the enhancement of thermal stability other than systems a (TGBAPB/MWCNTs) and b (TGBAPB). The improvement of thermal stability of system c is attributed to the

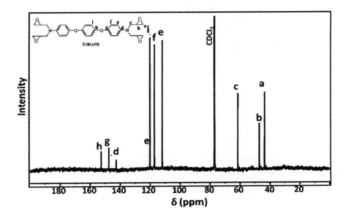

FIGURE 10.7 ¹³CNMR spectrum of TGBAPB epoxy resin. (From Duraibabu 2014 with permission.)

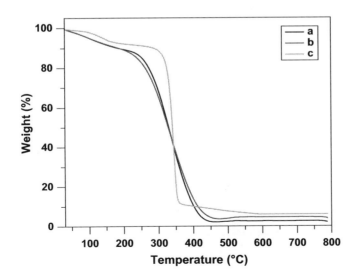

FIGURE 10.8 TGA of neat TGBAPB, TGBAPB/MWCNTs, and TGBAPB/A-MWCNTs.

uniform dispersion of A-MWCNTs with the TGBAPB epoxy matrix as well as the covalent bond between TGBAPAB epoxy/A-MWCNTs (Pourhashem et al. 2017; Dhanapal et al. 2018). In this case system b showed fewer thermal properties when compared to neat TGBAPB epoxy (system a). This may be due to the agglomeration of the MWNCTs inside the polymer matrix to reduce the thermal properties (Duraibabu et al. 2016, 2014).

10.3.2 Electrochemical Impedance Study

The Bode plots for neat TGBAPB, TGBAPB/MWCNTs, and TGBAPB epoxy reinforced with A-MWCNTs coating systems are depicted in Figure 10.9. The obtained values of the impedance of this coating system are between 105 and 106 ohm·cm$_2$. It is interesting to note that the A-MWCNTs reinforced TGBAPB epoxy coated system c specimens exhibited higher impedance values than the TGBAPB/MWCNTs (system b) and neat TGBAPB epoxy (system a) immersion in 3.5 wt.% NaCl solution for 7 days. This may be due to the nanostructured A-MWCNTs cross-linking between TGBAPB epoxy, and strong interfacial interaction exhibits stable corrosion resistance of the metal substrates (Zhou et al. 2019). TGBAPB/MWCNTs (system b) coatings show less corrosion resistance compared to the neat TGBAPB epoxy system a. This might be due to MWCNT particles coming close to one another and giving rise to an appreciable decrease in impedance value. A similar observation was made by Nayak et al. (2015) for CNT-poly sulfone nanocomposites coatings (Manjumeena et al. 2016; Saravanan et al. 2016).

10.3.3 Morphology Studies

The surface morphology of neat TGBAPB, TGBAPB/MWCNTs, and TGBAPB/A-MWCNTs coatings specimens were determined by SEM and illustrated in Figure 10.10. Figure 10.10a indicates that the smooth surface morphology was observed in the neat TGBAPB epoxy matrix (system a). However, the addition of nonfunctionalized MWCNTs (system b) and functionalized A-MWCNTs (system c) showed surface roughness increases. We observe that system c (TGBAPB/A-MWCNTs) exhibits excellent adhesion due to chemical bonding between A-MWCNTs and the TGBAPB epoxy resin (Yoonessi et al. 2014), as presented in Figure 10.10b. This confirms that the improvement of corrosion properties among the nonfunctionalized MWCNTs exhibited less impedance value due

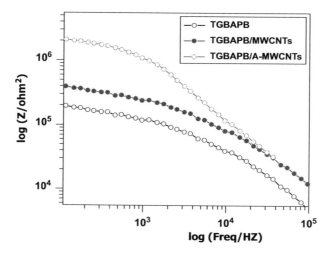

FIGURE 10.9 EIS Nyquist plot of neat TGBAPB, TGBAPB/MWCNTs, and TGBAPB/A-MWCNTs coated samples immersed in 3.5 wt.% NaCl solution for 7 days.

A-MWCNT Epoxy Coatings

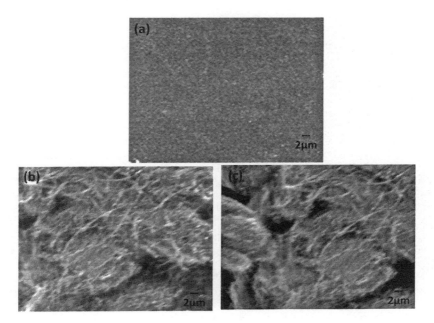

FIGURE 10.10 SEM image of systems (a) neat TGBAPB, (b) TGBAPB/MWCNTs, and (c) TGBAPB/A-MWCNTS epoxy coatings.

to poor dispersion of the MWCNTs within the TGBAPB epoxy matrix (Kim et al. 2016), as shown in Figure 10.10c.

10.4 CONCLUSION

Modified TGBAPB epoxy coating for corrosion was developed using the linkage of the intermediates 1,4′-bis(4-nitrophenoxy) benzene BHPB and 1,4′-bis(4-aminephenoxy)-benzene BAPB. The intermediates and TGBAPB epoxy resin were characterized by means of FTIR and NMR spectroscopic studies. The synthesized TGBAPB epoxy resin was coated on sandblasted mild steel specimens using the room temperature curing agent HY 951 in the presence and in the absence of an A-MWCNTs to a thickness of 100 microns by brush. The anticorrosive behavior of such coated specimens was evaluated by electrochemical impedance studies (EIS). It was interesting to note that the A-MWCNTs reinforced with TGBAPB epoxy coated specimens exhibited tremendous corrosion properties when compared to MWCNTs/TGBAPB and neat TGBAPB epoxy coatings, representing the good dispersion and better compatibility of A-MWCNTs with TGBAPB epoxy matrix. The nonfunctionalized MWCNTs/TGBAPB epoxy coating system displayed lesser impedance value than the neat TGBAPB epoxy coating system. This might be due to the continuous conducting network of nanocomposites coating just starting to form with many MWCNTs particles coming close to each other giving rise to an appreciable decrease in impedance indicating a conducting behavior of MWCNTs/TGBAPB epoxy coating. Thus, we concluded that the development of amine functionalized MWCNTs reinforced with TGBAPB epoxy coatings can be used for applications of corrosion protection of mild steel substrate for enhancement of longevity performance other than coatings systems in present usage.

ACKNOWLEDGMENTS

The authors thank facilities for operating of instruments provided by the Department of Chemistry, Anna University, Chennai, under the projects FIST-DST and DRS-UGC.

BIBLIOGRAPHY

Afzal, A., H. M. Siddiqi, N. Iqbal, and Z. Ahmad. 2013. "The effect of SiO$_2$ filler content and its organic compatibility on thermal stability of epoxy resin." *J. Therm. Anal. Calorim.*, 111, 247–252.

Behzadnasab, M., S. M. Mirabedini, K. Kabiri, and S. Jamali. 2011. "Corrosion performance of epoxy coatings containing silane treated ZrO$_2$ nanoparticles on mild steel in 3.5% NaCl solution." *Corros. Sci.*, 53, 89–98.

Bhattacharya, D., and R. K. Gupta. 2005. "Nanotechnology and potential of microorganisms." *Crit. Rev. Biotechnol.*, 25, 199–204.

Deyab, M. A., A. D. Riccardis, and G. Mele. 2016. "Novel epoxy/metal phthalocyanines nanocomposite coatings for corrosion protection of carbon steel." *J. Mol. Liq.*, 220, 513–517.

Dhanapal, D., G. Rebheka, S. Palanivel, and A. K. Srinivasan. 2015. "A comparative study on modified epoxy and glycidyl carbamate coatings for corrosion and fouling prevention." *Surf. Innovations*, 3, 127–139.

Dhanapal, Duraibabu, Alagar Muthukaruppan, and Ananda Kumar Srinivasan. 2018. "Tetraglycidyl epoxy reinforced with surface functionalized mullite fiber for substantial enhancement in thermal and mechanical properties of epoxy nanocomposites." *Silicon*, 10, 585–594.

Duraibabu, D., M. Alagar, and S. Ananda Kumar. 2014. "Studies on mechanical, thermal and dynamic mechanical properties of functionalized nanoalumina reinforced sulphone ether linked tetraglycidyl epoxy nanocomposites." *RSC Adv.*, 4, 40132–40140.

Duraibabu, D., M. Alagar, and S. Ananda Kumar. 2016. "Development and characterization of tetraglycidyl epoxy reinforced inorganic hybrid nanomaterials for high performance applications." *High Perfor. Polym.*, 28, 773–783.

Duraibabu, D., T. Ganeshbabu, R. Manjumeena, S. Ananda Kumar, and Priya Dasan. 2014, "Unique coating formulation for corrosion and microbial prevention of mild steel." *Prog. Org. Coat.*, 77, 657–664.

Hernandez, L. A., L. S. Hernandez, S. L. Rodriguez-Reyna. 2012. "Evaluation of corrosion behavior of galvanized steel treated with conventional conversion coatings and a chromate-free organic inhibitor." *Int. J. Corros.*, 2012, 1–8.

Hosseini, M. G., and K. Aboutalebi. 2019. "Enhancement the anticorrosive resistance of epoxy coatings by incorporation of CeO$_2$ @ polyaniline @ 2-mercaptobenzotiazole nanocomposites." *Synth. Met.*, 250, 63–72.

Jin, Fan-Long, Chang-Jie Ma, and Soo-Jin Park. 2011. "Thermal and mechanical interfacial properties of epoxy composites based on functionalized carbon nanotubes." *Mater. Sci. Eng. A*, 528, 8517–8522.

Kabeb, S. M., A. Hassan, Z. Mohamad, Z. Sharer, M. Mokhtar, and F. Ahmad. 2019. "Exploring the effects of nanofillers of epoxy nanocomposite coating for sustainable corrosion protection." *Chem. Eng. Trans.*, 72, 121–126.

Kadhum, A. A. H., A. B. Mohamad, L. A. Hammed, A. A. Al-Amiery, N. H. San, and A. Y. Musa. 2014. "Inhibition of mild steel corrosion in hydrochloric acid solution by new coumarin." *Materials*, 7, 4335–4348.

Khodair, Z. T., A. A. Khadom, and H. A. Jasim. 2018. "Corrosion protection of mild steel in different aqueous media via epoxy/nanomaterial coating: preparation, characterization and mathematical views." *J. Mater. Res. Technol.*, 8 (1), 424–435.

Kim, Sihwan, Woo I. Lee, Chung H. Park. 2016. "Assessment of carbon nanotube dispersion and mechanical property of epoxy nanocomposites by curing reaction heat measurement." *J. Reinf. Plast. Compos.*, 35, 71–80.

Kumar, Arun, P. K. Ghosh, K. L. Yadav, and Kaushal Kumar. 2017. "Thermo-mechanical and anti-corrosive properties of MWCNT/epoxy nanocomposite fabricated by innovative dispersion technique." *Composites Part B*, 113, 291–299.

Kumar, Arun, Kaushal Kumar, P. K. Ghosha, and K. L. Yadav. 2018. "MWCNT/TiO$_2$ hybrid nano filler toward high-performance epoxy composite." *Ultrason. Sonochem.*, 41, 37–46.

Lee, H. C., W. W. Liu, S. A. Chai, A. R. Mohamed, A. Aziz, C. S. Khe, N. M. S. Hidayah, and U. Hashim. 2017. "Review of the synthesis, transfer, characterization and growth mechanisms of single and multilayer grapheme." *RSC Adv.*, 7, 15644–15693.

Lou, C., R. Zhang, X. Lu, C. Zhou, and Z. Xin. 2019. "Facile fabrication of epoxy/polybenzoxazine based superhydrophobic coating with enhanced corrosion resistance and high thermal stability." *Colloids Surf. A*, 562, 8–15.

Majd, Mehdi Tabatabaei, Taghi Shahrabi, and Bahram Ramezanzadeh. 2019. "The role of neodymium based thin film on the epoxy/steel interfacial adhesion and corrosion protection promotion." *Appl. Surf. Sci.*, 464, 516–533.

Manjumeena, R., R. Venkatesan, D. Duraibabu, J. Sudha, N. Rajendran, and P. T. Kalaichelvan. 2016. "Green nanosilver as reinforcing eco-friendly additive to epoxy coating for augmented anticorrosive and antimicrobial behavior." *Silicon*, 8, 277–298.

Mohammadi, I., M. Izadi, T. Shahrabi, D. Fathib, and A. Fateha. 2019. "Enhanced epoxy coating based on cerium loaded Na-montmorillonite as active anti-corrosive nanoreservoirs for corrosion protection of mild steel: Synthesis, characterization, and electrochemical behavior." *Prog. Org. Coat.*, 131, 119–130.

Moitra, D., B. K. Ghosh, M. Chandel, R. K. Jani, M. K. Patra, S. R. Vadera, and N. N. Ghosh. 2016. "Synthesis of a Ni0.8Zn0.2Fe$_2$O$_4$–RGO nanocomposite: an excellent magnetically separable catalyst for dye degradation and microwave absorber." *RSC Adv.*, 6, 14090–14096.

Nayak, Lalatendu, Ranjan R. Pradhan, Dipak Khastgir, and Tapan K. Chaki. 2015. "Thermally stable electromagnetic interference shielding material from polysulfone nanocomposites: Comparison on carbon nanotube and nanofiber reinforcement." *Polym. Compos.*, 36, 3, 566–575.

Nayak, S. R., and K. N. S. Mohana. 2018. "Corrosion protection performance of functionalized graphene oxide nanocomposite coating on mild steel." *Surf. Interfaces*, 11, 63–73.

Neisiany, R. E., J. K. Y. Lee, S. Nouri Khorasani, R. Bagheri, and S. Ramakrishna. 2018. "Facile strategy toward fabrication of highly responsive self-healing carbon/epoxy composites via incorporation of healing agents encapsulated in poly(methylmethacrylate) nanofiber shell." *J. Ind. Eng. Chem.*, 59, 456–466.

Palanivelu, Saravanan, Duraibabu Dhanapal, Ananda Kumar Srinivasan. 2017. "Studies on silicon containing nano-hybrid epoxy coatings for the protection of corrosion and bio-fouling on mild steel." *Silicon*, 9, 447–458.

Park, S., and R. S. Ruoff. 2009. "Chemical methods for the production of graphenes." *Nat. Nanotechnol.*, 4, 217–219.

Pour, Zahra Sekhavat, Mousa Ghaemy, Sajjad Bordbar, and Hassan Karimi-Maleh. 2018. "Effects of surface treatment of TiO$_2$ nanoparticles on the adhesion and anticorrosion properties of the epoxy coating on mild steel using electrochemical technique." *Prog. Org. Coat.*, 119, 99–108.

Pourhashem, Sepideh, Mohammad Reza Vaezi, Alimorad Rashidi, and Mohammad Reza Bagherzadeh. 2017. "Exploring corrosion protection properties of solvent based epoxy-graphene oxide nanocomposite coatings on mild steel." *Corros. Sci.*, 115, 78–92.

Qiu, Shihui, Cheng Chen, Wenru Zheng, Wei Li, Haichao Zhao, and Liping Wang. 2017. "Long-term corrosion protection of mild steel by epoxy coating containing self-doped polyaniline nanofiber." *Synth. Met.*, 229, 39–46.

Ramezanzadeh, B., B. Karimi, M. Ramezanzadeh, and M. Rostami. 2019. "Synthesis and characterization of polyaniline tailored graphene oxide quantum dot as an advance and highly crystalline carbon-based luminescent nanomaterial for fabrication of an effective anti-corrosion epoxy system on mild steel." *J. Taiwan Inst. Chem. Eng.*, 95, 369–382.

Ramezanzadeh, B., and M. M. Attar. 2011. "Studying the effects of micro and nano sized ZnO particles on the corrosion resistance and deterioration behavior of an epoxy-polyamide coating on hot-dip galvanized steel." *Prog. Org. Coat.*, 71 (3), 314–328.

Ramezanzadeh, M., G. Bahlakeh, B. Ramezanzadeh, and M. Rostami. 2019. "Mild steel surface eco-friendly treatment by Neodymium-based nanofilm for fusion bonded epoxy coating anti-corrosion/adhesion properties enhancement in simulated seawater." *J. Ind. Eng. Chem.*, 72, 474–490.

Raotole, P. M., P. Koinkar, B. Joshi, and S. R. Patil. 2015. "Corrosion protective poly(aniline-co-oanisidine) coatings on mild steel." *J. Coat. Technol. Res.*, 12 (4), 757–766.

Sambyal, Pradeep, Gazala Ruhi, S. K. Dhawan, B. M. S. Bisht, and S. P. Gairola. 2018. "Enhanced anticorrosive properties of tailored poly(aniline-anisidine)/chitosan/SiO$_2$ composite for protection of mild steel in aggressive marine conditions." *Prog. Org. Coat.*, 119, 203–213.

Sanaei, Z., B. Ramezanzadeh, and T. Shahrabi. 2018. "Anti-corrosion performance of an epoxy ester coating filled with a new generation of hybrid green organic/inorganic inhibitive pigment; electrochemical and surface characterizations." *Appl. Surf. Sci.*, 454, 1–15.

Sanaei, Zahra, Ghasem Bahlakeh, and Bahram Ramezanzade. 2017. "Active corrosion protection of mild steel by an epoxy ester coating reinforced with hybrid organic/inorganic green inhibitive pigment." *J. Alloys Compd.*, 728, 1289–1304.

Saravanan, P., D. Duraibabu, K. Jayamoorthy, S. Suresh, S. Ananda Kumar. 2018. "Twin applications of tetrafunctional epoxy monomers for anticorrosion and antifouling studies." *Silicon*, 10, 555–565.

Saravanan, P., K. Jayamoorthy, and S. Ananda Kumar. 2016. "Design and characterization of non-toxic nano-hybrid coatings for corrosion and fouling resistance." *J. Sci.: Adv. Mater. Devices*, 1, 367–378.

Sørensen, P. A., S. Kiil, K. Dam-Johansen, and C. E. Weinell. 2009. "Anticorrosive coatings: a review." *J. Coat. Technol. Res.*, 6, 135–176.

Tasis, D., N. Tagmatarchis, A. Bianco, and M. Prato. 2006. "Chemistry of carbon nanotubes." *Chem. Rev.*, 106, 1105–1136.

Toorani, M., M. Aliofkhazraei, R. Naderi, M. Golabadi, and A. Sabour Rouhaghdam. 2017. "Role of lanthanum nitrate in protective performance of PEO/epoxy double layer on AZ31 Mg alloy: electrochemical and thermodynamic investigations." *J. Ind. Eng. Chem.*, 53, 213–227.

Vakili, H., B. Ramezanzadeh, and R. Amini. 2015. "The corrosion performance and adhesion properties of the epoxy coating applied on the steel substrates treated by cerium-based conversion coatings." *Corros. Sci.*, 94, 466–475.

Wang, Binhua, Yuxuan Bai, Xiaozhi Hu, and Pengmin Lu. 2016. "Enhanced epoxy adhesion between steel plates by surface treatment and CNT/short-fibre reinforcement." *Compos. Sci. Technol.*, 127, 149–157.

Wang, Na, Wanlu Fu, Jing Zhang, Xuri Li, and Qinghong Fang. 2015. "Corrosion performance of waterborne epoxy coatings containing polyethylenimine treated mesoporous-TiO_2 nanoparticles on mild steel." *Prog. Org. Coat.*, 89, 114–122.

Wei, Y. H., L. X. Zhang, and W. Ke. 2007. "Evaluation of corrosion protection of carbon black filled fusion-bonded epoxy coatings on mild steel during exposure to a quiescent 3% NaCl solution." *Corros. Sci.*, 49 (2), 287–302.

Wu, T. H., A. Foyet, A. Kodentsov, L. G. J. van der Ven, R. A. T. M. van Benthem, and G. de With. 2014. "Wet adhesion of epoxy–amine coatings on 2024-T3 aluminum alloy." *Mater. Chem. Phys.*, 145 (3), 342–349.

Yeganeh, M., M. Omidi, and T. Rabizadeh. 2019. "Anti-corrosion behavior of epoxy composite coatings containing molybdateloaded mesoporous silica." *Prog. Org. Coat.*, 126, 18–27.

Yeganeha, M., N. Asadi, M. Omidi, and M. Mohammad. 2019. "An investigation on the corrosion behavior of the epoxy coating embedded with mesoporous silica nanocontainer loaded by sulfamethazine inhibitor." *Prog. Org. Coat.*, 128, 75–81.

Yeole, K. V., I. P. Agarwal, and S. T. Mhaske. 2016. "The effect of carbon nanotubes loaded with 2-mercaptobenzothiazole in epoxy-based coatings." *J. Coat. Technol. Res.*, 13 (1), 31–40.

Yoonessi, Mitra, Marisabel Lebrón-Colón, Daniel Scheiman, and Michael A. Meador. 2014. "Carbon nanotube epoxy nanocomposites: the effects of interfacial modifications on the dynamic mechanical properties of the nanocomposites." *ACS Appl. Mater. Interfaces*, 6, 16621–16630.

Yuan, X., Z. F. Yue, X. Chen, S. F. Wen, L. Li, and T. Feng. 2015. "EIS study of effective capacitance and water uptake behaviors of silicone-epoxy hybrid coatings on mild steel." *Prog. Org. Coat.*, 86, 41–48.

Zhang, X. F., R. J. Chen, and J. M. Hu. 2016. "Superhydrophobic surface constructed on electrodeposited silica films by two-step method for corrosion protection of mild steel." *Corros. Sci.*, 104, 336–343.

Zhou, Changlu, Miao Tao, Juan Liu, Tao Liu, Xin Lu, and Zhong Xin. 2019. "Effects of interfacial interaction on corrosion resistance of polybenzoxazine/SiO_2 nanocomposite coatings." *ACS Appl. Polym. Mater.*, 1, 381–391.

11 Electrochemical Oxidation Reaction for the Treatment of Textile Dyes

Eswaran Prabakaran and Kriveshini Pillay

CONTENTS

11.1 Introduction	181
11.2 Advantages of EAOP Anodes	183
11.3 Types of Electrochemical Advanced Oxidation Methods	184
11.3.1 EAOPs by Direct Method	184
11.3.2 EAOPs by Indirect Method	185
11.4 EAOPs of Textile Dyes from Wastewater	186
11.4.1 Electrochemical Advanced Oxidation of Methylene Blue Dye	186
11.4.2 Electrochemical Advanced Oxidation of Rhodamine B Dye	188
11.4.3 Electrochemical Advanced Oxidation of Congo Red	188
11.4.4 Electrochemical Advanced Oxidation of Indigo Carmine Dye	188
11.4.5 Electrochemical Advanced Oxidation of Yellow 3 (DY3) Dye	189
11.4.6 Electrochemical Advanced Oxidation of Acid Blue 113 Dye	189
11.4.7 Electrochemical Advanced Oxidation of Triphenylmethane Dye	189
11.5 Conclusion	190
References	190

11.1 INTRODUCTION

Textile dyes have been used in dyeing and converting fibers into commercial substances. These processes include the dyeing of both manmade and natural fibers to produce permanent coloration for the textile production industry. This coloration process has resulted in a massive amount of colored wastewater with excessive pH, COD, and high temperature (Pathania et al. 2016). The color in the wastewater cannot be easily removed and requires some energy-intensive techniques (Sharma et al. 2015). The specific conventional techniques that have been used for the treatment of fabric dyes include biological (Paprowicz and Slodczyk 1988), adsorption (McKay 1980), coagulation (Hicham et al. 2019), and ozonation techniques (Lin and Lin 1992). These methods have, however, demonstrated drawbacks such as low efficiency and high cost, and additionally have not been employed for fabric dye remediation because of recalcitrant to biodegradation. The disadvantage of the adsorption technique is that the adsorbent cannot be regenerated and excessive costs arise depending on the type of adsorbent used. The other chemical strategies that have been employed have also generated a lot of sludge and these are difficult to eliminate.

Textile dyes contain a whole lot of chemical compounds and therefore it is difficult to isolate and remove the dyes and additionally these methods are not suitable (Anastasios et al. 2005). Consequently, low cost and environmentally friendly strategies are required for the removal of fabric dyes. Superior oxidation methods have been used for the removal of fabric dyes. These include electrochemical methods, ozonation, wet oxidation, and photocatalytic oxidation. Among these, electrochemical processes have exhibited greater oxidation of textile dyes because of their lower cost and less time consumption, and these are also less problematic to use (Lorimer et al. 2001). The electrochemical technique is a critical analytical technique that has been used to detect poisonous pollutants, which include inorganic and organic pollutants commonly contained in wastewater (Rajeshwar et al. 1994; Rao and Venkatarangaiah 2014; Martınez-Huitle et al. 2015). These exclusive electrochemical methods have been used to degrade toxic pollutants by electrocoagulation, electrodialysis, electroreduction, electroflotation, and electrochemical oxidation with energetic oxidation. These strategies have additionally included the electro-Fenton, photoelectro-Fenton, solar photoelectro-Fenton, and solar photoelectrocatalysis processes (Brillas et al. 2015). Electrochemical advanced oxidation processes (EAOPs) have been used for the degradation and removal of textile dyes from wastewater, as shown in Figure 11.1. In recent times, electrochemical anodic oxidation has been used for the elimination of fabric dyes and has manifested itself as a one of a kind methodology.

Electrochemical advance oxidation processes have been implemented for efficient anodic oxidation of textile dyes (Moreira et al. 2017). This method has the distinct advantage of not requiring any chemical reagents as the oxidizing agent is provided on the anodic electrode surface. The oxidizing agents are hydroxyl radicals, which are generated by the usage of water molecules at the anodic electrode. The hydroxyl radical is an incredible oxidizing agent for fabric dye remediation from wastewater (Buxton et al.1988). These hydroxyl radicals are also produced from H_2O_2 during EAOPs. This reaction takes place on the anodic surface. This anode therefore generates a lot of oxidizing agents and these are used for the treatment of textile dyes from contaminated water (Cañizares et al. 2009; Lan et al. 2017; Barazesh et al. 2018; Jasper et al. 2016; Martinez-Huitle et al. 2005). The ·OH radical is reacted with chlorine Cl_2 and $S_2O_8^{2-}$ to give sulfate radicals ($SO_4·^-$) and perchlorate (ClOH·$^-$) radicals as shown in Figure 11.2. S_2O_8, H_2O_2, and Cl_2 were used as oxidizing agents for the treatment of disinfectants in wastewater and they had been applied in aquifer remediation (Tsitonaki et al. 2010). $S_2O_8^{2-}$ has been explored at a standard oxidation potential of $E_0 = 2.01$ V and H_2O_2 at $E_0 = 1.8$ V and Cl_2 at $E_0 = 1.48$ V. Among them $S_2O_8^{2-}$ has demonstrated extraordinary potential as an oxidizing agent for the treatment of wastewater and as a disinfectant against bacterial action (Polcaro et al. 2007). The Cl·, ClOH^{2-}·, and SO_4^{2-}· radicals have been proven to be high-quality oxidants for the treatment of toxic pollutants from wastewater samples and by virtue of their immediate electron transfer to pollutants within the water samples. The pollutants in wastewater became directly carried out by the ·OH radicals and additional oxidants are applied to degrade the organic textile dyes with EAOPs. However, these do not completely degrade the textile dyes and

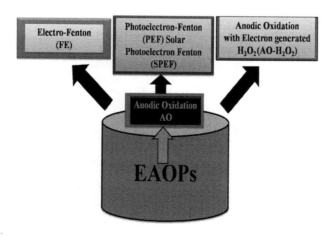

FIGURE 11.1 Different types of EAOPs employed for the removal of organic textile dyes.

Electrochemical Oxidation for the Treatment of Textile Dyes

$$H_2O \longrightarrow \cdot OH + H^+ + e^-$$
$$2 \cdot OH \longrightarrow H_2O_2$$
$$Cl^- \longrightarrow Cl \cdot + e^-$$
$$2Cl \cdot \longrightarrow Cl_2$$
$$SO_4^{2-} \longrightarrow SO_4^- + e^-$$
$$2SO_4^- \longrightarrow S_2O_8^{2-}$$
$$SO_4^{2-} + \cdot OH \longrightarrow SO_4^- + OH^-$$
$$Cl^- + OH \cdot \longrightarrow ClOH^-$$

FIGURE 11.2 Mechanism of oxidant generation.

need additional anodic electrode materials for the oxidation of textile pollutants with EAOP anodes (Carter et al. 2008; Zhuo et al. 2012; Zhuo et al. 2011).

11.2 ADVANTAGES OF EAOP ANODES

The electrochemical cell containing an anode and a cathode has been used for the electrochemical remediation of textile dyes from wastewater samples. Both anodes and cathodes have been separated via the usage of a membrane within the electrochemical cell, which maintains the flow of liquids within both compartments as shown in Figure 11.3. Therefore, the current response measures the flow of charge in the electrochemical cell while the ions are shifting among the membrane separators. The ion exchange membranes that are generally used as separators inside the electrochemical

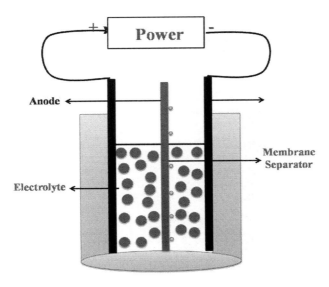

FIGURE 11.3 General setup of EAOPs.

cells include Nafion and fluoropolymer. In these, the membrane was extensively explored in both acidic and alkali environments with excellent selectivity of H⁺ ions. Despite this fact, these types of membranes are not used within electrochemical cells for the electrochemical treatment of textile dyes from wastewater because of excessive cost, effortless fouling, high electricity intake, excessive maintenance, and a reduction in overall performance (Sala et al. 2014). This results in the oxidation of textile dyes in wastewater without any problems provided that the samples are on the surface of the electrode, and the cathode is not affecting the anodic procedures. Electrochemical oxidation tactics have been carried out for the remediation of textile dyes in wastewater samples. This method can be utilized at a regular temperature. Other advantages presented by this method include the fact that the materials used to fabricate the electrode have a long-term lifespan and have a high resistance to acidic and basic media. The substances that constitute the anodic electrode in the electrochemical oxidation of textile dyes have shown improved remediation of dyes (Guo et al. 2016; Yoon et al. 2013). Oppositely charged materials were used in order to constitute the anode for the EAOPS inside the electrochemical mobile systems.

11.3 TYPES OF ELECTROCHEMICAL ADVANCED OXIDATION METHODS

A lot of these methods have been applied for the treatment of textile dyes from contaminated water and they have also shown some disadvantages. These processes are dependent on physical, chemical, and biological techniques. These methods also demonstrate different demerits due to the improper discharge of waste textile dye into the environment. In this sense, the problems posed include sludge formation, adsorbent problems, and complex dye formation. The complete mineralization of textile dyes results in the formation of carbon dioxide, water, and other products with photocatalytic processes. Advance oxidation processes (AOPs) have demonstrated better degradation of textile dyes from wastewater due to using powerful oxidants such as $^-\cdot OH$, F^-, HO_2, and H_2O_2, and these radicals have resulted in complete mineralization of the textile dyes from wastewater samples (Oturan et al. 2014; Sires et al. 2014). EAOPs were therefore used to complete the mineralization of organic textile dye pollutants by creating the hydroxyl radicals inside the supporting electrolyte. Two types of advanced oxidation procedures, namely direct EAOPs and indirect EAOPs have been explored. In direct EAOPs, hydroxyl radicals are produced inside the electrolyte system at the anode floor with direct techniques. This hydroxyl radical is produced at the anode surface, which is dependent on the anode substances, the active site of anode substances, and current density. This method has therefore eliminated the need for external reagents for the creation of ·OH radicals (Panizza and Cerisola 2009; Miled et al. 2010). However, the oblique technique of EAOPs has required external reagents for the production of hydroxyl radicals together with electro-Fenton, H_2O_2, and ferrous iron (Fered-Fenton). Hydrogen peroxide is used to produce cathode floor materials. The cathode surface then creates the hydroxyl radical electrolytically with H_2O_2 as an external reagent, and more hydroxyl radicals are also produced at the anode surface with ferrous iron as an external reagent (Brillas et al. 2009). The indirect EAOP method generates hydroxyl radicals at the cathode floor and produces H_2O_2 radicals and ferrous ions electrocatalytically. The EAOP is therefore an electrochemical method that offers the advantages of flexible nature, high removal performance, operational protection, and automation (Peralta-Hernandez et al. 2009; Jüttner et al. 2000; Anglada et al. 2009).

11.3.1 EAOPs BY DIRECT METHOD

This EAOPs method generates the hydroxyl radicals at the anode surface of materials in electrochemical systems without any external chemical reagents. This system also removes textile dyes from wastewater at the anodic surface. This direct electrochemical advanced oxidation method is therefore an important technique for generating the oxidants of hydroxyl radicals at the anode surface (Panizza and Cerisola 2009). The hydroxyl radicals are coated on the surface of the anode and these are actually chemisorbed on platinum (Pt) and Ti/RuO_2-IrO_2 DSA anodes. The formation of these types of anodes leads to the demand for the production of hydroxyl radicals. The oxygen evolution and hydroxyl radical formation anodes are known as boron-doped diamond (BDD). This anode is referred to as a thin film anode and it produces the hydroxyl radicals for the oxidation of textile dyes. The AO method results in the degradation of textile dyes at anodes surface (Comninellis 1994; Scialdone 2009). Two types of organic textile dyes have been degraded by using the AO method. The electrochemical conversion is conducted by degrading the persistent textile dyes into biodegradable by-products such as short-chain carboxylic acids, electrochemical combustion and incineration (Comninellis and De Battisti 1996). AO method is defined as the direct EAOPs in which the degradation of the organic pollutants occurs mainly by the adsorbed hydroxyl radicals produced by water oxidation in the presence of high O_2 overvoltage anodes. The organic pollutant degradation mechanism of AO process is shown in Figure 11.4.

11.3.2 EAOPs by Indirect Method

The manufacturing of hydroxyl radicals is accomplished with the addition of external chemicals for degrading the organic textile pollutants. Chlorine and hypochlorite ions have been used to degrade textile pollutants in the water medium, which may be generated by the anodic surface electrode as shown in Figure 11.5. Those processes are suitable for the electrochemical oxidation of inorganic

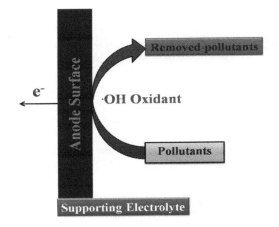

FIGURE 11.4 Electrochemical advance oxidation method with direct processes.

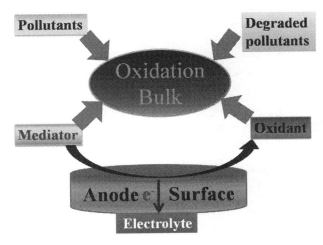

FIGURE 11.5 Electrochemical advance oxidation method with indirect processes.

and organic pollutants from wastewater samples (Hicham et al. 2019; He et al. 2016; Ines et al. 2014; Jalife-Jacobo et al. 2016). In this process the chlorine salts of sodium chloride and potassium chloride are brought into contact with organic and inorganic wastewater for good conductivity and the production of hypochlorite ions at the anodic surface. At the anodic surface chloride ions are also created and these form chlorine as shown in Figure 11.6.

11.4 EAOPS OF TEXTILE DYES FROM WASTEWATER

EAOPs had been conducted on different textile dyes with different anode substances because effective oxidizing agents were produced, and these include chlorine, oxygen, hydroxyl radicals, and other oxidants. These are generated at the anodic surface and serve as materials which result in better decolorization of the textile dyes due to the complete mineralization into carbon dioxide and the water molecule.

11.4.1 Electrochemical Advanced Oxidation of Methylene Blue Dye

Electrochemical oxidation of methylene blue (MB) was achieved with the electro-Fenton approach using stainless steel mesh as an electrode. This electrode was used for MB decolorization as a function of the following parameters, namely concentration of dye, pH and Fe:H_2O_2 reagent ratio, current density, and the diversity of the electrode. Satisfactory decolorization of MB was achieved at pH 3 after 20 minutes and also achieved the removal of MB dye at the highest Fe^{2+}:H_2O_2 ratio of 1:4. The current density resulted in a growing oxidation rate of MB dye and fast decolorization. The size of the mesh pores also resulted in MB dye degradation and additionally increased the proportion of elimination of MB dye with the optimum removal of MB dye of 94.5% within 25 minutes at the two electrodes instead of the metal plate. The metal mesh resulted in twice the rate of removal of MB dye solution than the steel plate with decreasing power consumption. Consequently, it is advised that steel mesh is used as an electrode for electrochemical techniques (Loloei and Rezaee 2016).

This has led to direct decolorization by the electrochemical oxidation of MB in aqueous solution with electroactive polymers (EAPs), wherein graphite becomes the anode and stainless steel 316 grade is used as the cathode. The electrochemical oxidation of MB dyes was reduced because of specific parameters, which include current density, electrolyte temperature, pH of the solution, and preliminary dye concentration. As a result, the MB dye decolorization was rapidly achieved with the graphite anode. Decolorization of MB of 99.75% and 99.60% was achieved. The energy consumption was increased from 2.055 to 11.714 Kw.h/m³ at the applied voltages 14.6 V and 28.5 V, respectively (Jawad and Najim 2018). These approaches verified that 10 mA/cm² current density, 0.5 g/L Na_2SO_4, 60 min electrolysis time, and concentrations of MB of 50 mg/L and 400 mg/L resulted in 72.51% and 77.94% removal of color at 20°C, respectively. An electrochemical oxidation approach

$2Cl \longrightarrow Cl_2 + 2e- \longrightarrow$ Anode reaction

Formation of hypochlorous from chlorine

$Cl_2 + H_2O \longrightarrow HOCl + H^+ + Cl$

$Cl_2 + H_2O \longrightarrow HOCl + H^+ + Cl$

Dissociation of hypochloride ion

$HOCl \rightarrow H^+ + OCl^-$

$MB + OCl^- \rightarrow$ dye intermediate $\rightarrow CO_2 + H_2O + Cl^-$

$MB + HOCl \rightarrow CO_2 + H_2O + HCl$

FIGURE 11.6 Electrochemical advance oxidation reaction on anode surface with indirect method.

Electrochemical Oxidation for the Treatment of Textile Dyes

is therefore useful for mineralizing methylene blue dye dissolved in water. For this reason, an electrochemical cell was used. This includes electrodes crafted from chrome steel and graphite. Different factors that influenced dye elimination were studied and these included current density, treatment time, and preliminary concentrations of dye solution. The increase in dye elimination was found to be in direct proportion with the initial dye concentration. However, the dye elimination was found to be improved with an increase in treatment time as high as 30–40 min and a contemporary density as high as 0.06 A cm^{-2}. The electrochemical oxidation of methylene blue was additionally conducted in an electrochemical cell consisting of an anode that was made of graphite and a cathode that was made from a graphite sheet of stainless steel. Each of the electrodes had been separated with the aid of setting a polyethylene-based membrane (Daramic 350). The cathode and anode were filled with electrolyte and methylene blue solution, respectively. The methylene blue solution was completely mineralized and the remaining concentration was used as a colorimetric agent in a spectrophotometric method. After filling the respective solutions in each of the compartments, the electrodes were linked to a DC current supply. The effect of preliminary attention, contemporary density, and remediation was studied with reference to the percentage elimination of methylene blue from its aqueous solution (Asghar et al. 2015). The cell was constituted of both anode and cathode as shown in Figure 11.7.

Electrochemical treatments of MB become indirectly decolorized by using a Pt electrode under a strong electrolyte where numerous parameters have been optimized to achieve maximum elimination of MB. The entire decolorization of MB was dependent on the presence of chloride as first-order reaction and bromide as a second-order reaction. The electrochemical oxidation of the MB decolorization rate was investigated as a function of various parameters, which include current density, the effect of chloride and bromide ions, supporting electrolyte, and initial pH of the solution. Although MB dye showed no effective decolorization with concentration, temperature, and chloride and bromide ions in MB solution, the electrochemical oxidation of MB was not complexly decolorized with iodide, fluoride, chloride, and bromide either. No effective decolorization coloration was also observed in the presence of sulfate ions in the MB solution (El Hajj Hassan and El Jamal 2012). BDD

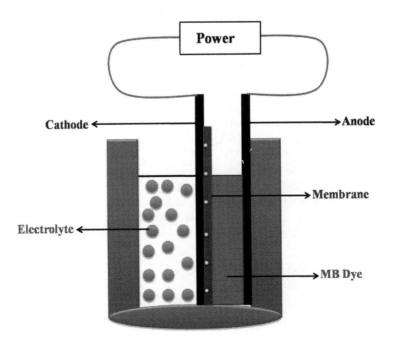

FIGURE 11.7 Electrochemical advance oxidation of methylene blue in the electrochemical cell.

electrodes act as anodes for the electrochemical oxidation of methylene dye in aqueous solution. In this case the level of MB dye decolorization was low because of the low rate of electrooxidation due to mass-transfer limitation. The adsorption techniques have resulted in a decrease in the interest of electrochemical oxidation of MB dye. The adsorbent–adsorbate combination of sawdust and methylene blue dye gave good results for the investigation of adsorption, kinetic rate activity, current density and regenerated time. MB decolorization was evaluated by the usage of electrochemical oxidation on sawdust, and the sawdust was regenerated for the next adsorption of MB dye. This cyclic process of adsorption and electrochemical regeneration has moved forward the decolorization of and enhanced adsorption of MB dye solution from wastewater samples (Bouaziz et al. 2014).

11.4.2 Electrochemical Advanced Oxidation of Rhodamine B Dye

Environmental pollution is the main problem that is created by surface waters contaminated with different types of dye industrial wastewater. Dye wastewater samples have been degraded with the electrochemical anodic oxidation method. Rhodamine B (RhB) dye was electrochemically degraded by using Ti/RuO$_2$-IrO$_2$ (DSA) and SnO$_2$ anodes. These individual electrode efficiencies have been studied for the treatment of dye wastewater. The effect of decolorization and chemical oxygen demand (COD) were optimized with distinctive parameters including current density, different initial pH, and the concentration of electrolyte and temperature on the electrochemical oxidation. The decolorization of RhB was achieved in the presence of chloride ions with Ti/RuO$_2$-IrO$_2$ (DSA) and SnO$_2$ anodes electrodes at various time intervals. DSA electrode was optimized for the decolorization of RhB dye including pH 6.5, t=25°C, the contemporary density of 40 mA cm^{-2} in NaCl 0.05 mol L^{-1} + Na$_2$SO$_4$ 0.1 mol L^{-1} mixture of supporting electrolyte solution. As a result, 100% color removal and 61.7% chemical oxygen demand removal after 90 min of electrolysis was achieved. The DSA anode exhibited better performance than the SnO$_2$ anode due to efficient activity and greater cost-effectiveness. The DSA anode resulted in good electrochemical anodic oxidation of RhB dye because of chloride ions within the mixture of electrolyte, which provided good oxidants such as Cl$_2$ and ClO$^-$, and these are used to facilitate the treatment of RhB at the electrode of Ti/RuO$_2$-IrO$_2$ (DSA) and SnO$_2$ anodes (Baddouh et al. 2018).

11.4.3 Electrochemical Advanced Oxidation of Congo Red

The electrochemical oxidation method was also used to degrade the diazo dye Congo red (CR) with anodic oxidation using the boron-doped diamond anode (AOx/BDD) with various supporting electrolytes. This treatment was performed by using the BDD anode and cathode of stainless steel (AISI-304) with exceptional current densities consisting of 7.5, 15, 30, and 50 mA/cm^2 in the 3L flow reactor. EAOPs of CR dye were additionally tested with distinctive supporting electrolytes (HClO$_4$, NaCl, Na$_2$SO$_4$, and H$_2$SO$_4$) with a 50 mM concentration and a flow rate of 7 L/min. The NaCl supporting electrolyte medium resulted in faster degradation of CR dye with EAOPs BDD anode when compared to HClO$_4$, Na$_2$SO$_4$, and H$_2$SO$_4$. The supporting electrolytes of HClO$_4$, Na$_2$SO$_4$, and H$_2$SO$_4$ resulted in lower oxidation of CR due to the presence of SO$_4^{2-}$ and ClO$_4^-$ in the bulk solution. EAOPs of CR degraded products have also shown evidence of mineralization with dissolved natural carbon ((DOC) and HPLC analysis (Jalife-Jacobo et al. 2016).

11.4.4 Electrochemical Advanced Oxidation of Indigo Carmine Dye

Electrochemical oxidation of textile dye wastewater remediation has not been directly performed inside the temporary mathematical model of FM01-LC reactor with the continuous stirring tank in the presence of chloride ions. IrO$_2$-SnO$_2$ doped with Sb$_2$O$_5$ was used as an anode with titanium multiplied meshed and stainless steel elevated meshes as the cathode. This reactor was operated as a batch recirculation system with the galvanostatic condition and current densities of 132 Am^{-2} and

200 Am^{-2}; 1.0 mM of indigo carmine dye; 0.05 M of NaCl supporting electrolyte; and flow rates of 0.9 L min^{-1}, 1.8 L min^{-1}, 2.7 L min^{-1}, and 3.6 L min^{-1}. FM01-LC reactor performance was studied with the aid of the liquid float pattern through axial dispersion coefficient (D_{ax}), the electrooxidation of chloride ions within the anode side (C_{Cl}^-), the electrochemical discount of chlorine oxidizing species (C_{ox}) under transport control in cathode, while inside the continuous stirring tank (CST) the chemical oxidation reactions between active chlorine species and organic matter through homogeneous second-order kinetics were considered. This had been evaluated through the consumption of the chemical oxygen demand (COD) and the color decay (C_{col}). The mathematical model was explained properly by the experimental investigation of different chemical species together with C_{Cl}^- ($k_{Cl} = (3.2 \times 10^3$ to 4.9×10^3 m^3 mol^{-1} s^{-1}), C_{COD} (1.16×10^5 to 4.7×10^5 m^3 mol^{-1} s^{-1}), and C_{col} (1.2×10^3 to 9.98×10^3 m^3 mol^{-1} s^{-1}), and these results were correlated with previous reported data (Cruz-Díaz et al. 2018).

11.4.5 Electrochemical Advanced Oxidation of Yellow 3 (DY3) Dye

The electrochemical oxidation method for the degradation of the industrial textile disperse yellow 3 (DY3) dye in aqueous solutions was studied with the use of specific electrocatalytic substances, BDD, Ti/Ru 0.3Ti 0.7O$_2$, and Ti/Pt anodes. The various current densities of 40 mA cm^{-2} and 60 mA cm^{-2} at 40°C with diverse supporting electrolytes Na$_2$SO$_4$ 50 mM and NaCl 50 mM under specific pH values 2.3, 7.0, and 10.0 were employed. The best degradation results achieved were 90% (TOC) in Na$_2$SO$_4$ supporting electrolyte, BDD, anode, and pH. However, the other anodic materials had simply resulted in 50% removal of the dye in Na$_2$SO$_4$ supporting electrolyte. Complete mineralization was achieved in the following order of anode usage: Ti/Ru0.3Ti0.7O$_2$ > BDD > Ti/Pt in a NaCl electrolyte medium. Chloride ions had been produced at the surface of Ti/Ru0.3Ti0.7O$_2$, BDD, and Ti/P at varying pHs. Active chloride oxidants were dependent on the pH and electrode substances. Ultimately, a rate of estimation for every electrocatalytic textile dye below various experimental conditions was realized exhibiting less energy consumption and electrolysis time in NaCl medium. Based totally on the results acquired, the electrochemical elimination of dyes and the profile of the carboxylic by-products depend on the nature of the material, pH, and supporting electrolyte (Salazar et al. 2018).

11.4.6 Electrochemical Advanced Oxidation of Acid Blue 113 Dye

In one of the studies reported, the degradation of a commercial textile dye by using the electrochemical oxidation (EO) technique was performed in an aqueous medium. Galvanostatic electrolysis, via the use of platinum supported on Ti (Ti/Pt), lead dioxide (Pb/PbO$_2$), and TiO$_2$-nanotubes decorated with PbO$_2$ supported on Ti (Ti/TiO$_2$-nanotubes/PbO$_2$) anodes, was carried out in an electrochemical flow cell with 1.0 L of solution containing 250 mg dm^{-3} of the textile dye acid blue 113 (AB113) using Na$_2$SO$_4$ as supporting electrolyte applying 20, 40, and 60 mA cm^{-2}. Large disk electrodes of Ti/TiO$_2$-nanotubes/PbO$_2$ (65 cm^2 of the geometrical area) were successfully synthesized by anodization and electrodeposition procedures. The electrolytic process was monitored by the ultraviolet (UV)-visible spectrometry and COD. The effects confirmed that AB113 was correctly degraded by using hydroxyl radicals electrogenerated from water discharge on the Ti/Pt and Pb/PbO$_2$; the Ti/TiO$_2$-nanotubes/PbO$_2$ anode showed better performance at removing AB113. In comparison with Ti/Pt and Pb/PbO$_2$, the Ti/TiO$_2$-nanotubes/PbO$_2$ anode showed a higher overall performance to recycle AB113. It provided a higher oxidation rate, higher current efficiency, and consumed less energy than the galvanostatic electrolysis using the other electrodes (Moura et al. 2016).

11.4.7 Electrochemical Advanced Oxidation of Triphenylmethane Dye

Electrochemical degradation and decolorization of triphenylmethane dye xylenol orange (XO) on BDD electrodes were also studied. A series of parameters, including electrode materials, current density, initial dye concentration, and electrolyte composition, was examined to discuss the effect of these factors in terms of COD removal, current efficiency, and decolorization rate. Complete decolorization and mineralization of XO could be realized on BDD electrodes while the oxidation is mass transfer controlled, and the degradation progress performs better in the case of the porous BDD electrode compared with the flat BDD electrode. The enhancement of effectiveness is highly correlated to the structure of the electrode material, where the porous BDD film provides more active sites for hydroxyl radicals. The presence of Cl⁻ in the solution promotes the COD and color removal due to the formation of active chlorine. The porous Ti/BDD electrode presents excellent potential in the electrochemical decolorization and mineralization of triphenylmethane dye (He et al. 2016).

11.5 CONCLUSION

Electrochemical advanced oxidation processes are an advanced oxidation method used to remove textile dyes from wastewater. These EAOPs have exhibited superior decolorization of textile dyes from wastewater samples. These processes have some demerits due to the cost of anodes as well as high oxygen overpotential. Different types of Fenton methods have resulted in effective decolorization of textile dye wastewater. EAOPs have shown varying effectiveness of different anode materials for the treatment of textile wastewater containing different types of organic textile dyes. The method of using different composite electrodes has been employed for the decolorization and chemical oxygen demand in textile wastewater in the presence of other chemicals and has shown promising potential. This book chapter focused on determining the effects of pH, inter-electrode distance, current intensity, reaction time, and applied voltage on decolorization and COD reduction of synthetic textile wastewater. Advanced electrochemical oxidation strategies are therefore used to improve the chemical oxidation of textile dyes from wastewater. This method helps to enhance the remediation of textile wastewater.

REFERENCES

Anastasios, Sakalis, Konstantinos Mpoulmpasakos, Ulrich Nickel, and Anastasios Voulgaropoulos. 2005. "Evaluation of a novel electrochemical pilot plant process for azodyes removal from textile wastewater". *Chem. Eng. J.* 111 (1): 63–70.

Anglada, A, A Urtiaga, and I Ortiz. 2009. "Contributions of electrochemical oxidation to waste-water treatment: fundamentals and review of applications". *J. Chem. Technol. Biotechnol.* 84 (12): 1747–1755.

Asghar, HMA, T Ahmad, SN Hussain, and H Sattar. 2015. "Electrochemical oxidation of methylene blue in aqueous solution". *Int. J. Chem. Eng. Appl.* 6 (5): 352–355.

Baddouh, A, GG Bessegato, MM Rguiti, B El Ibrahimi, L Bazzi, M Hilali, and MVB Zanoni. 2018. "Electrochemical decolorization of rhodamine B dye: influence of anode material, chloride concentration and current density". *J. Environ. Chem. Eng.* 6 (2): 2041–2047.

Barazesh, JM, C Prasse, and DL Sedlak. 2016. "Electrochemical transformation of trace organic contaminants in the presence of halide and carbonate ions". *Environ. Sci. Technol.* 50 (18): 10143–10152.

Bouaziz, I, C Chiron, R Abdelhedi, A Savall, and K Groenen Serrano. 2014. "Treatment of dilute methylene blue-containing wastewater by coupling sawdust adsorption and electrochemical regeneration". *Environ. Sci. Pollut. Res.* 21 (14): 8565–8572.

Brillas, E, and CA Martinez-Huitle. 2015. "Decontamination of wastewaters containing synthetic organic dyes by electrochemical methods. An updated review". *Appl. Catal. B Environ.* 166–167: 603–643.

Brillas, E, I Sirès, and MA Oturan. 2009. "Electro-Fenton Process and Related Electrochemical Technologies Based on Fenton's Reaction Chemistry". *Chem. Rev.* 109 (12): 6570–6631.

Buxton, GV, CL Greenstock, WP Helman, and AB Ross. 1988. "Critical review of rate constants for reactions of hydrated electrons, hydrogen atoms and hydroxyl radicals in aqueous, solution". *J. Phys. Chem. Ref. Data* 17 (2): 513–886.

Cañizares, P. R Paz, C Sáez, MA Rodrigo. 2009. "Costs of the electrochemical oxidation of wastewaters: a comparison with ozonation and Fenton oxidation processes". *J. Environ. Manage.* 90 (1): 410–420.

Carter, KE, and James Farrell. 2008. "Oxidative destruction of perfluorooctane sulfonate using boron doped diamond film electrodes". *Environ. Sci. Technol.* 42 (16): 6111–6115.

Comninellis, Christos. 1994. "Electrocatalysis in the electrochemical conversion/combustion of organic pollutants for waste water treatment". *Electrochim. Acta* 39 (11–12): 1857–1862.

Comninellis, C, and De Battisti A. 1996. "Electrocatalysis in anodic oxidation of organics with simultaneous oxygen evolution". *J. Chem. Phys.* 93: 673–679.

Cruz-Díaz, MR, EP Rivero, FA Rodríguez, and R Domínguez-Bautista. 2018. "Experimental study and mathematical modeling of the electrochemical degradation of dyeing wastewaters in presence of chloride ion with dimensional stable anodes (DSA) of expanded meshes in a FM01-LC reactor". *Electrochim. Acta* 260: 726–737.

El Hajj Hassan, MA, and MM El Jamal. 2012. "Kinetic study of the electrochemical oxidation of methylene blue with Pt electrode". *Portugaliae Electrochimica. Acta* 30 (5): 351–359.

Guo, L, K Ding, K Rockne, M Duran, and BP Chaplin. 2016. "Bacteria inactivation at a substoichiometric titanium dioxide reactive electrochemical membrane". *J. Hazard. Mater.* 319: 137–146.

Hicham, Zazoua, Hanane Afangaa, Siham Akhouairia, Hassan Ouchtaka, Abdelaziz Ait Addia, Rachid Ait Akboura, Ali Assabbanea, Jamaâ Doucha, Abdellah Elmchaourib, Joëlle Duplayc, Amane Jadad, and Mohamed Hamdania. 2019. "Treatment of textile industry wastewater by electrocoagulation coupled with electrochemical advanced oxidation process". *J. Water Proc. Eng.* 28: 214–221.

He, Y, X Wang, W Huang, R Chen, H Lin, and H Li. 2016. "Application of porous boron-doped diamond electrode towards electrochemical mineralization of triphenylmethane dye". *J. Electroanal. Chem.* 775: 292–298.

Ines, B, C Christelle, A Ridha, S André, and GS Karine. 2014. "Treatment of dilute methylene blue-containing wastewater by coupling sawdust adsorption and electrochemical regeneration". *Environ. Sci. Pollut. Res. Int.* 21 (14): 8565–8572.

Jalife-Jacobo, H, R Feria-Reyes, O Serrano-Torres, S Gutiérrez-Granados, and JM Peralta-Hernández. 2016. "Diazo dye Congo red degradation using a boron-doped diamond anode: an experimental study on the effect of supporting electrolytes". *J. Hazard. Mater.* 319: 78–83.

Jasper, JT, OS Shafaat, and MR Hoffmann. 2016. "Electrochemical Transformation of Trace Organic Contaminants in Latrine Wastewater." *Environ. Sci. Technol.* 50 (18): 10198–10208.

Jawad, Noor H, and Sarmad T Najim. 2018. "Removal of methylene blue by direct electrochemical oxidation method using a graphite anode". *Mater. Sci. Eng.* 454: 012023.

Jüttner, K, U Galla, and H Schmieder. 2000. "Electrochemical approaches to environmental problems in the process industry". *Electrochim. Acta* 45 (15–16): 2575–2594.

Lan, YD, C Coetsier, C Causserand, and KG Serrano. 2017. "On the role of salts for the treatment of wastewaters containing pharmaceuticals by electrochemical oxidation using boron doped diamond anode." *Electrochim. Acta* 231: 309–318.

Lin, SH, and CM Lin. 1992. "Decolorization of textile waste effluents by ozonation". *J. Environ. Syst.* 21: 143–153.

Loloei, Mahshid, and Abbas Rezaee. 2016. "Decolorization of methylene blue by the electroFenton process using stainless steel meshes electrodes". *Int. J. Env. Health Eng.* 5 (1): 27.

Lorimer, JP, TJ Mason, M Plattes, S Phull, and DJ Walton. 2001. "Degradation of dye effluent". *Pure Appl. Chem.* 73 (12): 1957–1968.

Martínez-Huitle, CA, MA Rodrigo, I Sires, and O Scialdone. 2015. "Single and coupled electrochemical processes and reactors for the abatement of organic water pollutants: a critical review". *Chem. Rev.* 115 (24): 13362–13407.

Martinez-Huitle, CA, S Ferro, and A De Battisti. 2005. "Electrochemical incineration in the presence of halides". *Electrochem. Solid. State Lett.* 8 (11): D35–D39.

McKay, G. 1980. "Color removal by adsorption". *Am. Dyest. Rep.* 69 (3): 38–44.

Miled, W, SH Said, and S Roudesli. 2010. "Decolorization of high polluted textile wastewater by indirect electrochemical oxidation process". *JTATM* 6 (3): 1–6.

Moreira, Francisca C, Rui AR Boaventura, Enric Brillas, and Vítor JP Vilar. 2017. "Electrochemical advanced oxidation processes: a review on their application to synthetic and real wastewaters". *Appl. Catal. Environ.* 202: 217–261.

Moura, DCd, MA Quiroz, DRD Silva, R Salazar, and CA Martínez-Huitle. 2016. "Electrochemical degradation of Acid Blue 113 dye using TiO2-nanotubes decorated with PbO2 as anode". *Environ. Nanotechnol. Monit. Manage.* 5: 13–20.

Oturan, MA, and JJ Aaron. 2014. "Advanced oxidation processes in water/wastewater treatment: principles and applications. A review". *Crit. Rev. Environ. Sci. Technol.* 44 (23): 2577–2641.

Panizza, M, and G Cerisola. 2009. "Direct and mediated anodic oxidation of organic pollutants". *Chem. Rev.* 109: 6541–6569.

Paprowicz, J, and Slodczyk S. 1988. "Application of biologically activated sorptive columns for textile wastewater treatment". *Envir. Technol. Lett.* 9 (4): 271–280.

Pathania, D, R Katwal, G Sharma, M Naushad. 2016. "Novel guar gum/Al2O3 nanocomposite as an effective photocatalyst for the degradation of malachite green dye." *Int. J. Biol. Macromol.* 87:366–374. doi: 10.1016/j.ijbiomac.2016.02.073

Peralta-Hernandez, JM, CA Martínez-Huitle, JL Guzman-Mar, and A Hernandez-Ramírez. 2009. "Recent advances in the application of electro-Fenton and photoelectro-Fenton process for removal of synthetic dyes in wastewater treatment". *J. Environ. Econ. Manag.* 19: 257–265.

Polcaro, AM, A Vacca, M Mascia, S Palmas, R Pompei, and S Laconi. 2007. "Characterization of a stirred tank electrochemical cell for water disinfection processes". *Electrochim. Acta* 52 (7): 2595–2602.

Rajeshwar, K, JG Ibanez, and GM Swain. 1994. "Electrochemistry and the environment". 24 (11): 1077–1091.

Sala, Mireia, Víctor López-Grimau, and Carmen Gutiérrez-Bouzán. 2014. "Photo-electrochemical treatment of reactive dyes in wastewater and reuse of the effluent: method optimization". *Materials (Basel).* 7 (11): 7349–7365.

Salazar, R, M Soledad Ureta-Zañartu, C Gonzalez-Vargas, C do Nascimento Brito, and CA Martinez-Huitle. 2018. "Electrochemical degradation of industrial textile dye disperse yellow 3: role of electrocatalytic material and experimental conditions on the catalytic production of oxidants and oxidation pathway". *Chemosphere* 198: 21–29.

Scialdone, O. 2009. "Electrochemical oxidation of organic pollutants in water at metal oxide electrodes: a simple theoretical model including direct and indirect oxidation processes at the anodic surface". *Electrochim. Acta* 54: 6140–6147.

Sharma, G, Naushad M, Pathania D, et al. 2015. Modification of *Hibiscus cannabinus* fiber by graft copolymerization: application for dye removal." *Desalin. Water Treat.* 54: 3114–3121. doi: 10.1080/19443994.2014.904822

Sires, I, E Brillas, MA Oturan, MA Rodrigo, and M Panizza. 2014. "Electrochemical advanced oxidation processes: today and tomorrow: a review". *Environ. Sci. Pollut. Res.* 21 (14): 8336–8367.

Subba Rao, AN, and VT Venkatarangaiah. 2014. "Metal oxide-coated anodes in wastewater treatment". *Environ. Sci. Pollut. Res. Int.* 21 (5): 3197–3217.

Tsitonaki, A, B Petri, M Crimi, H Mosbaek, RL Siegrist, and PL Bjerg. 2010. "In situ chemical oxidation of contaminated soil and groundwater using persulfate: a review". *Crit. Rev. Environ. Sci. Technol.* 40: 55–91.

Yoon, Y, Y Jung, M Kwon, E Cho, and JW Kang. 2013. "Alternative electrode materials and ceramic filter minimize disinfection byproducts in point-of-use electrochemical water treatment". *Environ. Eng. Sci.* 30 (12): 742–749.

Zhuo, Q, S Deng, B Yang, J Huang, B Wang, T Zhang, and G Yu. 2012. "Degradation of perfluorinated compounds on a boron-doped diamond electrode". *Electrochim. Acta* 77: 17–22.

Zhuo, Q, S Deng, B Yang, J Huang, and G Yu. 2011. "Efficient electrochemical oxidation of perfluorooctanoate using a Ti/SnO2-Sb-Bi anode". *Environ. Sci. Technol.* 45 (7): 2973–2979.

12 Materials Involved in Electrocoagulation Process for Industrial Effluents

Carlos Navas-Cárdenas, Herman Murillo, Maibelin Rosales, Cesar Ron, and Florinella Muñoz

CONTENTS

12.1 Introduction 193
12.2 Fundamentals of Electrocoagulation 194
    12.2.1 Colloidal Stability 194
    12.2.2 Electrocoagulation Principle 195
12.3 Effects of the Operating Parameters on the Performance of Electrocoagulation 196
    12.3.1 Electrode Material 196
    12.3.2 Initial pH 196
    12.3.3 Electrode Configuration 197
    12.3.4 Current Density 198
    12.3.5 Temperature 198
    12.3.6 Conductivity 199
12.4 Materials Employed for Electrocoagulation of Industrial Wastewater 200
    12.4.1 Textile and Tannery Industry 200
    12.4.2 Food Industry 200
    12.4.3 Oil Industry 201
    12.4.4 Pulp and Paper Industry 202
    12.4.5 Pharmaceutical Industry 202
    12.4.6 Wastewater Containing Heavy Metals 204
12.5 Future Perspectives 206
12.6 Conclusions 208
Acknowledgments 210
References 210

12.1 INTRODUCTION

The scarcity of hydric resources and industrial wastewater discharging have become worldwide threats. The environmental impact generated by industrial effluents has led to the development of suitable strategies to remove pollution in water and reuse the wastewater. In this sense, several wastewater treatment technologies have been proposed to reduce toxic pollutants from industrial effluents. However, most of these technologies require large amounts of nonrenewable energy, causing considerable carbon dioxide emissions into the environment (Garcia-Segura et al. 2017; Moussa et al. 2017).

Industrial wastewaters contain several pollutants (e.g., organic or inorganic components with complex structures, acid or alkaline constituents, toxic and biorefractory pollutants, dyes, etc.), which can cause thermal and biological impacts, scum formation, and a loss of biodiversity in the environment (Khansorthong and Hunsom 2009; Yavuz and Ogütveren 2018). Thus, conventional wastewater treatment methods (i.e., physical, chemical, and biological processes) have been proposed to eliminate toxic pollutants from industrial effluents. However, they are subjected to several drawbacks including low efficiency, high energy consumption, long time of treatment, generation of secondary pollutants, and high operating costs (An et al. 2017; Song et al. 2017). Therefore, research related to developing suitable methods for industrial wastewater treatment has nowadays intensified.

In this context, electrochemical methods such as electrocoagulation (EC), electrooxidation, electroflotation, and photo-assisted electrochemical methods have been emerging as alternatives of wastewater treatment, and are based on the use of electrons as the working agents to promote the pollutant removal (Brillas and Martínez-Huitle 2015; Moussa et al. 2017; Yavuz and Ogütveren 2018). Among the electrochemical methods, EC is one of the most promising water treatment technologies employed to remove refractory pollutants from industrial effluents. The EC process involves electrochemical reactions to lead the *in situ* formation of coagulating agents by the electrodissolution of a sacrificial anode due to the applied electric field to the electrodes (Aljaberi 2018; Chen 2004; Meas et al. 2010). The advantages of EC over conventional wastewater treatments include high removal efficiency, use of less or no chemicals, compact treatment facilities, relatively low cost, lower amount sludge production, and the possibility of complete automation (An et al. 2017; Jiang et al. 2017; Kobya et al. 2011).

In this chapter, we briefly review the fundamentals of EC, the factors that affect its performance on wastewater treatment, and the promising materials employed to treat effluents from several industries. The last part of this chapter describes future perspectives of the use of possible combinations of EC with other kinds of wastewater treatments, as conventional techniques or nonconventional technologies (e.g., advanced oxidation processes) in order to improve the pollutant removal performance and to exploit the properties of several materials that can be employed during the EC process.

12.2 FUNDAMENTALS OF ELECTROCOAGULATION

Electrocoagulation is a colloidal destabilization technique based on the perturbation of particle surface charge via applied electrical fields to the anodic–cathodic pair in order to promote coagulation, flocculation, and final precipitation of suspended particles (Ghernaout et al. 2011; Mollah et al. 2001; Shim et al. 2014).

12.2.1 COLLOIDAL STABILITY

A particle is considered as "colloidal" when its aggregation or sedimentation happens at such slow rates that it remains suspended in solution (Vepsäläinen et al. 2011). Colloidal particles (colloids) can be present in natural as well as in engineered systems, and typically their size ranges from ~1 nm to ~10 µm. The relatively small size of colloids leads to a very small mass-to-surface area ratio. Given that colloid mass and size are small compared to its surface area, gravitational effects are negligible as compared to surface phenomena (e.g., colloidal surface forces) (Ghernaout et al. 2011; Moussa et al. 2017).

The stability of colloids depends on the balance of surface forces described, physically and mathematically, in the classic DLVO theory (Derjaguin and Landau 1941; Verwey and Overbeek 1948). In the DLVO theory, the total interaction between colloid surfaces can be represented as the sum of van der Waals (vdW) and electric double layer (EDL) forces. vdW force emanates from fleeting and/or permanent dipoles appearing in molecular bonds, is attractive for most surfaces, and is independent of solution chemistry (Gregory 1981; Israelachvili 2011). EDL force arises from electroosmotic effects appearing when intrinsic surface charge interacts with ions present in a solution, is repulsive for like-charged surfaces (e.g., in nature, most surfaces are negatively charged), and relates inversely to solution ionic strength (Hogg et al. 1966; Wiese and Healy 1970).

When repulsive EDL forces overwhelm the attractive vdW forces, colloids repel each other and form a stable dispersion. In contrast, when attractive forces overcome repulsive forces, colloids attach and the coagulation process is triggered leading to the destabilization of a dispersion (Moussa et al. 2017). One of the basic principles of any coagulation process is to destabilize a colloidal dispersion by reducing or neutralizing the colloid surface charge. After the surface charge is disturbed, EDL forces reduce allowing the colloids aggregate via vdW forces (Ghernaout et al. 2011; Vepsäläinen 2012).

In real-life systems, DLVO theory is a good starting point to understand colloidal stability, however, it cannot totally explain the stabilization/destabilization phenomena. In complex systems, it is always required to consider surface forces arising from steric, hydration and solvation effects, hydrophobic–hydrophilic interactions, Lewis acid–base interactions, and others (Israelachvili 2011; van Oss 2006).

12.2.2 Electrocoagulation Principle

In conventional coagulation processes, chemical species (coagulant agent) (e.g., metallic, ionic polymers, ionic surfactants) are added to the system in order to reduce or neutralize the colloid surface charge. In EC, similar effects to conventional coagulation can be produced by promoting the generation of coagulant agents (e.g., metallic cations as Fe^{3+}, Al^{3+}). Coagulating agents are delivered to a to-be-treated system via dissolution of a metallic electrode after application of DC electrical fields (Brillas and Martínez-Huitle 2015; Chen 2004; Mollah et al. 2001). Although EC is considered to be similar to conventional coagulation techniques in terms of the colloidal destabilization mechanism, it is still different in other aspects such as the delivery mechanism of the coagulant agent, as well as the side reactions occurring simultaneously at the electrodes (Brillas and Martínez-Huitle 2015; Ghernaout et al. 2011; Moussa et al. 2017).

A basic EC cell system consists of an electrolyte cell wherein anode and cathode metal electrodes are externally connected to an electric power supply (Figure 12.1). The electrodes are immersed in the solution to be treated. Iron (Fe) and aluminum (Al) are the most extensively used electrode materials given their availability, reduced toxicity, and reliability (Garcia-Segura et al. 2017). The anode corresponds to the sacrifice electrode, and it serves as a coagulant agent while it dissolves and delivers cations (e.g., Fe^{3+}, Al^{3+}) to the solution (Ghernaout et al. 2011; Moussa et al. 2017).

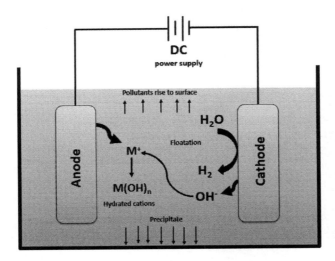

FIGURE 12.1 Schematic representation of a basic EC unit. (From Moussa et al. 2017.)

EC is a complicated process with several mechanisms operating in a coupling mode to destabilize colloids from solution. A strict description of the reaction mechanisms and physiochemical phenomena driving EC is out of the scope of this review. However, we herein provide a general description of the basic chemical reactions occurring simultaneously in an EC unit (Figure 12.1). In general, the EC process can be described considering the following steps (Chen 2004; Garcia-Segura et al. 2017; Ghernaout et al. 2011; Mollah et al. 2001):

1. Anodic reactions leading to *in situ* generation of metal cations and production of oxygen gas, as well as cathodic reactions producing hydrogen gas and releasing hydroxyl ions.

   At the anode:

   $$Al(s) \rightarrow Al^{3+}(aq) + n\,e^- \qquad (12.1)$$

   $$2H_2O \rightarrow 4H^+(aq) + O_2(g) + 4e^- \qquad (12.2)$$

   At the cathode:

   $$3H_2O + 3e^- \rightarrow \frac{3}{2}H_2(g) + 3OH^-(aq) \qquad (12.3)$$

   $$2Al(s) + 6H_2O + 2OH^-(aq) \rightarrow 2Al(OH)_4^-(aq) + 3H_2(g) \qquad (12.4)$$

2. Reduction or neutralization of the colloid surface charge via absorption of metal ions.
3. Formation of aggregates from destabilized colloids leading to coagulation and further flocculation.
4. Removal of coagulated colloids via sedimentation by flotation. Flotation occurs as a result of generated hydrogen and oxygen gases.

12.3 EFFECTS OF THE OPERATING PARAMETERS ON THE PERFORMANCE OF ELECTROCOAGULATION

The removal efficiency of the EC process depends on many factors that can affect the oxidation–reduction reactions in the system and the formation of the coagulating agents. The most important factors to be controlled in EC are detailed next.

12.3.1 Electrode Material

The electrode pair is the main component of an EC cell given that it will influence the overall performance of the EC process. In dependence of the system, both electrodes can be made of the same or different metal. The most common electrode materials employed for the EC process are aluminum and iron (Chen et al. 2000; Nidheesh and Singh 2017). Each electrode material exhibits different behavior (Linares-Hernández et al. 2009) since it determines the different reactions involved during this process (Moussa et al. 2017). Materials employed for EC process will be discussed further in Section 12.4.

12.3.2 Initial pH

Initial pH is an important parameter to be controlled in the EC process. It affects the conductivity of the aqueous medium, final pH, zeta potential, and the solubility of metal hydroxides (Khorram and Fallah 2018; Moussa et al. 2017; Trompette and Vergnes 2009). At an appropriate initial pH, the metal ions can form coagulating agents that remove contaminants from wastewater either by precipitation and adsorption of dissolved pollutants or by aggregation of suspended particles (Aoudj et al. 2010; Brillas and Martínez-Huitle 2015).

The pH effect on the EC process has been widely studied for the treatment of industrial effluents. The typical pH range for industrial effluents is between 5 and 7. However, it depends on the nature of pollutants and the material electrode. For textile wastewaters, higher removal efficiencies were found at neutral pH by using Al electrodes. On the other hand, when Fe is employed as the electrode, higher removal efficiencies, in terms of color and chemical oxygen demand (COD), were founded at basic pHs (Chen 2004). Different pH conditions were studied for EC treatment of industrial effluent with toxic metals (Gatsios et al. 2015), where the optimal pH was 6 and using iron electrodes and an applied current of 0.1 A. In addition, it has been found that an alkaline pH (between 7 to 9) can be the optimal condition for oily wastewater treatment, using Fe as the anode at a current density of 6 mA/cm^2 and a treatment time of 31 min (Zhao et al. 2014).

12.3.3 ELECTRODE CONFIGURATION

The electrode connection mode affects the removal performance of an EC cell, as well as the energy consumption and the total cost of the process. The electrode can be arranged in monopolar (in parallel, MP-P, or in serial connection, MP-S) or in bipolar (in serial connection, BP-S) configuration (Asselin, Drogui, Benmoussa et al. 2008; Garcia-Segura et al. 2017; Hakizimana et al. 2017). The different electrode arrangements are shown in Figure 12.2.

In MP-P configuration, all the sacrificial electrodes are directly connected to each other and to the power supply. The same configuration is employed for cathode electrodes. In this sense, the applied current to the system is distributed within each anode–cathode pair, and the voltage remains equal across the cell. In MP-S configuration, the power supply is only connected to the outermost electrodes (anode and cathode; Figure 12.2b), while the internal electrodes of each anode–cathode pair are connected to each other. In this case, the voltage is distributed in each anode–cathode pair and the applied current is the same for each one, resulting in a higher potential difference compared to the MP-P connection (Garcia-Segura et al. 2017; Moussa et al. 2017; Song et al. 2017).

On the other hand, the BS configuration can only be performed in serial connection. Here, the outermost electrodes are directly connected to the power supply, while the inner electrodes are not electrically connected to the external circuit. The electric current applied through the EC cell generates the polarization of the inner electrodes, which in turn leads to different polarities in

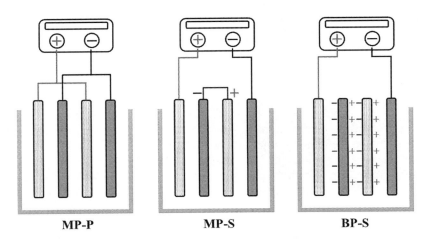

FIGURE 12.2 Different electrode arrangements employed for EC process. (Modified from Song et al. 2017.)

the opposite faces of them (Asselin, Drogui, Kaur et al. 2008; Khandegar and Saroha 2013). It is reported that the BP-S configuration is the most cost-effective electrode arrangement, however, the electric current can bypass the bipolar electrodes and flow in the electrolytic solution, generating a current loss in the system (Kobya et al. 2011; Song et al. 2017).

The choice of an appropriate electrode arrangement may be crucial in the EC process and depends on the kind of effluent. For example, Asselin and coworkers (Asselin, Drogui, Benmoussa, et al. 2008) analyzed the influence of electrode arrangement in the EC treatment of oily slaughterhouse wastewaters. They reported a system based on the use of iron (anode and cathode) or aluminum (anode and cathode) electrodes arranged in two different kinds of connections: MP-P and BP-S. An 82% COD removal performance was achieved using an MP-P configuration system with 0.3 A current density. In addition, a comparison between monopolar and bipolar electrode arrangements during the EC process employed to remove Cr^{3+} from aqueous solutions has been conducted. The optimal conditions to obtain complete Cr^{3+} removal were a current of 1 A, initial pH of 3.4, and using mild steel sheets as electrodes and a BP-S electrode configuration (Golder et al. 2007a). On the other hand, Kobya and coworkers studied the effect of different electrode connection modes on arsenic removal from potable water by EC. They compared MP-S and MP-P electrode arrangements, obtaining the highest arsenic removal (94%) at a current density of 2.5 A/m^2 for Al and Fe electrodes arranged at MP-S configuration (Kobya et al. 2011).

12.3.4 CURRENT DENSITY

The applied current density to the EC cell determines the number of metal ions released from the electrodes and, therefore, the amount of resulting coagulating agents for the EC process (Brillas and Martínez-Huitle 2015; Yavuz and Ogütveren 2018; Zodi et al. 2011). The effect of current density on the treatment of wastewater from textile industries by electrocoagulation has been widely studied. Verma (2017) investigated the pollutant removal efficiency by the EC process using Fe–Al electrodes and varying the current density. Here, the COD and color removal efficiencies increased when the current density increased from 5 to 25 A/m^2, due to the higher coagulant dosage and bubble generation rates during the EC process. A similar behavior was observed for the removal of textile dye (Remazol Red 133) from industrial wastewater by EC using three-dimensional steel wood electrodes and a cartridge-type reactor (López et al. 2017). A removal efficiency of 100% was obtained at current densities above 3.75 mA/cm^2, meanwhile 84% and 42% removal efficiencies were obtained at 2.5 and 1.25 mA/cm^2, respectively.

The current density depends on the initial concentration of pollutants in the wastewater (Hansen et al. 2019). For example, arsenic removal from industrial effluents was performed by EC at different current densities. It has been reported that for arsenic concentrations from 1 to 440 mg/L, the optimal current density may be between 6.5 and 50 A/m^2 (Ratna Kumar et al. 2004; Song et al. 2014), and for arsenic concentrations higher than 3000 mg/L, between 170 and 250 A/m^2 (Hansen et al. 2019).

12.3.5 TEMPERATURE

The EC process is usually performed at room temperature, however, some works have reported the effect of temperature on the removal efficiency of industrial wastewaters by EC. Higher temperatures o enhance the mass transport in the EC reactors, favoring a higher pollutant removal efficiency in the treatment (Vepsäläinen et al. 2009). In addition, the conductivity of the aqueous solution depends on the temperature. A higher temperature increases the conductivity of the wastewater, which in turn, decreases the power consumption during the EC process (An et al. 2017).

The effect of temperature on the EC process depends on the type of wastewater. For example, the effect of temperature on natural organic matter (NOM) removal from paper mill surface water by EC treatment was investigated (Vepsäläinen et al. 2009). A minor effect of the temperature of EC on the removal performance was determined using aluminum electrodes, a current density of 0.5 mA/m², initial pH of 5.0, and a temperature range of 2°C–22°C. At these conditions, the higher dissolved organic carbon removal was 80% at 22°C, while at 2°C, the dissolved organic carbon removal was 76%. This slight increase in the removal efficiency has been related to the higher dissolution of electrodes that occurs at high temperatures, however, the effect of temperature on the removal efficiency is minor for this type of wastewater.

On the contrary, Ulucan and coworkers (2014) reported a major effect of temperature on the COD removal from bilge wastewater by EC. The EC process was performed using an EC reactor with aluminum electrodes, at initial pH of water of 7.0, current density of 6 mA/m², and temperatures from 10°C to 55°C. At 10°C, the COD removal was only 38%, whereas COD removals of c.a. 90% were observed at 25°C. Similar COD removal was found for temperatures above 25°C. This indicates that 25°C is the optimal condition for this specific wastewater treatment. The same trend was observed for the removal of polyvinyl alcohol (PVA) from aqueous solution by the EC process using Fe and Al as the anode–cathode pair (Chou et al. 2010). Here, the effect of temperature and applied voltage on removal performance were studied at 15°C, 25°C, 35°C, and 45°C, and 5, 10 and 20 V, respectively. The highest PVA removal (85%) was achieved at 35°C and 45°C, whereas poor PVA removal (48%) was registered at 15°C. The difference in the removal performance is related to the higher destruction of iron oxide film on the anode and to the higher generation of hydroxyl radicals, which promote PVA removal from the wastewater (El-Ashtoukhy et al. 2009). However, it is reported that higher temperatures can increase the solubility of precipitates and promote the formation of unsuitable flocs in the system, generating an opposite effect on the removal efficiency. In this sense, it becomes necessary to determine the optimal temperature conditions to enhance the removal efficiency and to avoid technical problems during the EC process.

12.3.6 Conductivity

Solution conductivity is an important parameter that affects the removal performance and operating cost of EC treatment. The applied current passing through the circuit depends on the conductivity of the wastewater at a specified applied voltage (Chou et al. 2010). A high conductivity implies a higher generation of coagulating agents and a lower energy consumption during the EC process. The solution conductivity can be adjusted by adding supporting electrolytes (e.g., anions), which can also interact with the electrodes to promote the electrolytic reactions over their surface (Moussa et al. 2017; Trompette and Vergnes 2009).

Several studies have employed sodium chloride (NaCl) or sodium sulfate (Na_2SO_4) to obtain an appropriate conductivity for EC treatment of industrial effluents (Chou et al. 2010; Khandegar and Saroha 2013). Na_2SO_4 was used as supporting electrolyte at different concentrations (between 1 to 7 mM) to treat industrial estate wastewaters by EC with iron electrodes, wherein the maximum removal efficiency and the lowest power consumption were achieved at a concentration of 3 mM (Yavuz and Ogütveren 2018).

It has been reported that the use of NaCl as the supporting electrolyte can be more efficient compared to Na_2SO_4 (Ghernaout and Ghernaout 2011). The effect of the type of electrolyte on the EC system was analyzed for the treatment of oily effluents (Cañizarez et al. 2007). Here, higher COD removal efficiencies were obtained by using NaCl instead of Na_2SO_4. This fact was studied in a continuous EC reactor with Al electrodes at 25°C, at an initial pH of oily–wastewater of 8.5, and with a concentration of supporting electrolyte of 3000 mg/L. At an electrical charge passed of 0.15 Ah/L, ca. 90% COD removal was achieved with NaCl, whereas for Na_2SO_4, only ca. 60% COD removal was obtained. This difference has been related to the formation of different aluminum species depending on the type of supporting electrolytes. In this sense, the use of Na_2SO_4 can promotes the formation of amorphous Al hydroxide precipitates, which are not effective for the treatment of this kind of wastewater, while the chloride media (i.e., NaCl) can promote the formation of polymeric hydroxocations, which are active sites that favor the pollutant removal in the EC process (An et al. 2017; Cañizarez et al. 2007; Khorram and Fallah 2018).

12.4 MATERIALS EMPLOYED FOR ELECTROCOAGULATION OF INDUSTRIAL WASTEWATER

Different materials have been used as electrodes for the EC process in the treatment of industrial wastewaters and their selection depends on the nature of pollutant to be removed. The electrochemical reactions involved during EC are determined by the electrode materials employed in the process, considering the fact that their physicochemical properties are critical factors in the removal efficiency. Examples of some of the most effective employed materials in the EC process for different industrial wastewaters are discussed next.

12.4.1 Textile and Tannery Industry

In recent years, EC technology has become an effective tool for the treatment of textile wastewaters. High removal efficiency (>90%) has been typically obtained with the use of optimum material as the sacrificial electrode. In general, aluminum, iron, and stainless steel (SS) electrodes can be used as efficient electrode materials for the EC process (Brillas and Martínez-Huitle 2015; Khandegar and Saroha 2013; Kobya et al. 2016).

Khorram and Fallah reported that 97% of dye decolorization was obtained at 30 min of EC treatment, using Al electrodes (Khorram and Fallah 2018). A reduction in the initial dye concentration was obtained by Behin et al. (2015), when Fe electrodes were used as anodes in the EC process combined with ozonation. This improvement occurred due to the strongly oxidizing ozone that can react with wastewater compounds or with the oxidized iron to produce hydroxyl radicals. The latter react with organic molecules accelerating the decrease of dye concentration and consequently enhancing the efficiency of the decolorization. Linares–Hernández et al. (2009) studied the use of aluminum, iron, and a combination of them to decolorize textile wastewaters, and the dye removal efficiency was reported to be higher than 83% by employing a combination of electrodes (Al and Fe). Similar results were obtained when three different types of textile wastewaters were treated with an Fe–Al composite electrode; the treatment showed approximately 99% dye removal efficiency. Moreover, an anode of steel wool, also known as wire wool, was developed by A. López et al. (2017), and their results showed 99% of efficiency for removing the dye. Wei and coworkers (2012) observed higher color removal by employing iron plate as the anode and steel wool as the cathode. In that work, the authors investigated the efficiency of three types of cathodes (iron plate, stainless steel plate, and steel wool) and the results revealed higher dye removal using a steel wool cathode than by using the iron plate or the stainless steel plate cathodes. The higher efficiency was attributed to the higher surface area of the steel wool cathode. Finally, the literature indicates that electrocoagulation is an efficient process for removing dye from wastewater.

12.4.2 Food Industry

Several contaminants are found in wastewaters from the food industry. Common characteristics of food industry wastewater are highly degradability, nontoxic suspended solids, COD, and elevated biological oxygen demand (BOD) (Moussa et al. 2017). Color, oil, grease, dissolved organic compounds (e.g., proteins, blood), fat, hair, feather, flesh, manure, and grit are characteristic pollutants of wastewater from the meat processing industry and restaurants (Asselin et al. 2008; Chen et al. 2000). In the following, materials and removal efficiencies in specific food industry sectors are discussed.

Early attempts in the treatment of slaughterhouse wastewater with EC were performed employing Al and Fe electrodes. The process yielded 70% removal of COD, color, oil, and grease, working at acid or neutral pH. Al–Al electrodes performed better in COD and color removal, whereas Fe–Fe electrodes achieved higher efficiency in oil and grease removal (Kobya et al. 2006; Yetilmezsoy et al. 2009). Asselin et al. reported 82%, 100%, and 90% removal of COD, turbidity, and metals, respectively, after treating slaughterhouse wastewater with Fe–Fe electrodes at pH 6.1. In several other studies, the EC process allowed COD removal percentages ranging from 86% to 98% with both Al–Al and Fe–Fe electrodes (Bayramoglu et al. 2006; Chen et al. 2000; Kobya et al. 2006).

Wastewater from dairy production has been treated by EC with Fe–Fe, Al–Al, and Fe–Al electrodes. Treatment at pH 7.0 yielded COD removal efficiencies within 61% and 98%, and turbidity removal of 100% employing Al–Al electrodes (Şengil and Ozacar 2006; Tchamango et al. 2010; Yavuz et al. 2011). In effluents from the egg processing sector, the use of Al–Al, Fe–Fe, and SS–SS electrodes allowed removal above 92% of COD and 96% of turbidity at pH 7.0 (Xu et al. 2002). Treatment of gelatin production wastewater with Al–Al electrodes removed 54.7% COD at pH 5.9 (Lakshmi Kruthika et al. 2013).

Regarding the olive oil industry, wastewater from the milling process was treated with the EC process with Al–Al and Fe–Fe electrodes. Here, the wastewater treatment at a pH range from 4.2 and 6.0 removed the color up to 93%–100%, while the COD removal ranged from 42% to 70% (Coskun et al. 2012; Hanafi et al. 2010; Inan et al. 2004). As alternate materials, Zn–SS electrodes were tested to treat olive oil milling wastewater showing low removal efficiencies at pH 3.2 (<21% COD removal) (Fajardo et al. 2015). The EC treatment of wastewater from palm oil production with Al–Al electrodes showed a reduction of 30.4% in COD, 70% in turbidity, and 100% of metals (Agustin et al. 2008). While in the almond industry, Al–Fe electrodes at pH 6.1 allowed the reduction of 81% of COD, 99.6% of color, and 99.7% of turbidity removal (Valero et al. 2011).

Distilleries correspond to a sector of the food industry in which EC has been studied utilizing electrodes made of materials other than Al and Fe. At values of neutral pH, Al–SS electrodes proved to remove 80.1% of COD, 100% of color, and 99% of turbidity (Kannan et al. 2006). EC performed with SS–SS electrodes, yielded 61.6% COD removal, as well as reduced the color in 98.4% (Thakur et al. 2009). COD removal within 41.9% and 81.3% were achieved using Al–Al and Fe–Fe electrodes, respectively, at a pH ranging from 3.0 to 6.0 (Davila et al. 2011; Khandegar and Saroha. 2012; Kirzhner et al. 2008; Krishna et al. 2010).

12.4.3 Oil Industry

EC has been conducted for removing oil from petroleum refineries and petrochemical industries, being aluminum and iron the most commonly used electrodes for this purpose (An et al. 2017; Emamjomeh and Sivakumar 2009; Safari et al. 2016).

Studies about removing diesel from oil petroleum refinery wastewater were carried out by Safari et al. (2016), who employed Al and Fe as electrode materials. They reported that a higher capacity for removing diesel was achieved using Al anodes and cathodes, compared to that obtained with Fe electrodes. This is attributed to Al electrodes generating better stability during the EC process. However, Martínez–Delgadillo et al. (2010) investigated oil removal from similar oily wastewater also using Al and Fe electrodes. They reported that the use of Fe electrodes improves the removal efficiency compared to Al electrodes.

Chavalparit and Ongwandee (2009) showed that Al anodes and graphite cathodes can be an effective electrode combination to remove residual oil and grease from biodiesel wastewater. In addition, they developed a model for the prediction of oil and grease removal efficiency, which was verified with experimental results.

Due to the growing interest in improving oil removal efficiency from oil petroleum refinery wastewater, some modifications have been studied in the electrochemical processes. For example, dimensionally stable anodes (DSA) of nominal composition $Ti/Ru_{0.34}Ti_{0.66}O_2$ were employed in electrochemical remediation of oily wastewaters from the petroleum industry (Santos et al. 2006). The results showed changes on the catalytic surface of the DSA due to the possible adsorption of organic material at the electrode surface by the oil concentration, however, a high removal efficiency was reported.

On the other hand, the combination of carbon steel and aluminum as electrodes is efficient for removing oil and grease from bilge water using EC treatment (Rincón and La Motta 2014). The removal efficiency of oil and grease from bilge water was studied by Körbahti and Artut (2010), who proposed the use of platinum/iridium (Pt/Ir) anodes in electrochemical processes in order to enhance the resistance to possible chemical attacks and improve electrical conductivity and mechanical stability.

Oil removal from bio-oils and synthetic oils was evaluated using SS and Al anodes. The SS employed in the study was austenitic steel with molybdenum and 67% of iron. The results did not reveal a significant difference using an SS or Al anode. However, the SS anode showed a slightly higher efficiency for synthetic oil removal, while the Al anodes presented slightly better performance at reducing bio-oils (Karhu et al. 2015). The results presented in this section have demonstrated that the EC process is a feasible technique for the treatment of oily wastewaters.

12.4.4 Pulp and Paper Industry

In the pulp and paper industry, pollutants are released into water during wood bleaching, digestion, pulp washing, pulp bleaching, and papermaking processes (Vepsäläinen et al. 2011). For this reason, effluents are commonly blackish in color and contain high concentrations of lignin, cellulose, COD, BOD, organics, suspended solids, and arsenic (Moussa et al. 2017). In this sense, Al, Fe, and combinations of both electrodes have been employed in the treatment of this type of wastewater by EC. Al proved to be more effective, although the combination of electrodes has also shown high removal efficiencies (Vepsäläinen 2012).

Al–Al electrodes have been used in the treatment of black wastewater by the EC process. This black effluent is usually loaded with about 50% of lignin. Zaied and Bellkhal (2009) reported that COD reduction was about 98% with a color removal of 99% at pH 7.0. An effluent generated during the bleaching process has also been treated by EC, using Al–Al electrodes. Lignin and inorganic chloride are common constituents of this wastewater. After the EC treatment, COD was reduced by 90%, with a color reduction percentage of 94% (Sridhar et al. 2011); the EC process was performed at pH 7.0.

Another highly contaminated effluent coming from the pulp and paper industry is wastewater from the paper milling process. It commonly has high concentrations of lignin and phenol (Uğurlu et al. 2008). EC treatment with Al–Al electrodes at neutral pH ranges proved to be effective in the reduction of DOC with removal as high as 80%; however, the process yielded low removal efficiencies for COD (4.8%) (Uğurlu 2004). Further studies were performed in order to enhance the efficiency of the EC process in COD removal specifically. EC performed with Fe–Fe electrodes allowed COD reduction by about 55% at pH 7.5. Under the same conditions but with the use of Al–Al electrodes, COD removal reached 75% (Uğurlu et al. 2008).

12.4.5 Pharmaceutical Industry

Electrocoagulation has been successfully performed on the treatment of wastewaters from the pharmaceutical industry. Some reports can be found in this area of interest. One of them reported EC as a way to remove oxytetracycline hydrochloride by using Fe and Al as anodes, and SS as the cathode at 20 mA/cm^2. Iron was found better than aluminum on removal efficiency (Nariyan et al. 2017). Another report mentioned EC to effectively remove diclofenac, ibuprofen, and ketoprofen from water. Herein, Al electrodes were used along with cationic surfactants in the EC cell (Liu et al. 2015). Likewise, iron and aluminum plates were used as electrodes in the removal of salicylic acid from aqueous solutions. Some electrode combinations were tested, and the best removal efficiency was achieved by the Al–Al electrode pair, reaching 90% removal (Chou et al. 2011).

In terms of materials involved in EC, the goal must be focused on the electrode materials. As expected, most of the electrodes are made of iron and aluminum because both Fe^{2+} and Al^{3+} ions generated from these electrodes can work as coagulants. However, some authors have researched other materials since conventional materials have shown some drawbacks, such as electrode dissolution or low removal efficiencies (Sirés and Brillas 2012). In addition, it has been found that the aqueous system has turned turbid because of the Fe^{2+} ions, which can oxidize to Fe^{3+} and then $Fe(OH)_3$ is produced after a redox reaction with OH^- ions. This turns water to yellow and turbidity is increased. In this case, Al is chosen as electrode material (Linares-Hernández et al. 2009). Novel materials other than Fe and Al that are being used in EC are stainless steel (Olmez-Hanci et al. 2012) and titanium (Shon et al. 2010).

In Chou et al.'s (2011) report regarding salicylic acid removal, a comparison between Fe–Fe and Al–Al electrode pairs was made. Under the same operating conditions, both electrode pairs were tested and very different results were obtained. On the one hand, the EC involving the Al–Al pair achieved 92% removal by reaching steady-state conditions in about 20 minutes. On the other hand, the Fe–Fe pair allowed only a 68% removal, including a longer reaction time as 80 min were necessary to finally reach the steady state in the process.

A colorless effluent is obtained when EC was operated using the Al electrodes, whereas a highly turbid residue resulted from the usage of the Fe electrodes. A better performance of Al electrodes is related to the pH variation. A pH increase is observed for both electrode combinations, but in the case of Al, that increment is higher because of the different behavior between Al and Fe dominant species during the process. In other words, some dominant aluminum species like Al^{3+} and $Al(OH)^{2+}$ generated at the 2–3 pH range, and Al^{3+} and OH^- generated by electrodes at a pH range of 4–9 can form some intermediate species, which in turn form $Al(OH)_3$ that could lead to very efficient pollutant removal. $Fe(OH)_3$ was dominant in the pH range of 6–10, but this is not good enough since pH operating conditions for salicylic acid were in a pH range of 3.4–5.8. Consequently, no Fe-based coagulant agent is completely produced under these conditions, whereas these are optimal for $Al(OH)_3$ production (Chou et al. 2011).

In a report regarding removal of the antibiotic oxytetracycline hydrochloride, biological treatments were found to be not useful at all, as microorganism activity is inhibited by antibiotics (Nariyan et al. 2017). Therefore, EC is considered to be a more effective technique for antibiotic removal from wastewater. The electrode materials tested were Fe and Al as anodes, and SS as the cathode. Thus, the electrode pairs Fe–SS and Al–SS were assessed. Under optimum conditions, removal efficiencies of 93.2% and 87.8% were achieved by Fe–SS and Al–SS, respectively. In the two cases, pH was found to increase up to 12, where iron and aluminum were in the forms of hematite and gibbsite, according their E_h-pH diagrams, respectively. These species are capable of absorbing oxytetracycline through some mechanisms like hydrogen bonding or electrostatic attraction (Nariyan et al. 2017).

Other parameters such as current density and oxygen produced by the anodes are supposed to affect removal efficiency. Moreover, dissolved oxygen was also analyzed due to its relevance in terms of showing the self-cleaning capability of water. For instance, it has been demonstrated that removal of organic matter is feasible when oxygen in water is exhausted (Kabdaşlı et al. 2009). Dissolved oxygen as well as redox potential decreased in all experiments performed. Production of hematite and gibbsite is also related to this trend in both cases, except for lower current density values when using an aluminum anode, where no pH increase occurred because of repulsion between oxytetracycline (in its positive form) and aluminum-dominant species (Al^{3+} ions). The initial concentration of oxytetracycline was found to have a nonsignificant effect over removal efficiency, no matter the electrode material. Regarding the power consumption, the Al–SS had a higher level of consumption; therefore, Al electrodes demand more energy than Fe. This is an issue for the operation, considering that removal was even lower using the Al–SS electrode pair as well (Nariyan et al. 2017).

A rarely used material is titanium. For example, laboratory-scale research was performed with synthetic pharmaceutical wastewater. In this study, titanium was used as the anode in order to generate Ti⁴⁺ ions. Since most of the colloids to be treated by EC are negatively charged, Ti⁴⁺ ions are more likely to work better than Fe³⁺ or Al³⁺ ions as coagulants. Like Fe and Al, Ti⁴⁺ ions are hydrolyzed to form hydroxides and polyhydroxides capable of producing the precipitation/flotation of the destabilized colloids as aggregated flocs (Moreno-Casillas et al. 2007). In this research, the optimum current density was found to be higher than the iron or aluminum values, due to less dissolution of ions from the titanium-based anode. In terms of pH, an optimum condition for titanium-based EC (Ti-EC) is supposed to be a pH value of 4, but when OH⁻ ions are generated, pH is expected to increase, and removal efficiency might decrease during the first stages of the process. Regarding chemical oxygen demand, under optimum parameters, Ti-EC reached 70% removal. It is also suggested that using Ti-EC could be more effective than chemical coagulation when removing compounds when a much smaller molecular weight is required. Finally, this research includes a preparation route of titanium dioxide nanoparticles from the Ti-EC sludge, so that the environmental impact of the sludge can be mitigated (Shon et al. 2010).

When comparing the last detailed reports, it can be concluded that there is no optimal electrode material for all types of pharmaceutical wastewaters, i.e., every pollutant may react in a different way according to the electrode materials used and the operating conditions performed. Therefore, several preliminary tests must be done to find out the optimum electrode materials as well as the operating conditions.

Table 12.1 shows a comparative overview of different reports regarding EC for pharmaceutical wastewater treatment.

12.4.6 Wastewater Containing Heavy Metals

EC has been successfully applied to remove heavy metals and metalloids from wastewaters. Generally, high pollutant removal can be achieved by choosing the appropriate electrode material. In literature, Al, Fe, and combination of the two have been reported as electrode material for heavy metals removal via EC (Balasubramanian et al. 2009; Hansen et al. 2019; Ratna Kumar et al. 2004; Song et al. 2017).

Studies regarding arsenic (As) removal from industrial effluent by EC have reported strong As adsorption on hydrous metal oxides such as Al and Fe hydroxides/oxyhydroxides. These species are formed in electrochemical reactions during EC (Song et al. 2017). Several authors have reported Fe electrodes in As removal by the EC process (Balasubramanian et al. 2009; Ratna Kumar et al. 2004; Wan et al. 2010). Balasubramanian et al. (2009) employed mild steel as the sacrificial anode, thus EC yielded 94% of arsenic removal. Ratna Kumar et al. (2004) evaluated three types of electrodes: Fe, Al, and Ti. The highest efficiency achieved (99%) was obtained with Fe electrodes. Studies on EC performed with Fe electrode, have reported As(III) oxidation to As(V), with As(V) adsorbing into hydrous ferric oxides (Ratna Kumar et al. 2004; Wan et al. 2010). Alternative materials to Fe and Al have been proposed to improve removal efficiency with EC technology. For example, high removal efficiencies were reported for EC performed with zinc (Zn), copper (Cu), and Cu–Zn electrodes (Ali et al. 2012; Maldonado-Reyes et al. 2007). Cu electrodes were more efficient than Al and Fe for this type of industrial effluent (Gomes et al. 2010).

Fe electrodes have been used in Se removal from oil refinery wastewater, as well as Cu production wastewater. For both, above 90% of selenium was removed. Researchers suggested that Fe hydroxides were the chemical species capable of adsorbing Se (Hansen et al. 2019; Mavrov et al. 2006).

Fe and Fe–Cu electrodes have shown high Cu, Cr, Ag, Zn, and Ni removal when EC was applied to metal plating and automotive industry wastewater (Beyazit 2014; Gatsios et al. 2015; Golder et al. 2007b; Hanay and Hasar 2011; Heidmann and Calmano 2008; Mercado-Martínez et al. 2013). Beyazit (Beyazit 2014) reported a set of possible electrode pair combinations using Al, Fe, and SS electrodes. The author concluded that Fe–SS electrode pairs were optimal for removing the aforementioned metals. Al electrodes have been used in heavy metal ion (e.g. Zn^{2+}, Cu^{2+}, Ni^{2+}, Ag^+, and

TABLE 12.1
Overview of Pharmaceutical Wastewater Pollutants Treated by EC Process

Pollutant	Electrodes	Distance between Electrodes (cm)	Effective Area (cm²)	Reactor Volume (L)	Initial Pollutant Concentration	Current Density	% Removal	pH	EC Time (min)	Reference
Dexamethasone (DEX)	Al	—	61	1.0	100 µg/L	—	DEX (38)	—	45	Arsand et al. 2013
Hydrolyzed peptone	Fe	2.0	31	0.5	COD: 1753 mg/L	763.0 A/m²	Turb (91) COD (86)	6.0	90	Boroski et al. 2009
Salicylic acid (SA)	Fe, Al	2.0	33	0.5	100 mg/L	—	Al/Al: SA (92) Fe/Fe: SA (68)	7.0	20–80	Chou et al. 2011
Ciprofloxacin (CIP)	Al	2.3	450	1.0	25 mg/L	18.0 A/m²	TOC (87) CIP (98)	9.0	40	Espinoza-Quiñones et al. 2016
Oxytetracycline hydrochloride	Fe, Al, SS	5.0	35	0.2	50 mg/L	20.0 mA/cm²	Fe: SS (93.17) Al: SS (87.68)	—	60, 40	Nariyan et al. 2017
Naphthalene sulfonates	316 SS	0.3	286	3.0	—	—	COD (98) COD (91) TOC (88)	11.0	150	Olmez-Hanci et al. 2012
Salicylic acid (SA)	Hybrid Al–Fe	0.2	35	1.0	100 mg/L	50.0 A/m²	COD (85)	6.0	10	Ozyonar and Aksoy 2016
Hospital wastewater	Fe, Al, SS	—	176	2.0	TOC: 276 mg/L	4.9, 4.5, 2.7 mA/cm²	TOC: Fe (99), Al (100), SS (100)	5.5–7.8	—	Veli et al. 2016
Synthetic wastewater	Ti	—	207	5.0	10 mg/L	8.3 mA/cm²	COD (60–70)	4.0	4	Shon et al. 2010
Flurbiprofen	Fe	0.5	81	0.6	—	2.5 mA/cm²	TOC (66.3)	6.5	20	Barısçı et al. 2015
Ayurveda pharmaceutical wastewater	SS, Al	—	82	1	—	Al: 99.9 A/m², SS: 125.8 A/m²	Al: COD (58), SS: COD (79)	Al: 6.0, SS: 7.5	120	Singh et al. 2016
Azithromycin	Fe	0.2	48	1	190 mg/L	20.0 mA/cm²	COD (96)	3.0	60	Yazdanbakhsh et al. 2015

$Cr_2O_7^{2-}$) removal. In this application, Zn, Cu, Ni, and Ag ions were first hydrolyzed and then coprecipitated as hydroxides. Cr(VI) is reduced at the cathode to Cr(III) to then precipitate as hydroxide (Golder et al. 2007b; Heidmann and Calmano 2008). Al alloy anodes (including Zn, In, Fe, and Si in the alloy) have been also utilized to remove Cr from water (Vasudevan et al. 2011).

Complete removal of indium from water was achieved in studies considering Fe–Al electrode pair (Chou et al. 2009). Al and Fe electrodes have been reported to effectively remove mercury from wastewater (Nanseu-Njiki et al. 2009). Al electrodes showed to improve the removal of manganese (Gatsios et al. 2015; Hanay and Hasar 2011; Shafaei et al. 2010).

12.5 FUTURE PERSPECTIVES

Even though new materials in EC are being investigated, there are still some problems in the process. Namely, it should not be focused only on electrode materials and their dependence on the operating conditions such as pH range, current density, or the type of the pollutant to be removed, but also other drawbacks regardless of the materials used as electrodes. Few weaknesses have been detected in EC like the inexorable production of sludge, electrode dissolution and subsequent replacement, or power consumption (Sirés and Brillas 2012).

One of the possibilities to reach a process enhancement might be combining EC with other wastewater treatment techniques or modifying the electrode materials by coating them or adding some compounds to remediate some of the mentioned problems.

For example, in a study about urban wastewater treatment, a process called electrodisinfection/electrocoagulation (ED-EC) was performed employing boron-doped diamond as the anode and stainless steel as the cathode. Additionally, a perforated iron plate was settled between the electrodes to work as a bipolar electrode. The aim of this work was to achieve simultaneously the removal of turbidity and microbiological content. In this context, total disinfection and more than 90% of turbidity removal were obtained. Compounds formed during the ED-EC process such as hypochlorite and chloramines were found responsible for disinfection, and its efficiency mostly depends on the anode material. Meanwhile, turbidity removal is attributed to the electrochemical cell arrangement, reaching a better performance when using iron as a bipolar electrode, mainly, at low current densities when the bipolar electrode dissolution is inhibited (Llanos et al. 2014). In addition, novel anodic materials like boron-doped conductive diamond electrodes appear to be more stable and efficient (Linares-Hernández et al. 2010).

Given that electro-Fenton (EF) treatment is not good enough for the treatment of cyanotoxins removal due to its high energy consumption and some issues about the use of chemicals, EC has been coupled to the Fenton processes (An et al. 2019). Hence, EC-Fenton appears to be an improved EF where no chemicals need to be added, and the energy consumption is reduced. The EF process is an advanced oxidation process (AOP), which consists in producing highly oxidizing species, e.g., hydroxyl radicals (·OH). These radicals can completely oxidize some organic compounds into CO_2 and H_2O. In EF, Fe^{2+} ions are added to the continuous generation of H_2O_2 (which is formed through oxygen reduction on the cathode). This interaction between Fe^{2+} and H_2O_2 can produce hydroxyl radicals. In this case, EC could enhance the EF process by providing coagulating agents to the hydroxyl radicals, improving the pollutant removal efficiency. Iron is the key element in this process because both EC and EF use Fe ions in their wastewater treatment mechanisms.

The experimental stage consisted of a three-electrode setup, where iron was used as the anode. A graphite cathode and a Ti/IrO_2 counter electrode were also employed. The aim is to alternate EC and EF in a single reactor by only switching the electrode connection, where Fe ions and ·OH were produced *in situ*, so that the addition of chemicals was not required. A final comparison was made in terms of the single EF process and the new EC-Fenton. Results have shown an improvement of the 30% for TOC removal efficiency by using this new technique, and energy consumption decreased by up to 92%. More research efforts need to be done in this field of interest as new electrode materials can be discovered (An et al. 2019).

Materials Involved in Electrocoagulation Process

Among the applications of EC, wastewater treatment from the textile, food, oil, pulp and paper, and pharmaceutical industries were discussed so far. However, there are other uncommon types of wastewater, for instance, cooling tower blowdown water could also possibly be treated by EC, due the presence of scale-forming species. Specifically, hardness ions (Ca^{2+} and Mg^{2+}) and dissolved silica are considered as the scale-forming species mentioned in this report. These pollutants are generated when water evaporates from the cooling tower, when solids concentrate in water, causing scale deposits over heat transfer surfaces. Therefore, hardness and silica ions must be removed in order to avoid decreasing efficiencies in terms of heat transfer. Sampling was made from a urea fertilizer industry in Egypt (Hafez et al. 2018). In this case, three different electrode materials were tested (Al, Fe, and Zn) and aluminum appeared to be more efficient than the other two materials, as removal efficiencies of 55.4% and 99.5% for both hardness and silica ions were obtained, respectively. Additionally, at optimum conditions, power consumption is also lower for the Al electrodes (Hafez et al. 2018).

Future perspectives for electrode materials in EC processes should be focused to treat other types of wastewater, so that more industries can employ EC as a cost-effective water treatment technique. In consequence, new materials could be developed as electrodes.

Considering a recent report including wastewater from the olive mill industry, EC was combined with a novel technique called the catalytic sonoperoxone process (Khani et al. 2019). Here, EC was performed to improve the biological degradability of the mentioned wastewater. In this case, traditional EC was performed using Fe and Al electrodes followed by several processes involving advanced oxidation processes. In other words, EC was carried out as a pretreatment, whereas the other processes were assessed to determine which one was better for removal efficiency. These techniques are the following: single ozonation process (SOP), catalytic ozonation process (COP), ultrasonic process (US), H_2O_2, COP/US, a peroxone process (H_2O_2/O_3), and a catalytic sonoperoxone process (H_2O_2/COP/US). This last one is a combination of the others. In terms of total organic carbon removal, Figure 12.3 shows the efficiency of each AOP.

A previous EC pretreatment was performed in order to evaluate the removal performance of both iron and aluminum electrodes. Here, the optimal conditions were reached using the Fe electrodes. Then, the AOPs were performed over the pretreated water from EC at optimal conditions. As can be seen in Figure 12.3, the catalytic sonoperoxone process showed better performance since a 75% TOC removal was obtained with this technique (Khani et al. 2019).

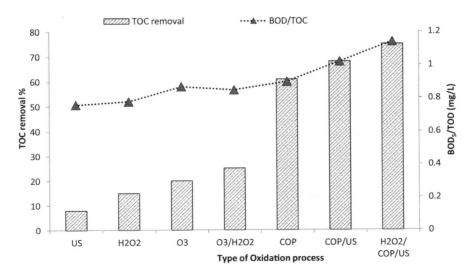

FIGURE 12.3 Wastewater degradability for different AOPs. (From Khani et al. 2019.)

FIGURE 12.4 Comparison of the textile wastewater sample before and after the treatments. (From Tavangar et al. 2019.)

On the other hand, the combination of EC with a nanofiltration (NF) process was performed to treat wastewater from the textile industry (Tavangar et al. 2019). Here, an individual comparison between EC and NF was made, and then their hybrid (EC-NF) was assessed. Figure 12.4 shows a comparison of the samples after the treatment.

For the EC process, Al, Fe, and Ti electrodes were used. Al has appeared as the best of these three materials, reaching 64% and 94% of COD and color removal, respectively. On the other hand, NF was made using an NF membrane, which achieved more than 87% color removal. In the end, EC-NF is expected to improve certain drawbacks of the individual processes. For instance, NF can remove the color of the solutions remaining from EC, thus improving the whole process performance. Once again, EC is intended to be a pretreatment that works at decreasing the membrane contamination, so that permeate flux can increase. Overall results led to a 95% color removal and 99% turbidity removal for the EC-NF hybridization (Tavangar et al. 2019).

Table 12.2 provides an overview of some recent studies about EC-based processes.

12.6 CONCLUSIONS

In this work, we have outlined the approach of using the EC process to treat wastewaters from different industries that contain toxic and nonreadily biodegradable pollutants (e.g., textile, pulp and paper mill, food, oil, and pharmaceutical industries). As detailed in Section 12.3, many operational factors play an important role in the removal efficiency of the EC process, including the electrode material, pH, electrode arrangement, current density, temperature, and conductivity. The choice of an appropriate electrode material is a crucial issue to ensure the success of EC through the *in situ* formation of coagulating agents that can remove several pollutants from effluents.

According to the results of this chapter, it has been detailed that both aluminum and iron work better as electrode materials for the EC process. However, in the last years, many studies have focused on developing new electrode materials for EC, such as the use of stainless steel, zinc, copper, and titanium-based anodes, as indicated in Section 12.4. On the other hand, different combinations among the EC process and other wastewater treatment technologies have been employed to remove some pollutants from industrial effluents that are not easily removed by conventional methods. Particularly, it has been detailed in Section 12.5 that the use of EC with other electrochemical processes and advanced oxidation processes allow removal of refractory compounds from many types of effluents, reaching removal efficiencies close to 100%, and therefore improving the water quality of the mentioned wastewaters.

TABLE 12.2
Overview of Recent Reports about Future Perspectives for EC-Based Wastewater Treatment Processes

Process	Pollutant/Wastewater	Electrodes	Removal Efficiency (%)	Reference
EC-Fenton	Cyanobacteria/cyanotoxins	Fe, graphite and a Ti/IrO$_2$ counter electrode	Cyanobacterial cells (98), Microcystins (100), TOC (96)	An et al. 2019
EC	Hardness (H) and dissolved silica (DS) from cooler tower blowdown water	Al, Fe, Zn	Al: H (55), DS (100); Fe: H (37), DS (99); Zn: H (39), DS (96)	Hafez et al. 2018
EC-H$_2$O$_2$/COP/US	Olive mill wastewater	Fe, Al	TOC (75)	Khani et al. 2019
EC-Nanofiltration	Textile wastewater	Al, Fe, Ti	Color (95), Turbidity (99)	Tavangar et al. 2019
EC-AOP (photoelectro–Fenton and electrochemical oxidation)	Raw cheese whey wastewater	Fe, Al, SS (AISI 304 and AISI 316 L)	TOC (49)	Tirado et al. 2018
EC-Fenton	Oil mill effluent	Fe	COD (100)	Chairunnisak et al. 2018
EC-AOP (peroxi-coagulation, anodic oxidation, and electro–Fenton)	Mixture of reactive dyes (mainly methylene blue)	Fe, Boron doped diamond (BDD)	Optimal EC-Fenton: TOC (97), Turbidity (100), Color (100)	Zazou et al. 2019
EC-Electrooxidation (EC–EF), Electrochemical peroxidation (ECP)	Canola oil refinery wastewater	Al, BDD, SS	EC-EF: Soluble COD (99)	Sharma and Simsek 2019

ACKNOWLEDGMENTS

This work was financially supported by a grant from CONICYT (Scholarship Programs No. 21150520 and 21151390). The authors acknowledge the support of Escuela Politécnica Nacional, Ecuador, through the Internal Project PII-DCN-2012.

REFERENCES

Agustin, Melissa, Waya Sengpracha, and Weerachai Phutdhawong. 2008. "Electrocoagulation of Palm Oil Mill Effluent." *International Journal of Environmental Research and Public Health* no. 5 (3):177–80.

Ali, Imran, Mohd Asim, and Tabrez Alam Khan. 2012. "Arsenite Removal from Water by Electro-Coagulation on Zinc–zinc and Copper–copper Electrodes." *International Journal of Environmental Science and Technology* no. 10 (2):377–84.

Aljaberi, Forat Yasir. 2018. "Studies of Autocatalytic Electrocoagulation Reactor for Lead Removal from Simulated Wastewater." *Journal of Environmental Chemical Engineering* no. 6:6069–6078.

An, Chunjiang, Gordon Huang, Yao Yao, and Shan Zhao. 2017. "Emerging Usage of Electrocoagulation Technology for Oil Removal from Wastewater: A Review." *Science of the Total Environment* no. 579:537–556.

An, Jingkun, Nan Li, Shu Wang, Chengmei Liao, Lean Zhou, Tian Li, Xin Wang, and Yujie Feng. 2019. "A Novel Electro-Coagulation-Fenton for Energy Efficient Cyanobacteria and Cyanotoxins Removal without Chemical Addition." *Journal of Hazardous Materials* no. 365:650–658.

Aoudj, Salaheddine, Abdellah Khelifa, Nadjib Drouiche, Mouna Hecini, and Houria Hamitouche. 2010. "Electrocoagulation Process Applied to Wastewater Containing Dyes from Textile Industry." *Chemical Engineering & Processing: Process Intensification* no. 49:1176–1182.

Arsand, Daniel R., Klaus Kümmerer, and Ayrton F. Martins. 2013. "Removal of Dexamethasone from Aqueous Solution and Hospital Wastewater by Electrocoagulation." *Science of the Total Environment* no. 443:351–357.

Asselin, Mélanie, Patrick Drogui, Hamel Benmoussa, and Jean-François Blais. 2008. "Effectiveness of Electrocoagulation Process in Removing Organic Compounds from Slaughterhouse Wastewater Using Monopolar and Bipolar Electrolytic Cells." *Chemosphere* no. 72:1727–1733.

Asselin, Mélanie, Patrick Drogui, Satinder Kaur, and Hamel Benmoussa. 2008. "Organics Removal in Oily Bilgewater by Electrocoagulation Process." *Journal of Hazardous Materials* no. 151:446–455.

Balasubramanian, Natesan, Toshinori Kojima, Chiya Ahmed Basha, and Chand Srinivasakannan. 2009. "Removal of Arsenic from Aqueous Solution Using Electrocoagulation." *Journal of Hazardous Materials* no. 167:966–969.

Barışçı, Sibel, Feride Ulu, Mika Sillanpää, and Anatholy Dimoglo. 2015. "Evaluation of Flurbiprofen Removal from Aqueous Solution by Electrosynthesized Ferrate(VI) Ion and Electrocoagulation Process." *Chemical Engineering Journal* no. 262:1218–1225.

Bayramoglu, Mahmut, Mehmet Kobya, Murat Eyvaz, and Elif Senturk. 2006. "Technical and Economic Analysis of Electrocoagulation for the Treatment of Poultry Slaughterhouse Wastewater." *Separation and Purification Technology* no. 51 (3):404–408.

Behin, Jamshid, Negin Farhadian, Mojtaba Ahmadi, and Mehdi Parvizi. 2015. "Ozone Assisted Electrocoagulation in a Rectangular Internal-Loop Airlift Reactor: Application to Decolorization of Acid Dye." *Journal of Water Process Engineering* no. 8:171–178.

Beyazit, Nevzat. 2014. "Copper (II), Chromium (VI) and Nickel (II) Removal from Metal Plating Effluent by Electrocoagulation." *International Journal of Electrochemical Science* no. 9:4315–4330.

Boroski, Marcela, Angela Cláudia Rodrigues, Juliana Carla Garcia, Luiz Carlos Sampaio, Jorge Nozaki, and Noboru Hioka. 2009. "Combined Electrocoagulation and TiO2 Photoassisted Treatment Applied to Wastewater Effluents from Pharmaceutical and Cosmetic Industries." *Journal of Hazardous Materials* no. 162:448–454.

Brillas, Enric and Carlos A. Martínez-Huitle. 2015. "Decontamination of Wastewaters Containing Synthetic Organic Dyes by Electrochemical Methods. An Updated Review." *Applied Catalysis B: Environmental Environmental* 166–167:603–643.

Cañizarez, Pablo, Fabiola Martínez, Justo Lobato, and Manuel Andrés Rodrigo. 2007. "Break-up of Oil-in-Water Emulsions by Electrochemical Techniques." *Journal of Hazardous Materials* no. 145:233–240.

Chairunnisak, Aula, Bastian Arifin, Hizir Sofyan, Mirna R. Lubis, and Darmandi Darmadi. 2018. "Comparative Study on the Removal of COD from POME by Electrocoagulation and Electro-Fenton Methods: Process Optimization." *IOP Conference Series: Materials Science and Engineering* no. 334:1–12.

Chavalparit, Orathai and Maneerat Ongwandee. 2009. "Optimizing Electrocoagulation Process for the Treatment of Biodiesel Wastewater Using Response Surface Methodology." *Journal of Environmental Sciences* no. 21 (11):1491–1496.

Chen, Guohua. 2004. "Electrochemical Technologies in Wastewater Treatment." *Separation and Purification Technology* no. 38:11–41.

Chen, Xueming, Guohua Chen, and Po Lock Yue. 2000. "Separation of Pollutants from Restaurant Wastewater by Electrocoagulation." *Separation and Purification Technology* no. 19:65–76.

Chou, Wei-Lung, Chi-Ta Wang, Te-Chao Liu, and Li-Chen Chou. 2011. "Effect of Process Parameters on Removal of Salicylic Acid from Aqueous Solutions via Electrocoagulation." *Environmental Engineering Science* no. 28 (5):365–372.

Chou, Wei-Lung, Chih-Ta Wang, and Kai-Yu Huang. 2009. "Effect of Operating Parameters on Indium (III) Ion Removal by Iron Electrocoagulation and Evaluation of Specific Energy Consumption." *Journal of Hazardous Materials* no. 167:467–474.

Chou, Wei-Lung, Chih-Ta Wang, and Kai-Yu Huang. 2010. "Investigation of Process Parameters for the Removal of Polyvinyl Alcohol from Aqueous Solution by Iron Electrocoagulation." *Desalination* no. 251:12–19.

Coskun, Tamer, Fatih İlhan, Neslihan Manav Demir, Eyup Debik, and Uğur Kurt. 2012. "Optimization of Energy Costs in the Pretreatment of Olive Mill Wastewaters by Electrocoagulation." *Environmental Technology* no. 33 (7):801–807.

Davila, Javier A., Fiderman MacHuca, and Nilson Marrianga. 2011. "Treatment of Vinasses by Electrocoagulation-Electroflotation Using the Taguchi Method." *Electrochimica Acta* no. 56 (22):7433–7436.

Derjaguin, Boris and Lev Landau. 1941. "Theory of the Stability of Strongly Charged Lyophobic Sols and of the Adhesion of Strongly Charged Particles in Solutions of Electrolytes." *Progress in Surface Science* no. 43:30–59.

El-Ashtoukhy, El Sayed Zakaria, Nevine Kamal Amin, and Ola Abdelwahab. 2009. "Treatment of Paper Mill Effluents in a Batch-Stirred Electrochemical Tank Reactor." *Chemical Engineering Journal* no. 146:205–210.

Emamjomeh, Mohammad M. and Muttucumaru Sivakumar. 2009. "Review of Pollutants Removed by Electrocoagulation and Electrocoagulation/Flotation Processes." *Journal of Environmental Management* no. 90 (5):1663–1679.

Espinoza-Quiñones, Fernando R., Ariádine R. C. de Souza, Aparecido N. Módenes, Daniela E. G. Trigueros, Aline R. de Pauli, Patrícia S. C. de Souza, and Alexander D. Kroumov. 2016. "Removal Performance, Antibacterial Effects, and Toxicity Assessment of Ciprofloxacin Treated by the Electrocoagulation Process." *Water, Air, & Soil Pollution* no. 227:460–472.

Fajardo, Ana S., Raquel F. Rodrigues, Rui C. Martins, Luis M. Castro, and Rosa M. Quinta-Ferreira. 2015. "Phenolic Wastewaters Treatment by Electrocoagulation Process Using Zn Anode." *Chemical Engineering Journal* no. 275:331–341.

Garcia-Segura, Sergi, Maria Maesia S. G. Eiband, Jailson Vieira de Melo, and Carlos Alberto Martínez-Huitle. 2017. "Electrocoagulation and Advanced Electrocoagulation Processes: A General Review about the Fundamentals, Emerging Applications and Its Association with Other Technologies." *Journal of Electroanalytical Chemistry* no. 801:267–299.

Gatsios, Evangelos, John N. Hahladakis, and Evangelos Gidarakos. 2015. "Optimization of Electrocoagulation (EC) Process for the Purification of a Real Industrial Wastewater from Toxic Metals." *Journal of Environmental Management* no. 154:117–127.

Ghernaout, D., Wahib Mohamed Naceur, and Badiaa Ghernaout. 2011. "A Review of Electrocoagulation as a Promising Coagulation Process for Improved Organic and Inorganic Matters Removal by Electrophoresis and Electroflotation." *Desalination and Water Treatment* no. 28 (1–3):287–320.

Ghernaout, Djamel and Badiaa Ghernaout. 2011. "On the Controversial Effect of Sodium Sulphate as Supporting Electrolyte on Electrocoagulation Process: A Review." *Desalination and Water Treatment* no. 27 (1–3):243–254.

Golder, Animes K., Amar Nath Samanta, and Subhabrata Ray. 2007a. "Removal of Cr3+ by Electrocoagulation with Multiple Electrodes: Bipolar and Monopolar Configurations." *Journal of Hazardous Materials* no. 141:653–661.

Golder, Animes K., Amar Nath Samanta, and Subhabrata Ray. 2007b. "Removal of Trivalent Chromium by Electrocoagulation." *Separation and Purification Technology* no. 53:33–41.

Gomes, Jewel Andrew, Md Sanoar Rahman, Kamol Das, Srikanth Varma, and David L. Cocke. 2010. "A Comparative Electrochemical Study on Arsenic Removal Using Iron, Aluminum, and Copper Electrodes." *ECS Transactions* no. 25 (28):59–68.

Gregory, John. 1981. "Approximate Expressions for Retarded van Der Waals Interaction." *Journal of Colloid and Interface Science* no. 83 (1):138–145.

Hafez, Omar M., Madiha A. Shoeib, Mohamed A. El-Khateeb, Hussein I. Abdel-Shafy, and Ahmed O. Youssef. 2018. "Removal of Scale Forming Species from Cooling Tower Blowdown Water by Electrocoagulation Using Different Electrodes." *Chemical Engineering Research and Design* no. 136:347–357.

Hakizimana, Jean Nepo, Bouchaib Gourich, Mohammed Cha, Youssef Stiriba, Christophe Vial, Patrick Drogui, and Jamal Naja. 2017. "Electrocoagulation Process in Water Treatment: A Review of Electrocoagulation Modeling Approaches." *Desalination* no. 404:1–21.

Hanafi, Fatiha, Omar Assobhei, and Mohamed Mountadar. 2010. "Detoxification and Discoloration of Moroccan Olive Mill Wastewater by Electrocoagulation." *Journal of Hazardous Materials* no. 174 (1–3):807–812.

Hanay, Özge and Halil Hasar. 2011. "Effect of Anions on Removing Cu^{2+}, Mn^{2+} and Zn^{2+} in Electrocoagulation Process Using Aluminum Electrodes." *Journal of Hazardous Materials* no. 189:572–576.

Hansen, Henrik K., Sebastián Franco, Claudia Gutiérrez, Andrea Lazo, Pamela Lazo, and Lisbeth M. Ottosen. 2019. "Selenium Removal from Petroleum Refinery Wastewater Using an Electrocoagulation Technique." *Journal of Hazardous Materials* no. 364:78–81.

Heidmann, Ilona and Wolfgang Calmano. 2008. "Removal of Zn (II), Cu (II), Ni (II), Ag (I) and Cr (VI) Present in Aqueous Solutions by Aluminium Electrocoagulation." *Journal of Hazardous Materials* no. 152:934–941.

Hogg, Richard, Tom W. Healy, and Douglas W. Fuerstenau. 1966. "Mutual Coagulation of Colloidal Dispersions." *Transactions of the Faraday Society* no. 62 (615):1638–1651.

Inan, Hatice, Anatoly Dimoglo, and M. Karpuzcu, H. Şimşek. 2004. "Olive Oil Mill Wastewater Treatment by Means of Electro-Coagulation." *Separation and Purification Technology* no. 36:23–31.

Israelachvili, Jacob N. 2011. *Intermolecular and Surface Forces*. 3rd ed. Burlington, MA: Academic Press.

Jiang, Wenming, Mingcan Chen, Jie Yang, Zhanfei Deng, Yang Liu, Jiang Bian, Shilin Du, and Danyang Hou. 2017. "Dynamic Experimental Study of a New Electrocoagulation Apparatus with Settlement Scheme for the Removal Process in Oil Field." *Journal of Electroanalytical Chemistry* no. 801:14–21.

Kabdaşlı, Işık, Burcu Vardar, Idil Arslan-Alaton, and Olcay Tünay. 2009. "Effect of Dye Auxiliaries on Color and COD Removal from Simulated Reactive Dyebath Effluent by Electrocoagulation." *Chemical Engineering Journal* no. 148:89–96.

Kannan, Nagarethinam, G. Karthikeyan, and N. Tamilselvan. 2006. "Comparison of Treatment Potential of Electrocoagulation of Distillery Effluent with and without Activated Areca Catechu Nut Carbon." *Journal of Hazardous Materials* no. 137 (3):1803–1809.

Karhu, Mirjam, Ville Kuokkanen, Toivo Kuokkanen, and Jaakko Rämö. 2015. "Bench Scale Electrocoagulation Studies of Bio Oil-in-Water and Synthetic Oil-in-Water Emulsions." *Separation and Purification Technology* no. 96:296–305.

Khandegar, Vinita and Anil Kumar Saroha. 2012. "Electrochemical Treatment of Distillery Spent Wash Using Aluminum and Iron Electrodes." *Chinese Journal of Chemical Engineering* no. 20 (3):439–443.

Khandegar, Vinita and Anil K. Saroha. 2013. "Electrocoagulation for the Treatment of Textile Industry Effluent - A Review." *Journal of Environmental Management* no. 128:949–963.

Khani, Mohammad Reza, Hadi Kuhestani, Laleh R. Kalankesh, Bahram Kamarehei, Susana Rodríguez-Couto, Mohammad Mehdi Baneshi, and Yousef Dadban Shahamat. 2019. "Rapid and High Purification of Olive Mill Wastewater (OMV) with the Combination Electrocoagulation-Catalytic Sonoproxone Processes." *Journal of the Taiwan Institute of Chemical Engineers* no. 97:47–53.

Khansorthong, Sittichok and Mali Hunsom. 2009. "Remediation of Wastewater from Pulp and Paper Mill Industry by the Electrochemical Technique." *Chemical Engineering Journal* no. 151:228–234.

Khorram, Atousa Ghaffarian and Narges Fallah. 2018. "Treatment of Textile Dyeing Factory Wastewater by Electrocoagulation with Low Sludge Settling Time: Optimization of Operating Parameters by RSM." *Journal of Environmental Chemical Engineering* no. 6:635–642.

Kirzhner, Felix, Yoram Zimmels, and Yossi Shraiber. 2008. "Combined Treatment of Highly Contaminated Winery Wastewater." *Separation and Purification Technology* no. 63 (1):38–44.

Kobya, Mehmet, Elif Senturk, and Mahmut Bayramoglu. 2006. "Treatment of Poultry Slaughterhouse Wastewaters by Electrocoagulation." *Journal of Hazardous Materials* no. 133 (1–3):172–176.

Kobya, Mehmet, Erhan Gengec, and Erhan Demirbas. 2016. "Operating Parameters and Costs Assessments of a Real Dyehouse Wastewater Effluent Treated by a Continuous Electrocoagulation Process." *Chemical Engineering & Processing: Process Intensification* no. 101:87–100.

Kobya, Mehmet, Feride Ulu, Ugur Gebologlu, Erhan Demirbas, and Mehmet S. Oncel. 2011. "Treatment of Potable Water Containing Low Concentration of Arsenic with Electrocoagulation: Different Connection Modes and Fe–Al Electrodes." *Separation and Purification Technology* no. 77:283–293.

Körbahti, Bahadir and Kahraman Artut. 2010. "Electrochemical Oil/Water Demulsification and Purification of Bilge Water Using Pt/Ir Electrodes." *Desalination* no. 258:219–28.

Krishna, B. M., Usha N. Murthy, B. Manoj Kumar, and K. S. Lokesh. 2010. "Electrochemical Pretreatment of Distillery Wastewater Using Aluminum Electrode." *Journal of Applied Electrochemistry* no. 40 (3):663–673.

Lakshmi Kruthika, N., S. Karthika, G. Bhaskar Raju, and S. Prabhakar. 2013. "Efficacy of Electrocoagulation and Electrooxidation for the Purification of Wastewater Generated from Gelatin Production Plant." *Journal of Environmental Chemical Engineering* no. 1 (3):183–188.

Linares-Hernández, Ivonne, Carlos Barrera-Díaz, Bryan Bilyeu, Pablo Juárez-GarcíaRojas, and Eduardo Campos-Medina. 2010. "A Combined Electrocoagulation – Electrooxidation Treatment for Industrial Wastewater." *Journal of Hazardous Materials* no. 175:688–694.

Linares-Hernández, Ivonne, Carlos Barrera-Díaz, Gabriela Roa-Morales, Bryan Bilyeu, and Fernando Ureña-Núñez. 2009. "Influence of the Anodic Material on Electrocoagulation Performance." *Chemical Engineering Journal* no. 148:97–105.

Liu, Yu-Jung, Shang-Lien Lo, Ya-Hsuan Liou, and Ching-Yao Hu. 2015. "Removal of Nonsteroidal Anti-Inflammatory Drugs (NSAIDs) by Electrocoagulation-Flotation with a Cationic Surfactant." *Separation and Purification Technology* no. 152:148–154.

Llanos, Javier, Salvador Cotillas, Pablo Cañizares, and Manuel A. Rodrigo. 2014. "Effect of Bipolar Electrode Material on the Reclamation of Urban Wastewater by an Integrated Electrodisinfection/Electrocoagulation Process." *Water Research* no. 53:329–338.

López, Ainhoa, David Valero, Leticia García-cruz, Alfonso Sáez, Vicente García-garcía, Eduardo Expósito, and Vicente Montiel. 2017. "Characterization of a New Cartridge Type Electrocoagulation Reactor (CTECR) Using a Three-Dimensional Steel Wool Anode." *Journal of Electroanalytical Chemistry* no. 793:93–98.

Maldonado-Reyes, Araceli, Cecilia Montero-Ocampo, and Omar Solorza-Feria. 2007. "Remediation of Drinking Water Contaminated with Arsenic by Electro Removal Process Using Different Metal Electrodes Remediation of Drinking Water Contaminated with Arsenic by the Electro-Removal Process Using Different Metal Electrodes." *Journal of Environmental Monitoring* no. 9:1241–1247.

Martínez-Delgadillo, Sergio Alejandro, Miguel Angel Morales Mora, and I. D. Barceló-Quintal. 2010. "Electrocoagulation Treatment to Remove Pollutants from Petroleum Refinery Wastewater." *Sustainable Environment Research* no. 20 (4):227–231.

Mavrov, Valco, Stefan Stamenov, Ekaterina Todorova, Horst Chmiel, and Torsten Erwe. 2006. "New Hybrid Electrocoagulation Membrane Process for Removing Selenium from Industrial Wastewater." *Desalination* no. 201:290–296.

Meas, Yunny, José A. Ramirez, Mario A. Villalon, and Thomas W. Chapman. 2010. "Industrial Wastewaters Treated by Electrocoagulation." *Electrochimica Acta* no. 55:8165–8171.

Mercado-Martínez, Iván Darío, Germán González-Silva, and Sergio Humberto Valencia-Hurtado. 2013. "Removal of Nickel and COD Present in Wastewaters from Automotive Industry by Electrocoagulation." *Revista EIA* no. 10 (19):13–21.

Mollah, M. Yousuf A., Robert Schennach, José R. Parga, and David L. Cocke. 2001. "Electrocoagulation (EC)-Science and Applications." *Journal of Hazardous Materials* no. 84 (1):29–41.

Moreno-Casillas, Hector A., David L. Cocke, Jewel A. G. Gomes, Paul Morkovsky, J. R. Parga, and Eric Peterson. 2007. "Electrocoagulation Mechanism for COD Removal." *Separation and Purification Technology* no. 56:204–211.

Moussa, Dina T., Muftah H. El-Naas, Mustafa Nasser, and Mohammed J. Al-Marri. 2017. "A Comprehensive Review of Electrocoagulation for Water Treatment: Potentials and Challenges." *Journal of Environmental Management* no. 186:24–41.

Nanseu-Njiki, Charles Péguy, Serge Raoul Tchamango, Philippe Claude Ngom, André Darchen, and Emmanuel Ngameni. 2009. "Mercury (II) Removal from Water by Electrocoagulation Using Aluminium and Iron Electrodes." *Journal of Hazardous Materials* no. 168:1430–1436.

Nariyan, Elham, Ailin Aghababaei, and Mika Sillanpää. 2017. "Removal of Pharmaceutical from Water with an Electrocoagulation Process; Effect of Various Parameters and Studies of Isotherm and Kinetic." *Separation and Purification Technology* no. 188:266–281.

Nidheesh, P. V. and T. S. Anantha Singh. 2017. "Arsenic Removal by Electrocoagulation Process: Recent Trends and Removal Mechanism." *Chemosphere* no. 181:418–432.

Olmez-Hanci, Tugba, Zeynep Kartal, and İdil Arslan-Alaton. 2012. "Electrocoagulation of Commercial Naphthalene Sulfonates: Process Optimization and Assessment of Implementation Potential." *Journal of Environmental Management* no. 99:44–51.

Ozyonar, Fuat and Sümeyye Aksoy. 2016. "Removal of Salicylic Acid from Aqueous Solutions Using Various Electrodes and Different Connection Modes by Electrocoagulation." *International Journal of Electrochemical Science* no. 11:3680–3696.

Ratna Kumar, Pankaj, Sanjeev Chaudhari, Kartic C. Khilar, and Sapna P. Mahajan. 2004. "Removal of Arsenic from Water by Electrocoagulation." *Chemosphere* no. 55:1245–1252.

Rincón, Guillermo J. and Enrique J. La Motta. 2014. "Simultaneous Removal of Oil and Grease, and Heavy Metals from Artificial Bilge Water Using Electro-Coagulation/Flotation Oxidation e Reduction Potential Society of Automotive Engineers." *Journal of Environmental Management* no. 144:42–50.

Safari, Sania, Mojtaba Azadi Aghdam, and Hamid-Reza Kariminia. 2016. "Electrocoagulation for COD and Diesel Removal from Oily Wastewater." *International Journal of Environmental Science and Technology* no. 13:231–242.

Santos, Marcos R. G., Marilia O. F. Goulart, Josealdo Tonholo, and Carmem L. P. S. Zanta. 2006. "The Application of Electrochemical Technology to the Remediation of Oily Wastewater." *Chemosphere* no. 64:393–399.

Şengil, I. Ayhan and Mahmut Ozacar. 2006. "Treatment of Dairy Wastewaters by Electrocoagulation Using Mild Steel Electrodes." *Journal of Hazardous Materials* no. 137 (2):1197–1205.

Shafaei, Ashraf, Maryam Rezayee, Mokhtar Arami, Manouchehr Nikazar, and Al Al. 2010. "Removal of Mn^{2+} Ions from Synthetic Wastewater by Electrocoagulation Process." *Desalination* no. 260 (1–3):23–28.

Sharma, Swati and Halis Simsek. 2019. "Treatment of Canola-Oil Refinery Effluent Using Electrochemical Methods: A Comparison between Combined Electrocoagulation + Electrooxidation and Electrochemical Peroxidation Methods." *Chemosphere* no. 221:630–639.

Shim, Ho Y., Kyo S. Lee, Dong S. Lee, Dae S. Jeon, Mi S. Park, Ji S. Shin, Yun K. Lee, Ji W. Goo, Soo B. Kim, and Doug Y. Chung. 2014. "Application of Electrocoagulation and Electrolysis on the Precipitation of Heavy Metals and Particulate Solids in Washwater from the Soil Washing." *Journal of Agricultural Chemistry and Environment* no. 3 (4):130–138.

Shon, Ho Kyong, Sherub Phuntsho, S. Vigneswaran, Jaya Kandasamy, Long D. Nghiem, Geon-Joon Kim, J. B. Kim, and J. H. Kim. 2010. "Preparation of Titanium Dioxide Nanoparticles from Electrocoagulated Sludge Using Sacrificial Titanium Electrodes." *Environmental Science and Technology* no. 44:5553–5557.

Singh, Shriom, Seema Singh, Shang Lien, and Navneet Kumar. 2016. "Electrochemical Treatment of Ayurveda Pharmaceuticals Wastewater: Optimization and Characterization of Sludge Residue." *Journal of the Taiwan Institute of Chemical Engineers* no. 67:385–396.

Sirés, Ignasi and Enric Brillas. 2012. "Remediation of Water Pollution Caused by Pharmaceutical Residues Based on Electrochemical Separation and Degradation Technologies: A Review." *Environment International* no. 40:212–229.

Song, Peipei, Zhaohui Yang, Guangming Zeng, Xia Yang, Haiyin Xu, Like Wang, and Rui Xu. 2017. "Electrocoagulation Treatment of Arsenic in Wastewaters: A Comprehensive Review." *Chemical Engineering Journal* no. 317:707–725.

Song, Peipei, Zhaohui Yang, Haiyin Xu, Jing Huang, Xia Yang, and Like Wang. 2014. "Investigation of Influencing Factors and Mechanism of Antimony and Arsenic Removal by Electrocoagulation Using Fe – Al Electrodes." *Industrial & Engineering Chemistry Research* no. 53 (33):12911–12919.

Sridhar, Ramasamy, Venkatachala Sivakumar, V. Prince Immanuel, and J. Prakash Maran. 2011. "Treatment of Pulp and Paper Industry Bleaching Effluent by Electrocoagulant Process." *Journal of Hazardous Materials* no. 186:1495–1502.

Tavangar, Tohid, Kamran Jalali, Mohammad Amin Alaei Shahmirzadi, and Mohammad Karimi. 2019. "Toward Real Textile Wastewater Treatment: Membrane Fouling Control and Effective Fractionation of Dyes/Inorganic Salts Using a Hybrid Electrocoagulation – Nanofiltration Process." *Separation and Purification Technology* no. 216:115–125.

Tchamango, Serge, Charles P. Nanseu-Njiki, Emmanuel Ngameni, Dimiter Hadjiev, and André Darchen. 2010. "Treatment of Dairy Effluents by Electrocoagulation Using Aluminium Electrodes." *Science of the Total Environment* no. 408 (4):947–952.

Thakur, Chandrakant, Vimal Chandra Srivastava, and Indra Deo Mall. 2009. "Electrochemical Treatment of a Distillery Wastewater: Parametric and Residue Disposal Study." *Chemical Engineering Journal* no. 148 (2–3):496–505.

Tirado, Lydia, Ömür Gökkuş, Enric Brillas, and Ignasi Sirés. 2018. "Treatment of Cheese Whey Wastewater by Combined Electrochemical Processes." *Journal of Applied Electrochemistry* no. 48 (12):1307–1319.

Trompette, Jean-Luc and Hugues Vergnes. 2009. "On the Crucial Influence of Some Supporting Electrolytes during Electrocoagulation in the Presence of Aluminum Electrodes." *Journal of Hazardous Materials* no. 163:1282–1288.

Uğurlu, Mehmet. 2004. "The Removalof Some Inorganic Compounds from Paper Mill Effluents by the Electrocoagulation Method." *G.U. Journal of Science* no. 17 (3):85–99.

Uğurlu, Mehmet, Ahmet Gürses, Çetin Doğar, and Mehmet Yalçın. 2008. "The Removal of Lignin and Phenol from Paper Mill Effluents by Electrocoagulation." *Journal of Environmental Management* no. 87 (3):420–428.

Ulucan, Kubra, Harun Akif Kabuk, Fatih Ilhan, and Ugur Kurt. 2014. "Electrocoagulation Process Application in Bilge Water Treatment Using Response Surface Methodology." *International Journal of Electrochemical Science* no. 9:2316–2326.

Valero, David, Juan M. Ortiz, Vicente García, Eduardo Expósito, Vicente Montiel, and Antonio Aldaz. 2011. "Electrocoagulation of Wastewater from Almond Industry." *Chemosphere* no. 84 (9):1290–1295.

van Oss, Carel J. 2006. *Interfacial Forces in Aqueous Media*. 2nd. Ed. Boca Raton, FL: Taylor & Francis.

Vasudevan, Subramanyan, Jothinathan Lakshmi, and Ganapathy Sozhan. 2011. "Studies on the Al–Zn–In-Alloy as Anode Material for the Removal of Chromium from Drinking Water in Electrocoagulation Process." *Desalination* no. 275:260–268.

Veli, Sevil, Ayla Arslan, and Deniz Bingöl. 2016. "Application of Response Surface Methodology to Electrocoagulation Treatment of Hospital Wastewater." *Clean-Soil, Air, Water* no. 44 (11):1516–1522.

Vepsäläinen, Mikko. 2012. "Electrocoagulation in the Treatment of Industrial Waters and Wastewaters." *VTT Science* no. 19:1–96.

Vepsäläinen, Mikko, Heli Kivisaari, Martti Pulliainen, Aimo Oikari, and Mika Sillanpää. 2011. "Removal of Toxic Pollutants from Pulp Mill Effluents by Electrocoagulation." *Separation and Purification Technology* no. 81:141–150.

Vepsäläinen, Mikko, Mohammad Ghiasvand, Jukka Selin, Jorma Pienimaa, Eveliina Repo, Martti Pulliainen, and Mika Sillanpää. 2009. "Investigations of the Effects of Temperature and Initial Sample PH on Natural Organic Matter (NOM) Removal with Electrocoagulation Using Response Surface Method (RSM)." *Separation and Purification Technology* no. 69:255–261.

Verma, Akshaya Kumar. 2017. "Treatment of Textile Wastewaters by Electrocoagulation Employing Fe-Al Composite Electrode." *Journal of Water Process Engineering* no. 20:168–172.

Verwey, E. J. W. and J. Th G. Overbeek. 1948. *Theory of the Stability of Lyo-Phobic Colloids*. Amsterdam, Holland: Elsevier.

Wan, Wei, Troy J. Pepping, Tuhin Banerji, Sanjeev Chaudhari, and Daniel E. Giammar. 2010. "Effects of Water Chemistry on Arsenic Removal from Drinking Water by Electrocoagulation." *Water Research* no. 45:384–392.

Wei, Ming-Chi, Kai-Sung Wang, Chin-Lin Huang, Chih-Wei Chiang, Tsung-Jen Chang, Shiuan-Shinn Lee, and Shih-Hsien Chang. 2012. "Improvement of Textile Dye Removal by Electrocoagulation with Low-Cost Steel Wool Cathode Reactor." *Chemical Engineering Journal* no. 192:37–44.

Wiese, G. R. and Thomas W. Healy. 1970. "Effect of Particle Size on Colloid Stability." *Transactions of the Faraday Society* no. 66 (13):490–499.

Xu, Le Jia, Brian W. Sheldon, Duan K. Larick, and Roy E. Carawan. 2002. "Recovery and Utilization of Useful By-Products from Egg Processing Wastewater by Electrocoagulation." *Poultry Science* no. 81 (6):785–792.

Yavuz, Yusuf, Eren Öcal, Ali Savaş Koparal, and Ülker Bakir Öğütveren. 2011. "Treatment of Dairy Industry Wastewater by EC and EF Processes Using Hybrid Fe-Al Plate Electrodes." *Journal of Chemical Technology and Biotechnology* no. 86 (7):964–969.

Yavuz, Yusuf and Ülker Bakir Ogütveren. 2018. "Treatment of Industrial Estate Wastewater by the Application of Electrocoagulation Process Using Iron Electrodes." *Journal of Environmental Management* no. 207:151–158.

Yazdanbakhsh, Ahmad Reza, Mohammad Reza Massoudinegad, Sima Eliasi, and Amir Sheikh Mohammadi. 2015. "The Influence of Operational Parameters on Reduce of Azithromyin COD from Wastewater Using the Peroxi-Electrocoagulation Process." *Journal of Water Process Engineering* no. 6:51–57.

Yetilmezsoy, Kaan, Fatih Ilhan, Zehra Sapci-Zengin, Suleyman Sakar, and M. Talha Gonullu. 2009. "Decolorization and COD Reduction of UASB Pretreated Poultry Manure Wastewater by Electrocoagulation Process: A Post-Treatment Study." *Journal of Hazardous Materials* no. 162 (1):120–132.

Zaied, Mourad and Nizar Bellakhal. 2009. "Electrocoagulation Treatment of Black Liquor from Paper Industry." *Journal of Hazardous Materials* no. 163 (2–3):995–1000.

Zazou, Hicham, Hanane Afanga, Siham Akhouairi, Hassan Ouchtak, Abdelaziz Ait Addi, Rachid Ait Akbour, Ali Assabbane, Jamaâ Douch, Abdellah Elmchaouri, Joëlle Duplay, Amane Jada, and Mohamed Hamdani. 2019. "Treatment of Textile Industry Wastewater by Electrocoagulation Coupled with Electrochemical Advanced Oxidation Process." *Journal of Water Process Engineering* no. 28:214–221.

Zhao, Shan, Guohe Huang, Guanhui Cheng, Yafei Wang, and Haiyan Fu. 2014. "Hardness, COD and Turbidity Removals from Produced Water by Electrocoagulation Pretreatment Prior to Reverse Osmosis Membranes." *Desalination* no. 344:454–462.

Zodi, Salim, Jean-noël Louvet, Clémence Michon, Olivier Potier, Marie-noëlle Pons, François Lapicque, and Jean-pierre Leclerc. 2011. "Electrocoagulation as a Tertiary Treatment for Paper Mill Wastewater: Removal of Non-Biodegradable Organic Pollution and Arsenic." *Separation and Purification Technology* no. 81:62–68.

13 Role of Organic Materials in Electrochemical Applications

S. Ganesan

CONTENTS

13.1 Introduction 217
13.2 Photoelectrochemical Measurements 219
    13.2.1 Preparation of Solar Cells 220
13.3 Organic Materials as Photosensitizers in Dye-Sensitized Solar Cells 220
    13.3.1 Organic Molecules as Additives in Redox Electrolyte in DSSCs 220
    13.3.2 Conductivity Studies of Organic Compounds as Additives in Polymer Electrolytes 221
    13.3.3 Differential Scanning Calorimetry (DSC) with Organic Molecules as Additives in Polymer Electrolytes 223
    13.3.4 X-Ray Diffraction (XRD) Studies 224
    13.3.5 Scanning Electron Microscope (SEM) Analysis of Organic-Compound-Doped Polymer Electrolytes 225
    13.3.6 Photoelectrochemical Analysis 225
13.4 Conclusions 228
Acknowledgments 229
References 229

13.1 INTRODUCTION

In the context of environmental concerns, solar energy systems offer more advantages in that they are clean and emit no pollutants into the atmosphere. Solar energy is renewable, whereas fossil fuels are not. In recent years, an ever-increasing interest has been directed toward photoelectrochemical processes as applied for solar energy to electric energy conversion (Harris and Wilson 1978; Memming 1980; Fujishima and Honda 1972; Maruthamuthu and Ashokkumar 1988; Chapin et al. 1954). In dye-sensitized solar cells, a di-tetrabutylammonium cis-bis(isothiocyanato)bis(2,2′-bipyridyl-4,4′-dicarboxylato)ruthenium(II) (N719 dye)-coated TiO_2 on fluorinated tin oxide (FTO) was employed as the working electrode, the KI and I_2 in acetonitrile solvent as redox electrolyte, and platinum-coated FTO as the counter electrode (O'Regan and Gratzel 1991). Dye-sensitized solar cells (DSSCs) are assured as next-generation solar cells due to their potentially high conversion efficiency (Nazeeruddin et al. 1993). However, the liquid electrolyte in DSSCs creates low stability, due to easy evaporation of solvent and sublimation of iodine. For the past two decades, most research groups focused on replacing of liquid electrolyte by polymer electrolyte, gel polymer electrolyte, incorporation of inorganic fillers such as Al_2O_3 and ZnO, and organic compounds as plasticizers (Kubo et al. 2001; Tennakone et al. 1998; Kruger et al. 2002; Wang et al. 2003; Nogueira et al. 2002; McFarland and Tang 2003; O'Regan and Schwartz 1996) with polymer electrolytes.

    To make a dye-sensitized solar cell, a solution containing nanoparticles of TiO_2 is pasted on an FTO-coated conducting glass plate, then dried and heated to form a porous, high surface area TiO_2 structure that when magnified would look like a thin membrane. The prepared TiO_2 film on FTO plate is dipped into a solution of a cis-dithiocyanato-bis(2,2′bipyridyl-4,4′-dicarboxy) ruthenium (II) (N3 dye). To complete the device, a drop of liquid electrolyte containing potassium iodide and iodine as electrolyte placed on the TiO$_2$ film and the platinum-coated FTO plate as a counter electrode forms sandwich-type solar cells that are illuminated through the TiO$_2$ film. The possibility for charge transfer within the dye layer is low, and hence it is reasonable to assume that only a very thin layer is interconnected in the case of nanocrystalline DSSCs. The electrons lost by the N3 dye through light absorption are replaced by the mediator, which is the iodide ion within the redox electrolyte. The oxidized mediator forms I$^-$ and I$_3^-$, which in turn obtains an electron at the counter electrode after it has moved through the circuit.

The energy producing reactions in the dye-sensitized solar cells are

$$\text{Dye} + \text{light} \rightarrow \text{Dye}^* \tag{13.1}$$

$$\text{Dye}^* + \text{TiO}_2 \rightarrow e^-\left(\text{TiO}_2\right) + \text{oxidized dye} \tag{13.2}$$

$$\text{Oxidized dye} + 3/2\ \text{I}^- \rightarrow \text{Dye} + \tfrac{1}{2}\ \text{I}_3^- \tag{13.3}$$

$$1/2\ \text{I}_3^- + e^-\left(\text{counter electrode}\right) \rightarrow 3/2\ \text{I}^- \tag{13.4}$$

The voltage produced by the DSSC is the difference in energy levels between the TiO$_2$ and the electrolyte, and depends on the electrolyte and solvent used as well as the conduction of TiO$_2$ apart from the intensity of illumination (Gratzel 2003; O'Regan and Gratzel 1991; Smestad et al. 1994; Nazerruddin et al. 1993). The DSSC is a photoelectrochemical cell that resembles natural photosynthesis in two respects: (1) the organic dye molecule acts like chlorophyll to absorb light and produce a flow of electrons, and it uses multiple layers to enhance both the light absorption and electron collection efficiency (Equation 13.2). The choice of the charge mediator, the electrolyte material, and its composition are also crucial in making efforts to obtain optimized overall efficiencies of DSSC. For more than a decade most research groups focused on modifying dye-sensitized solar cells. The presence of redox couple in the electrolyte is of crucial importance for the stable operation of a DSSC, and in transporting the charge between the photoanode and the counter electrode during regeneration of the dye. The ionic liquid electrolyte has been actively pursued as a nonvolatile electrolyte for DSSCs (Oskam et al. 2001; Papageorgiou et al. 1996). Molten salts based on imidazolium iodides have revealed very attractive stability features (Murai et al. 2003; Matsumoto et al. 2001; Wand et al. 2002). Owing to their high viscosity, linear photocurrent response up to full light intensities have been observed. The redox electrolyte (I$^-$/I$_3^-$) system in DSSCs provides contact between the solar cell electrodes (working and counter), and it exists in different forms such as solid or liquid. The redox electrolytes typically working in photoelectrochemical solar cells are dissolved in organic solvents, as the dye degenerates in the presence of solvent. For long-term operation, liquid electrolytes cause stability problems due to leakage of solvent, sensitivity to air and water, as well as elevated temperatures (Wang et al. 2003; Kuciauskas et al. 2001; Alebbi et al. 1998). The organic molecules are like dendrimers with regular and highly branched three-dimensional structure with variety of applications in light harvesting systems.

In recent years organic dyes such as BODIPY and carbazole-based dendrimers with or without metal complexes were analyzed. Organic sensitizers/dyes (Yen et al. 2012; Mishra et al. 2009) gained popularity in recent decades and are replacing metal-based dyes due to their easy structural design, high absorbance value, and low cost when compared to the expensive Ru metal. Organic sensitizers consist of a D-π-A type molecular structure (Liang and Chen 2013; Wu and Zhu 2013). When the sensitizer absorbs the incident light, a push–pull mechanism operates between the donor (D) molecules and acceptor (A) molecules through π conjugation, which results in the intramolecular charge transfer. The synthesized dendrimer macromolecules with highly disperse structure and excellent material are made more attractive by adding organic molecules of some functionality at the margin to create versatile macromolecules for solar to electric energy conversion. The carbazole-based dendrimer is an excellent hole transporting material with high ability for donating electrons and is broadly used in DSSCs (Degbia et al. 2014). Some research groups focused on modifying and replacing N3 dye with organic molecules particularly BODIPY. Hattori and coworkers (2005) first introduced the synthesized BODIPY as a photosensitizer for dye-sensitized solar cells. Zhang et al. (2016) reported 4.42% efficiency with 9H-carbazol-9-yl group-based BODIPY core molecules. Other than dye molecules, the electrolyte in DSSCs plays a vital role in determining efficiency (Wu and Zhu 2013). The redox couple is more responsible for electron transfer and regeneration of the Ru metal complex dye molecules. Leakage and evaporation of electrolyte are the major problems faced when we use the liquid electrolyte and thus reduced the durability of the cells, which results in very low performance in harvesting solar energy. When I^-/I_3^- redox mediator in the electrolyte is used, the results in the sublimation of iodine from the cell also drastically affect the performance of the DSSCs. Satoh et al. (2005) reported the use of a fifth generation hole transporting phenylazomethine-based dendrimer and triphenylamine core as an additive to improve the efficiency of solar energy harvesting systems with Ru (II)-based dye molecules. Similarly, nitrogen- and sulfur-based heterocycles are employed as additives in redox couple electrolyte solutions to improve the solar to electric energy conversion efficiency of DSSCs (Kusama et al. 2008).

In the polymer electrolyte field, polymer blending techniques are used to increase the conductivity of the polymer and enhance the efficiency of DSSCs (Nogueira et al. 2001; Stathatos et al. 2001; Wang et al. 2004; Katsaros et al. 2002; Masamoto et al. 1996). The presence of polar moiety such as ether linkage (-C-O-C-) in PEO, ester linkage (-COO-) in PMMA, and high electronegative atoms (-CF$_2$-) in PVdF exhibit reasonable ionic conductivities and eliminate the problems of sealing and solvent leakage (Liu et al. 2004; Kim et al. 2005; Scully et al. 2001; Lee et al. 2008; Ganesan et al. 2013). The incorporation of organic compounds, namely 2,6-bis (N-pyrazolyl)isonicotinic acid (BNIN) (Ganesan et al. 2013), 2,6-bis(2-thiopyridyl) pyridine (BTPP) (Ganesan et al. 2016), and 2,6-bis (N-pyrazolyl) pyridine (BNPP) (Donald and Kenneth 1990) in polymer electrolyte reduce the sublimation of iodine in the redox couple and tend to shift the negative band of TiO$_2$ and improve open-circuit voltage (V_{oc}) (Kusama et al. 2008; Ganesan et al. 2013; Ganesan et al. 2016).

13.2 PHOTOELECTROCHEMICAL MEASUREMENTS

When the circuit system containing DSSC is switched on by light, a voltage is developed, which under infinite load resistance is called the open-circuit voltage, V_{oc}, and then by connecting the terminals directly, thus closing the circuit, the so-called short-circuit current, J_{sc}, can be obtained. To compare the current–voltage (J-V) characteristics for each dye-sensitized solar cell, the illuminated areas were measured separately, and the J_{sc} is introduced and most often chosen. The efficiency of the solar energy to electric energy conversion is generally distinct by the ratio between the maximum power output and the power of the incident sunlight. The performance of the solar cell also depends on the fill factor, representing the ratio of the maximum power output to the product of the short-circuit current and the open-circuit voltage. The photocurrent–photovoltage diagrams were monitored and recorded using a BAS 100A electrochemical analyzer unit.

The short-circuit current density (J_{sc}) and the open-circuit voltage (V_{oc}) were calculated from the J-V curve. The fill factor (FF) and the solar to electric energy conversion efficiency (η) were determined using the following equations, where P^s means the power of the incident (solar) light (W m^{-2}).

$$FF = \frac{(JV)_{MAX}}{J_{sc}V_{oc}} \tag{13.5}$$

$$\eta = \frac{J_{SC} \cdot V_{OC} \cdot FF}{P^s} \tag{13.6}$$

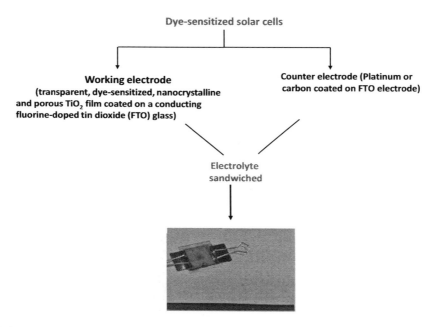

SCHEME 13.1 The schematic diagram for the preparation of dye-sensitized solar cells.

The value of the maximum power point, $(JV)_{MAX}$, was received from the largest area of a rectangle possible under the J-V curve. The product of J_{sc} and V_{oc}, being the point of connection with their axes, is the area of a rectangle formed outside the J-V curve. The solar cells were irradiated by a tungsten halogen lamp (OSRAM, Germany) providing AM 1.5.

13.2.1 Preparation of Solar Cells

Solar cells containing two electrodes merged into a sandwich-type construction have been prepared. The active area of the cell was 1 cm² (Scheme 13.1).

13.3 ORGANIC MATERIALS AS PHOTOSENSITIZERS IN DYE-SENSITIZED SOLAR CELLS

The organic compound carbazole dendrimer (Figure 13.1) based BODIPY (Janakiramanbabu et al. 2018) (Figure 13.2) was used as a photosensitizer in dye-sensitized solar cells with a good efficiency of 2.7%. The photosensitizing ability of the dendrimer with carbazole moiety acts as the electron donating unit, the BODIPY core acts as the electron acceptor, and the BODIPY is anchored to TiO_2 material and the transporting role is played by the triazole units in between them.

13.3.1 Organic Molecules as Additives in Redox Electrolyte in DSSCs

The performance of DSSCs mainly depends on (1) concise charge recombination fatalities at the interface of N3 dye coated TiO_2 and redox electrolyte and (2) the dye restoration efficiency. The charge recombination at the interface of the TiO_2 and redox electrolyte leads to loss in the short-circuit current (J_{sc}) and the open-circuit voltage (V_{oc}), resulting in a decrease in solar to electric energy conversion efficiency. In general, the addition of organic nitrogenous compounds in potassium iodide and iodine significantly improves the open-circuit voltage, which is commonly recognized by the shift of the conduction band edge of TiO_2 toward more negative potentials, and this is

Role of Organic Materials in Electrochemical Applications

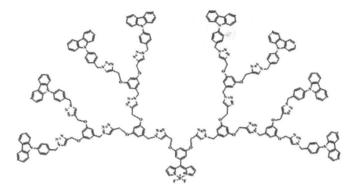

FIGURE 13.1 The structure of carbazole-dendrimer-based BODIPY as photosensitizer for DSSCs.

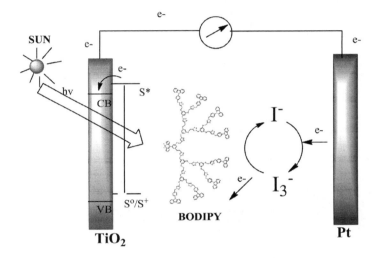

FIGURE 13.2 The schematic diagram of an organic molecule as a photosensitizer for dye-sensitized solar cells.

probably attributed to electron recombination between TiO$_2$ and redox electrolyte in dye-sensitized cells. The synthesized cyclohexadienone cored 3,6-ditertiary butyl carbazole decorated triazolyl bridged dendrimers (Figure 13.3) with I$^-$/I$_3^-$ (Babu et al. 2019) increased the current and V_{oc} and reported high efficiency and more stability up to 85% in one week (Figure 13.4).

13.3.2 Conductivity Studies of Organic Compounds as Additives in Polymer Electrolytes

Figure 13.5 shows the conductivity of polymer electrolyte based on PEO-PVdF/KI/I$_2$ with organic compounds such as 2,6, bis (N-pyrazolyl) methyl isonicotinate (BNMI), 2,2'-((4-methyl pyridine-2,6-diyl) bis (sulfanediyl)) dipyridine (MBSD), and 2,6-bis (N-pyrazolyl) 4-methyl pyridine (BNMP), and the conductivity values are listed in Table 13.1. The organic compounds (Figure 13.6) nitrogen, oxygen, and sulfur have a lone pair of electrons that can interact with iodine in redox electrolytes, which decreases the sublimation and enhances the conductivity of polymer (Kang et al. 2005; Wu et al. 2006; Jacob et al. 1997). The BNMI-doped polymer electrolyte reports high conductivity of 5.7×10^{-4} Scm^{-1}. The BNMI organic molecule has nitrogen and oxygen with resonance activity; they easily interact with iodine in polymer electrolytes, which increases the segmental motion and enhances the conductivity.

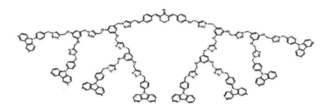

FIGURE 13.3 Structure of cyclohexadienone cored carbazole dendrimer.

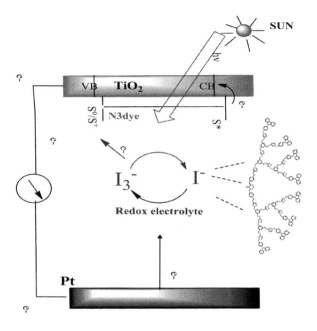

FIGURE 13.4 A schematic diagram of carbazole-dendrimer-doped additive for DSSC performance.

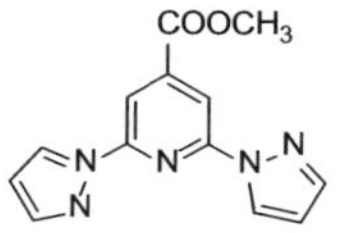

2,6, bis (N-pyrazolyl) methyl isonicotinate (BNMI)

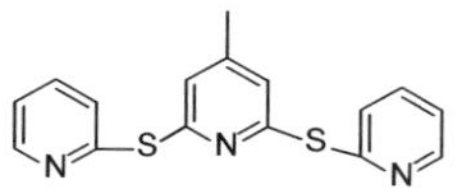

2 2'-((4-methyl pyridine-2,6-diyl) bis (sulfanediyl)) dipyridine (MBSD)

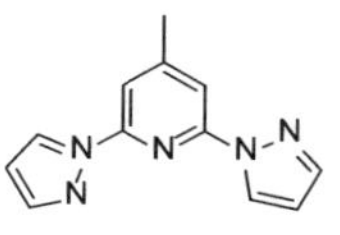

2,6-bis (N-pyrazolyl) 4-methyl pyridine (BNMP)

FIGURE 13.5 The structure of synthesized organic compounds as plasticizer for PEO-PVdF/potassium iodide/iodine.

Role of Organic Materials in Electrochemical Applications

TABLE 13.1
DSC and Conductivity Data of Polymer Electrolyte Systems

System	Melting Temperature (T_m) °C	Melting Enthalpy (ΔH_m) J/g	Degree of Crystallinity (χ_c)	Conductivity (σ) Scm^{-1}
PEO-PVdF/potassium × iodide/iodine	36	53	29	4.5×10^{-5}
PEO-PVdF/potassium × iodide/iodine/BNMI	−23	−24	13	5.7×10^{-4}
PEO-PVdF/potassium × iodide/iodine/MBSD	−19	−33	18	3.1×10^{-4}
PEO-PVdF/potassium × iodide/iodine/BNMP	−18	−38	21	2.6×10^{-4}

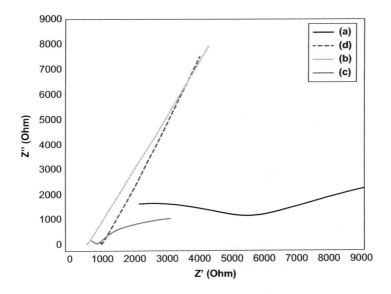

FIGURE 13.6 Complex impedance graph of (a) PEO-PVdF/potassium iodide/iodine, (b) PEO-PVdF/potassium iodide/iodine/BNMI, (c) PEO-PVdF/potassium iodide/iodine/MBSD, and (d) PEO-PVdF/potassium iodide/iodine/BNMP.

13.3.3 Differential Scanning Calorimetry (DSC) with Organic Molecules as Additives in Polymer Electrolytes

As seen in Figure 13.7a the melting temperature of 36°C and high enthalpy of 54 J/g resulted in the appearance of more crystallinity for the PEO-PVdF/potassium iodide/iodine system. However, the melting enthalpy and melting temperature were shifted by the incorporation of synthesized organic compounds, namely BNMI, MBSD, and BNMP to poly (ethylene oxide)/poly (vinylidene fluoride/potassium iodide/iodine, and these values are listed in Table 13.1. As seen in the Figure 13.7b–d, the melting temperatures observed for the synthesized organic compounds of BNMI-, MBSD-, and BNMP-doped polymer blends are −23°C, −19°C, and −18°C, respectively. The changes in melting temperature confirm the free segmental motion of polymers, which decreases the crystallinity of polymers and enhances the conductivity.

The degree of crystallinity (χ_c) in the polymer electrolyte was calculated by using the following relation:

$$\chi_c = \Delta H_m / \Delta H_m° \times 100$$

where $\Delta H_m°$ represents the pure melting enthalpy of PEO (~181.3).

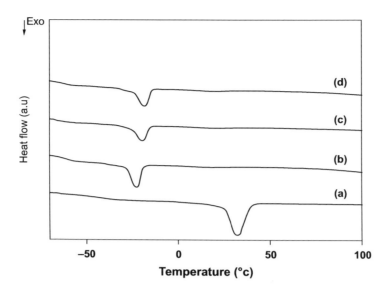

FIGURE 13.7 The differential scanning calorimetry curves of (a) PEO-PVdF/potassium iodide/iodine, (b) PEO-PVdF/potassium iodide/iodine/BNMI, (c) PEO-PVdF/potassium iodide/iodine/MBSD, and (d) PEO-PVdF/potassium iodide/iodine/BNMP.

13.3.4 X-Ray Diffraction (XRD) Studies

Figure 13.8 confirms the morphological studies of changes of the polymer electrolyte samples poly (ethylene oxide)/poly (vinylidene fluoride)/potassium iodide/iodine and incorporation of organic compounds (BNMI, MBSD, and BNMP). The figure shows the x-ray diffraction (XRD) peak is broader and less intense due to the incorporation of organic compounds with polymer electrolytes. The broadening of the XRD peak confirms the low crystallinity of the polymer, which increases the conductivity as well as the efficiency of DSSCs (Ganesan et al. 2011; Yang et al. 2008). This XRD again proves the important role of organic compounds with polymer electrolytes.

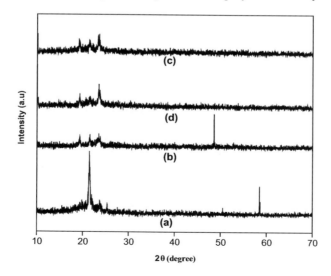

FIGURE 13.8 XRD patterns obtained for (a) PEO-PVdF/potassium iodide/iodine, (b) PEO-PVdF/potassium iodide/iodine/BNMI, (c) PEO-PVdF/potassium iodide/iodine/MBSD, and (d) PEO-PVdF/potassium iodide/iodine/BNMP.

Role of Organic Materials in Electrochemical Applications

13.3.5 Scanning Electron Microscope (SEM) Analysis of Organic-Compound-Doped Polymer Electrolytes

Figure 13.9 shows the scanning electron microscope (SEM) images at 30 μm of PEO-PVdF/potassium iodide/iodine, PEO-PVdF/potassium iodide/iodine/BNMI, PEO-PVdF/potassium iodide/iodine/MBSD, and PEO-PVdF/potassium iodide/iodine/BNMP. As seen in Figure 13.9a, there is very little space between the spherical grains, confirming that PEO-PVdF/potassium iodide/iodine possesses considerable crystallinity. The organic-compound-doped PEO-PVdF/potassium iodide/iodine system (Figure 13.9b–d) shows free space between the spherical grains and confirming more randomness in the polymer. It is well known that an increase in randomness decreases the degree of crystallinity. As seen in Figure 13.9b, the BNMI-doped polymer electrolyte shows more space between the spherical grains and it confirms that BNMI has more interaction with the redox couple due to its higher number of lone pairs of electrons from nitrogen and oxygen. This result also holds good with conductivity, DSC, and XRD results.

13.3.6 Photoelectrochemical Analysis

Figure 13.10 shows the current–voltage curve of the doped and undoped organic compound polymer electrolytes of PEO-PVdF/potassium iodide/iodine, PEO-PVdF/potassium iodide/iodine/BNMI, PEO-PVdF/potassium iodide/iodine/MBSD, and PEO-PVdF/potassium iodide/iodine/BNMP. The short-circuit current (J_{sc}), the open-circuit voltage (V_{oc}), fill factor, and conversion efficiency values for DSSCs employing the undoped and BNMI-, MBSD-, and BNMP-doped polymer electrolytes are listed in Table 13.2. The V_{oc} and J_{sc} values and efficiency were high with synthesized organic-compound-doped polymer blend electrolytes compared to undoped organic compound (PEO-PVdF/potassium iodide/iodine). The conversion efficiency of 9.5% and 8.9% under the illumination of 60 mW cm^{-2} was observed with the MBSD- and BNMP-doped polymer electrolyte, and the BNMI-added polymer electrolyte reported high efficiency of 9.9% in PEO-based DSSC field. Usually, organic nitrogenous compounds influence the redox couple to make it more stable and enhance the conversion efficiency of DSSC (Ganesan et al. 2008; Kusama et al. 2005; Kawano et al. 2004). The organic compounds MBSD, BNMP, and BNMI possess conjugated aromatic ring and a lone pair of electron from nitrogen, oxygen and sulfur, which interact with iodine in the redox electrolyte. The schematic electron transfer mechanism of organic nitrogenous compound-doped PEO-PV dF/potassium iodide/iodine polymer blend electrolyte is shown in Figure 13.11. Table 13.3 represents the results of research on particularly organic compounds as additives in redox electrolyte-based performance of dye-sensitized solar cells, and this result reveals that the synthesized organic compound-doped electrolyte is best in DSSCs. Table 13.4 reports the research on organic molecules as photosensitizer dye instead of Ru complex dye in dye-sensitized solar cells. These results confirm that the organic molecules work best in dye-sensitized solar cells.

TABLE 13.2
Current and Voltage Data of DSSCs Fabricated Using Synthesized Organic-Compound Doped PEO-PVdF/Potassium Iodide/Iodine under Illumination of 60 mW cm^{-2}

System	Voltage (mV)	Current (mA)	Fill Factor	Efficiency (η) %
TiO$_2$/N719dye/PEO-PVdF/potassium iodide/iodine/Pt	690	5.7	0.47	3
TiO$_2$/N719dye/PEO-PVdF/potassium iodide/iodine/BNMI/Pt	890	13.4	0.5	9.9
TiO$_2$/N719dye/PEO-PVdF/potassium iodide/iodine/MBSD/Pt	881	12.7	0.51	9.5
TiO$_2$/N719dye/PEO-PVdF/potassium iodide/iodine/BNMP/Pt	874	12	0.51	8.9

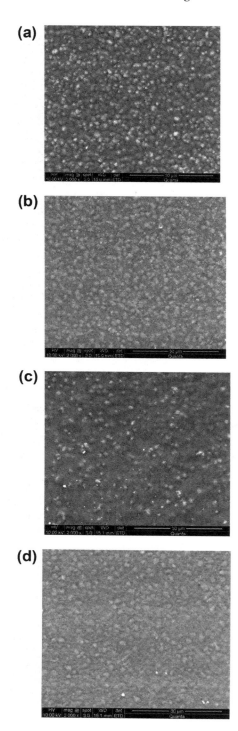

FIGURE 13.9 Scanning electron microscope (SEM) images of organic-compound-doped polymer electrolyte for (a) PEO-PVdF/potassium iodide/iodine, (b) PEO-PVdF/potassium iodide/iodine/BNMI, (c) PEO-PVdF/potassium iodide/iodine/MBSD, and (d) PEO-PVdF/potassium iodide/iodine/BNMP.

Role of Organic Materials in Electrochemical Applications 227

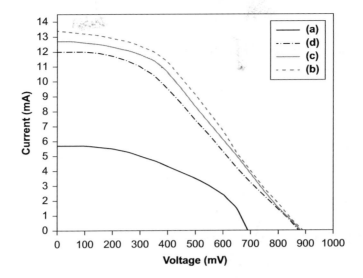

FIGURE 13.10 The photovoltaic (I-V) curves under 60 mWcm^{-2} illumination of DSSCs fabricated of (a) PEO-PVdF/potassium iodide/iodine, (b) PEO-PVdF/potassium iodide/iodine/BNMI, (c) PEO-PVdF/potassium iodide/iodine/MBSD, and (d) PEO-PVdF/potassium iodide/iodine/BNMP.

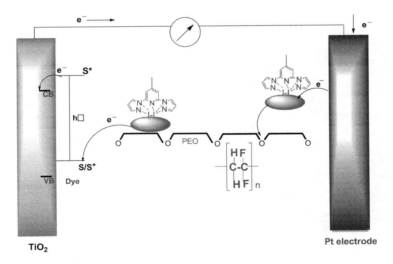

FIGURE 13.11 The electron transfer mechanism taking place in the DSSCs fabricated using organic compound doped as additive in PEO-PVdF/potassium iodide/iodine.

TABLE 13.3
Comparison of Efficiency of Organic Compound as Additive in Redox-Electrolyte-Based Dye-Sensitized Solar Cells

Organic Compound Used As Additive	Redox Couple	Dye Used in DSSC	Photo Anode	Photo Cathode	Efficiency Attained	Reference
(4,4'-((((oxybis(ethane-2,1-diyl))bis(oxy))bis(ethane-2,1diyl))bis(sulfanediyl))dipyridine))	I^-/I_3^-	N719	TiO_2	Platinum	9.2%	Karthika et al. 2018
1,1'-(3,3',5,5'-tetramethyl-[1,1'-biphenyl]-4,4'-diyl)bis(3-(3,4,5-trimethoxyphenyl)thiourea)	Co^{2+}/Co^{3+}	N719	TiO_2	Platinum	9.10%	Karthika et al. 2019
Cyclohexadienone cored 3,6-ditertiary butyl carbazole decorated triazolyl dendrimer	I^-/I_3^-	N3	TiO_2	Platinum	9%	Babu et al. 2018
Triazole-based phenothiazine dendrimer	I^-/I_3^-	N3	TiO_2	Platinum	8.6%	Rajakumar et al. 2013
Phenothiazine	I^-/I_3^-	N3	TiO_2	Platinum	8.5%	Ganesan et al. 2013

TABLE 13.4
Comparison of Efficiency of Organic Compound as Photosensitizer Dye in Dye-Sensitized Solar Cells

Organic Compound Used As Photosensitizer	Redox Couple	Photo Anode	Photo Cathode	Efficiency Attained	Reference
BODIPY dendrimer	I^-/I_3^-	TiO_2	Platinum	2.7%	Janakiraman Babu et al. 2018
9H-carbazol-9-yl group based BODIPY	I^-/I_3^-	TiO_2	Platinum	4.4%	Zhang et al. 2016
3,6-ditert-butyl-9H-carbazole	I^-/I_3^-	TiO_2	Platinum	2.66%	Barea et al. 2010
BODIPY with N-carbazole	I^-/I_3^-	TiO_2	Platinum	0.67%	Cheema et al. 2016
Benzothiadiazole- hexylthiophene	I^-/I_3^-	TiO_2	Platinum	1.96%	Yang et al. 2010

13.4 CONCLUSIONS

In the photoelectrochemical field, particularly dye-sensitized solar cells, the organic molecule plays a vital role. In dye-sensitized solar cells, the dendrimer-based BODIPY molecule reports good solar to electric energy conversion efficiency. The organic compounds were reported as the best additives, with high conversion efficiency with polymer electrolyte-based dye-sensitized solar cells. The influence of organic compounds, such as BNMI, MBSD, and BNMP as additives, with PEO-PVdF/potassium iodide/iodine–based polymer electrolyte systems were successfully analyzed. The organic compounds BNMI, MBSD, and BNMP have an aromatic ring and special atoms like N, S, and O, and have more lone pairs of electrons. These electrons interact well with iodine in the redox electrolyte to minimize the sublimation of iodine leading to enhanced stability and efficiency of DSSCs. XRD, SEM, DSC, and conductivity studies also proved the organic-doped polymer electrolytes decrease the degree of crystallinity of polymer sand improves the conductivity. High efficiencies for solar energy to electric energy conversion—9.9% for BNMI-, 9.5% for MBSD-, and 8.9% for BNMP-doped polymer electrolytes—were also reported.

ACKNOWLEDGMENTS

The authors are sincerely thankful to the Department of Science and Technology (DST) Fast Track, New Delhi, for financial support (SB/FT/CS-063/2013).

REFERENCES

Alebbi, M., Bignozzi, C. A., Heimer, T. A., Hasselmann, G. M., and Meyer, G. J. 1998. "The Limiting Role of Iodide Oxidation in cis-Os(dcb)$_2$(CN)$_2$/TiO$_2$Photoelectrochemical Cells". *J. Phys. Chem. B* 102: 7577–7581.

Babu, J., Ganesan, S., Karuppusamy, M., and Rajakumar, P. 2018. "Synthesis, photophysical, electrochemical properties, DFT studies and DSSC performance of BODIPY cored triazole bridged 3,6-ditertiary butyl carbazole decorated dendrimers". *Chemistry Select* no. 3: 9222–9231.

Babu, J., Ganesan, S., Karuppusamy, M., and Rajakumar, P. 2019. "Cyclohexadienone cored 3,6-ditertiary butyl carbazole decorated triazolyl bridged dendrimer – Synthesis and application in dye sensitized solar cells". *New J. Chem.* 43: 4036–4048.

Barea, E. M., Zafer, C., Gultekin, B., Aydin, B., Koyuncu, S., Icli, S., Santiago, F. F., and Bisquert, J. 2010. "Quantification of the effects of recombination and injection in the performance of dye-sensitized solar cells based on N-substituted carbazole dyes". *J. Phys. Chem. C* no. 114: 19840–198481.

Chapin, D. M., Fuller, C. S., and Pearson, G. L. 1954. "A new silicon p-n junction photocell for converting solar radiation into electrical power". *J. Appl. Phys.* no. 25: 676–677.

Cheema, H., Younts, R., Gautam, B., Gundogdu, K., and El-Shafei, A. 2016. "Design and synthesis of BODIPY sensitizers with long alkyl chains tethered to N-carbazole and their application for dye sensitized solar cells". *Mat. Phy. Chem.* no. 184: 57–63.

Degbia, M., Schmaltz, B., Boucle, J., Grazuleviciub, J. V., and Tran-Van, F. 2014. "Carbazole based hole transporting materials for solid state dye-sensitizer solar cells: role of the methoxy groups". *Polym. Int.* 63: 387–1393.

Donald, L. J., and Kenneth, A. G. 1990. "2,6-bis(N-pyrazolyl)pyridines: the convenient synthesis of a family of planar tridentate N3 ligands that are terpyridine analogs". *J. Org. Chem.* no. 55: 4992–4994.

Fujishima, A. and Honda, K. 1972. "Electrochemical photolysis of water at a semiconductor electrode". *Nature* no. 238: 37–38.

Ganesan, S., Karthika, P., Rajarathinam, R., Arthanareeswari, M., Mathew, V., and Maruthamuthu, P. 2016. "A poly (ethylene oxide), poly (vinylidene fluoride) and polycapro lactone polymer blend doped with an indigenous nitrogen–sulfur based organic compound as a novel electrolyte system for dye-sensitized solar cell applications". *Sol. Energy* no. 135: 84–91.

Ganesan, S., Mathew, V., Paul, B. J., Maruthamuthu, P., Suthanthiraraj, S. A. 2013. "Influence of organic nitrogenous compounds phenothiazine and diphenyl amine in poly(vinylidene fluoride) blended with poly(ethylene oxide) polymer electrolyte in dye-sensitized solar cells". *Electrochimica Acta* no. 102: 219–224.

Ganesan, S., Muthuraaman, B., Madhavan, J., Mathew, V., Maruthamuthu, P., and Suthanthiraraj, S. A. 2008. "The use of 2, 6-bis(N-pyrazolyl) pyridine as an efficient dopant in conjugation poly (ethylene oxide) for nanocrystalline dye-sensitized solar cells". *Electrochimica Acta* no. 53: 7903–7907.

Ganesan, S., Muthuraaman, B., Mathew, V., Vadivel, M. K., Maruthamuthu, P., Ashokkumar, M., and Suthanthiraraj, S. A. 2011. "Influence of 2, 6 (N-pyrazolyl) isonicotinic acid on the photovoltaic properties of a dye-sensitized solar cell fabricated using poly (vinylidene fluoride) blended with poly (ethylene oxide) polymer electrolyte". *Electrochimica Acta* no. 56: 8811–8817.

Gratzel, M. 2003. "Effects of ion doping on the optical properties of dye-sensitized solar cells". *J. Photochem. Photobiol. C Photochem. Rev.* 4: 145–153.

Harris, L. A. and Wilson, R. H. 1978. "Semiconductors for photoelectrolysis". *Ann. Rev. Mater. Sci.* no. 8: 99–134.

Hattori, S., Ohkubo, K., Urano, Y., Sunahara, H., Nagano, T., Wada, Y., Tkachenko, N. V., Lemmetyinen, H., and Fukuzumi, S. 2005. "Charge separation in a nonfluorescent donor–acceptor dyad derived from boron dipyrromethene dye, leading to photocurrent generation". *J. Phys. Chem. B* 109: 15368–15375.

Jacob, M. M. E., Prabaharan, S. R. S., and Radhakrishnan, S. 1997. "Effect of PEO addition on the electrolytic and thermal properties of PVDF-LiClO$_4$ polymer electrolytes". *Solid State Ionics* no. 104: 267–276.

Kang, M. S., Kim, J. H., Kim, Y. J., Won, J., Park, N. G., and Kang, Y. S. 2005. "Dye-sensitized solar cells based on composite solid polymer electrolytes". *Chem. Commun.* no. 2005: 889–891.

Karthika, P., Ganesan, S., and Arthanareeswari, M. 2018. "Low-cost synthesized organic compounds in solvent free quasi-solid state polyethyleneimine, polyethylene glycol based polymer electrolyte for dye-sensitized solar cells with high photovoltaic conversion efficiencies." *Sol. Energy* no. 160: 225–250.

Karthika, P., Ganesan, S., Thomas, A., Rani, T. M., and Prakash, M. 2019. "Influence of synthesized thiourea derivatives as a prolific additive with tris(1,10-phenanthroline)cobalt(II/III)bis/tris(hexafluorophosphate)/ hydroxypropyl cellulose gel polymer electrolytes on dye-sensitized solar cells." *Electrochimica Acta* no. 298: 237–247.

Katsaros, G., Stergiopoulos, T., Arabatzis, I. M., Papadokostak, K. G., and Falaras, P. 2002. "A solvent-free composite polymer/inorganic oxide electrolyte for high efficiency solid-state dye-sensitized solar cells". *J. Photochem. Photobiol. A: Chem.* no. 149: 191–198.

Kawano, R., Matsui, H., Matsuyama, C., Sato, A., Susan, M. A., Tanabe, N., and Watanabe, M. 2004. "High performance dye-sensitized solar cells using ionic liquids as their electrolytes". *J. Photochem. Photobiol. A: Chem.* no. 164: 87–92.

Kim, J. H., Kang, M. S., Kim, Y. J., Won, J., and Kang, Y. S. 2005. "Poly(butyl acrylate)/NaI/I$_2$ electrolytes for dye-sensitized nanocrystalline TiO$_2$ solar cells". *Solid State Ionics* no. 176: 579–584.

Krüger, J., Plass, R., Grätzel, M., and Matthieu, H. J. 2002. "Improvement of the photovoltaic performance of solid-state dye-sensitized device by silver complexation of the sensitizer cis-bis (4,4'-dicarboxy-2, 2' bipyridine) –bis (isothiocyanato) ruthenium (II)". *J. Appl. Phys. Lett.* no. 81: 367–370.

Kubo, W., Murakoshi, K., Kitamura, T., Yoshida, S., Haruki, M., Hanabusa, K., Shirai, H., Wada, Y., and Yanagida, S. 2001. "Quasi-solid-state dye-sensitized TiO$_2$ solar cells: effective charge transport in mesoporous space filled with gel electrolytes containing iodide and iodine". *J. Phys. Chem. B* no. 105: 12809–12815.

Kuciauskas, D., Freund, M. S., Gray, H. B., Winkler, J. R., and Lewis, N. S. 2001. "Electron transfer dynamics in nanocrystalline titanium dioxide solar cells sensitized with ruthenium or osmium polypyridyl complexes". *J. Phys. Chem. B* 105: 392–403.

Kusama, H., Kurashige, M., and Arakawa, H. 2005. "Influence of nitrogen-containing heterocyclic additives in I$^-$/I$_3^-$ redox electrolytic solution on the performance of Ru-dye-sensitized nanocrystalline TiO$_2$ solar cell". *J. Photochem. Photobiol. A: Chem.* no. 169: 169–176.

Kusama, H., Orita, H., and Sugihara, H. 2008. "TiO$_2$ band shift by nitrogen-containing heterocycles in dye-sensitized solar cells: a periodic density functional theory study". *Langmuir* no. 24: 4411–4419.

Lee, J. Y., Bhattacharya, B., Kim, D. W., and Park, J. K. 2008. "Poly(ethylene oxide)/poly(dimethylsiloxane) blend solid polymer electrolyte and its dye-sensitized solar cell applications". *J. Phys. Chem. C* no. 112: 12576–12582.

Liang, M., and Chen, J. 2013 "Arylamine organic dyes for dye-sensitized solar cells". *Chem. Soc. Rev.* 42: 3453–3488.

Liu, Y., Lee, J. Y., and Hong, L. 2004. "In situ preparation of poly(ethylene oxide)–SiO$_2$ composite polymer electrolytes". *J. Power Sources* no. 129: 303–311.

Maruthamuthu, P., and Ashokkumar, M. 1988. "Doping effects of transition metal ions on the photosensitization of WO3 particles". *Sol. Energy Mater. Sol. Cells* no. 17: 433–438.

Masamoto, M., Hiromitsu, H., Kikuo, K., Yoshimasa, Y., and Yoichi, Y. 1996. "A dye sensitized TiO$_2$ photoelectrochemical cell constructed with polymer solid electrolyte". *Solid State Ionics* no. 89: 263–267.

Matsumoto, H., Mtsuada, T., Tsuade, T., Hagiwara, R., Ito, Y., and Miyazaki, Y. 2001. "The application of room temperature molten salt with low viscosity to the electrolyte for dye-sensitized solar cell". *Chem. Lett.* 30: 26–27.

McFarland, E. W., and Tang, J. 2003. "A photovoltaic device structure based on internal electron emission". *Nature* no. 421: 616–618.

Memming, R. 1980. "Solar energy conversion by photoelectrochemical processes". *Electrochim. Acta* no. 25: 77–88.

Mishra, A., Fischer, M. K., and Bauerle, P. 2009. "Metal-free organic dyes for dye-sensitized solar cells: from structure: property relationships to design rules". *Angew. Chem. Int. Ed.* no. 48: 2474–2499.

Murai, S., Mikoshiba, S., Sumino, H., Kato, T., and Hayese, S. 2003. Quasi-solid dye sensitised solar cells filled with phase-separated chemically cross-linked ionic gels". *Chem. Commun.* 2003: 1534–1535.

Nazeeruddin, M. K., Kay, A., Rodicio, I., Humphry-Baker, R., Muller, E., Liska, P., Vlachppoulos, N., and Gratzel, M. 1993. "Conversion of light to electricity by cis-X2bis(2,2'-bipyridyl-4,4'-dicarboxylate) ruthenium(II) charge-transfer sensitizers (X = Cl−, Br−, I−, CN−, and SCN−) on nanocrystalline titanium dioxide electrodes". *J. Am. Chem. Soc.* no. 115: 6382–6390.

Nogueira, A. F., Durrant, J. R., and De Paoli, M. A. 2001. "Dye-sensitized nanocrystalline solar cells employing a polymer electrolyte". *Adv. Mater.* no. 13: 826–830.

Nogueira, A. F., Durrent, J. R., DePaoli, M. A., Boschloo, G. K., and Goossens, A. 2002. "Solid-state and flexible dye-sensitized TiO$_2$ solar cells: a study by electrochemical impedance spectroscopy". *J. Phys. Chem. B* no. 106: 5925–5930.

O'Regan, B., and Gratzel, M. 1991. "A low-cost, high-efficiency solar cell based on dye-sensitized colloidal TiO$_2$ films". *Nature* no. 353: 737–740.

O'Regan, B., and Schwartz, D. T. 1996. "Efficient dye-sensitized charge separation in a wide-band-gap *p-n* heterojunction". *J. Appl. Phys.* no. 80: 4749–4754.

Oskam, G., Bergeron, B. V., Meyer, G. J., and Searson, P. C. 2001. "Pseudohalogens for dye-sensitized TiO$_2$ photoelectrochemical cells". *J. Phys. Chem. B* 105: 6867–6873.

Papageorgiou, N., Athanassov, Y., Armand, M., Bonholte, P., Pettersson, H., Azam, A., and Gratzel, M. 1996. "The performance and stability of ambient temperature molten salts for solar cell applications". *J. Electrochem. Soc.* 143: 3099–3108.

Rajakumar, P., Satheeshkumar, C., Ravivarma, M., Ganesan, S., and Maruthamuthu, P. 2013. "Enhanced performance of dye-sensitized solar cell using triazole based phenothiazine dendrimers as additives". *J. Mater. Chem.* no. 1: 13941–13948.

Satoh, N., Nakashima, T., and Yamamoto, K. 2005. "Metal-assembling dendrimers with a triarylamine core and their application to a dye-sensitized solar cell". *J. Am. Chem. Soc.* 127: 13030–13038.

Scully, S. R., Lloyd, M. T., Herrera, R., Giannelis, E. P., and Malliaras, G. G. 2001. "Dye-sensitized solar cells employing a highly conductive and mechanically robustnanocomposite gel electrolyte". *Synth. Met.* no. 144: 291–296.

Smestad, G., Bignozzi, C., and Arazzi, R. 1994. "Testing of dye sensitized TiO$_2$ solar cells I: Experimental photocurrent output and conversion efficiencies". *Sol. Energy Mater. Sol. Cells* 32: 259–272.

Stathatos, E., Lianos, P., and Krontiras, C. 2001. "Dye-sensitized photoelectrochemical cell using a nanocomposite SiO$_2$/poly(ethylene glycol) thin film as electrolyte support. characterization by time-resolved luminescence and conductivity measurements". *J. Phys. Chem. B* no. 105: 3486–3492.

Tennakone, K., Kumara, G. R. R. A., Kottegoda, I. R. M., Wijayantha, K. G. U., and Perera, V. P. S. 1998. "A solid-state photovoltaic cell sensitized with a ruthenium bipyridyl complex". *J. Phys. D: Appl. Phys.* no. 31: 1492–1496.

Wang, G., Zhou, X., Lin, Y., Fang, S., and Xiao, X. 2004. "Gel polymer electrolytes based on polyacrylonitrile and a novel quaternary ammonium salt for dye-sensitized solar cells". *Mater. Res. Bull.* no. 39: 2113–2118.

Wang, P., Zakeeruddin, S. M., Comte, P., Exnar, I., and Gratzel, M. 2003. "Gelation of ionic liquid- based electrolytes with silica nanoparticles for quasi-solid-state dye-sensitized solar cells". *J. Am. Chem. Soc.* 125: 1166–1167.

Wang, P., Zakeeruddin, S. M., Exnar, I., and Gratzel, M. 2002. "High efficiency dye-sensitized nanocrystalline solar cells based on ionic liquid polymer gel electrolyte". *Chem. Commun.* 2002: 2972–2973.

Wang, P., Zakeeruddin, S. M., Moser, J. E., and Grätzel, M. 2003. "A new ionic liquid electrolyte enhances the conversion efficiency of dye-sensitized solar cells". *J. Phys. Chem. B* no. 107: 13280–13285.

Wu, J., Lan, Z., Wang, D., Hao, S., Lin, J., Huang, Y., Yin, S., and Sato, T. 2006. "Gel polymer electrolyte based on poly(acrylonitrile-co-styrene) and a novel organic iodide salt for quasi-solid state dye-sensitized solar cell". *Electrochimica Acta* no. 51: 4243–4249.

Wu, Y., and Zhu, W. 2013. "Organic sensitizers from D–π–A to D–A–π–A: effect of the internal electron-withdrawing units on molecular absorption, energy levels and photovoltaic performances". *Chem. Soc. Rev.* 42: 2039–2058.

Yang, Y., Zhang, J., Zhou, Y., Zhao, G., He, C., Li, Y., Andersson, M., Inganas, O., and Zhang, F. 2010. "Solution-processable organic molecule with triphenylamine core and two benzothiadiazole-thiophene arms for photovoltaic application". *J. Phys. Chem. C* no. 114: 3701–3706.

Yang, Y., Zhang, J., Zhou, C., Wu, S., Xu, S., Liu, W., Han, H., Chen, B., and Zhao, X.-Z. 2008. "Effect of lithium iodide addition on poly(ethylene oxide)–poly(vinylidene fluoride) polymer-blend electrolyte for dye-sensitized nanocrystalline solar cell". *J. Phys. Chem. B* no. 112: 6594–6602.

Yen, Y. S., Chou, H. H., Chen, Y. C., Hsu, C. Y., and Lin, J. T. 2012. "Recent developments in molecule-based organic materials for dye-sensitized solar cells". *J. Mater. Chem.* 2012: 8734–8747.

Zhang, J., Lu, F., Qi, S., Zhao, Y., Wang, K., Zhang, B., and Feng, Y. 2016. "Influence of various electron-donating triarylamine groups in BODIPY sensitizers on the performance of dye- sensitized solar cells". *Dyes Pigm.* 128: 296–303.

14 Advanced Conducting Polymers for Electrochemical Applications

R. Suresh, R. V. Mangalaraja, Paola Santander, and Jorge Yáñez

CONTENTS

14.1 Introduction 233
14.2 Examples of Conducting Polymers 234
    14.2.1 Polyaniline and Its Derivatives 234
    14.2.2 Polythiophene and Its Derivatives 235
    14.2.3 Polypyrrole 235
14.3 Electrochemical Applications of Conducting Polymers 235
    14.3.1 Solar Cells 235
    14.3.2 Batteries 238
    14.3.3 Fuel Cells 238
    14.3.4 Supercapacitors 240
    14.3.5 Electrochemical Sensors 240
14.4 Conclusions 245
Acknowledgments 245
References 245

14.1 INTRODUCTION

The polymer is a long-chain organic molecule that is made up of a number *n* of repeating units, usually known as "monomers." For example, polyethylene is composed of *n* ethylene monomers. Moreover, polymers such as polyvinyl chloride (PVC), polyethylene terephthalate (PET), polytetrafluroethylene (PTFE), and polymethylmethacrylate (PMMA) are widely used in our daily lives. Based on the electrical conductivity, polymers are classified as (a) nonconducting or insulating polymers, and (b) metallic or semiconducting or conducting polymers. As the name implies, nonconducting polymers don't conduct electricity. Polyethylene and polyvinyl chloride are examples of insulating polymers. On the other hand, conducting polymers (CPs) conduct electricity due to the presence of conjugated double bonds in their structure. Poly(ethylenedioxythiphene), polypyrrole, polyfuran, polyaniline, polyindole, polyacetylene, poly(thieno[3,2-b]thiophene), and poly(phenylene-vinylene) are some of the typical examples of widely used CPs (Figure 14.1). CPs have high electrical conductivity due to delocalization of π electrons through the polymer chain. They also exhibit magnetic, optical, mechanical, and redox properties. They have high chemical sensitivity, easy processability, are lightweight, and microwave shielding properties when compared to insulating polymers (Das and Prusty 2012).

CPs in nanoscale with different morphologies are of particular interest. Like inorganic nanoparticles, the nanostructures of CPs significantly differ from their bulk form (Ghosh et al. 2016). CP

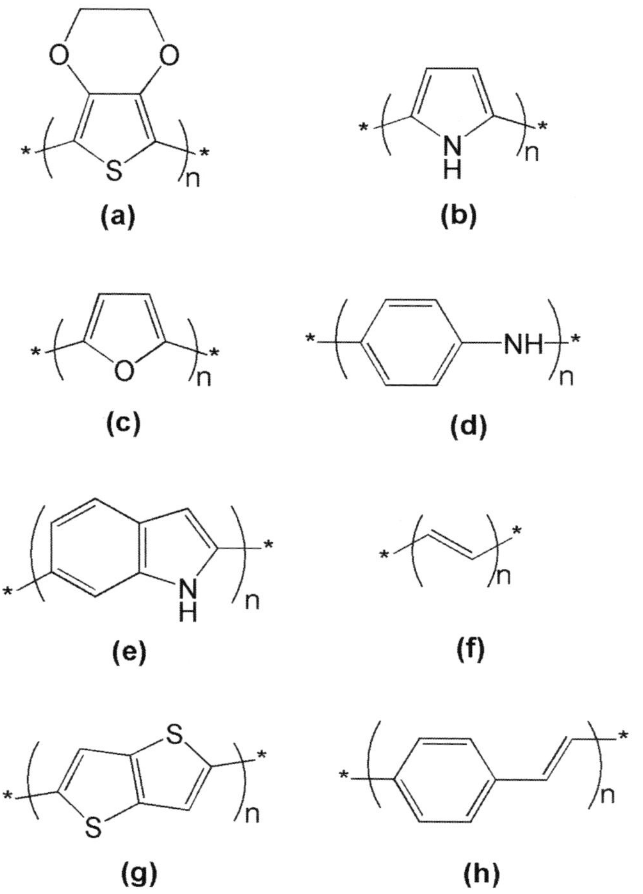

FIGURE 14.1 Structure (undoped form) of (a) poly(ethylenedioxythiphene), (b) poly(pyrrole), (c) poly(furan), (d) poly(aniline), (e) poly(indole), (f) poly(acetylene), (g) poly(thieno[3,2-b]thiophene) and (h) poly(phenylene-vinylene).

nanomaterials have served in catalysts; photocatalysts; drug delivery agents; electroheological fluids; EMI shielding; light emitting diodes; lithium sulfur batteries; electrodes in solar cells; chemical, gas, and biosensors; field-effect transistors; field emission devices; electrochromic display devices; tissue engineering; and supercapacitors (Das and Prusty 2012; Ghosh et al. 2016; Malinauskas et al. 2005). Among many applications, the vital role of CPs in electrochemical devices such as solar cells, batteries, fuel cells, supercapacitors, and electrochemical sensors are discussed in this book chapter.

14.2 EXAMPLES OF CONDUCTING POLYMERS

14.2.1 POLYANILINE AND ITS DERIVATIVES

In the 19th century, Henry Lethe discovered polyaniline (PANI) during his experiments on "oxidation of aniline" (Bhandari 2018) (Figure 14.2). Later, it gained much attention from researchers worldwide due to its low weight, easy synthesis, flexibility, low cost, and good stability. The optical absorption coefficient of PANI is also very high. Therefore, it can trap vast amounts of light from solar radiation. It can be easily doped with inorganic or organic acids, thereby tuning its electrical conductivity. However, it suffers in practical applications due to its poor processability.

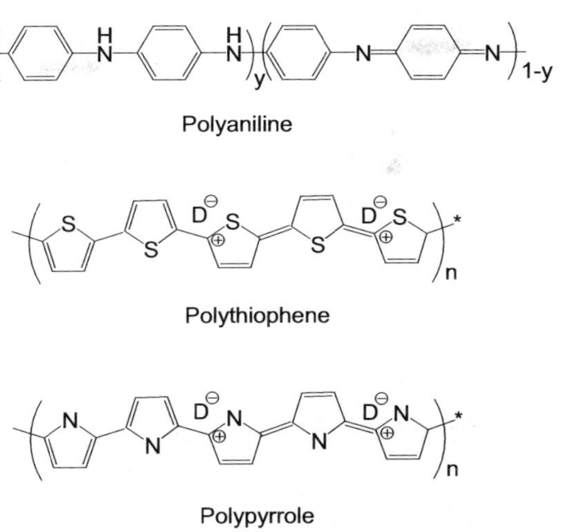

Advanced Conducting Polymers for Electrochemical Applications

FIGURE 14.2 Structure of conducting polymers in doped form. D⁻ represents anionic dopant.

To solve this issue, researchers introduced polar functional groups (-COOH, -OH, -OCH$_3$) in PANI, i.e., substituted PANI such poly(2-methoxyaniline) and poly(anthranilic acid) have been prepared.

14.2.2 Polythiophene and Its Derivatives

In the early 1980s, polythiophene (PT) (Jaymand et al. 2015) was first synthesized via oxidative polymerization of thiophene monomer. Nowadays, PT (Figure 14.2) and its derivatives are the most attractive conducting polymer nanomaterials in electrochemical devices. Like PANI, unsubstituted PT is also insoluble in many solvents, which made it poor in processability. This restriction can be invoked by functionalization in the side chain of PT. The numerous derivatives of thiophene have been synthesized by chemical and electrochemical methods, resulting in good processability.

14.2.3 Polypyrrole

In 1968, polypyrrole (PPY) was shown to be a conducting polymer (Vernitskaya and Efimov 1997). It is a p-type organic semiconductor (Figure 14.2) that has broad absorption from ultraviolet (UV) to the visible light region, easy preparation, superior electrochemical properties, electrical conductivity, water solubility, and flexibility. Hence, PPY is also widely studied for its various electrochemical applications.

14.3 ELECTROCHEMICAL APPLICATIONS OF CONDUCTING POLYMERS

14.3.1 Solar Cells

A human population explosion leads to increasing demand for energy. At the same time, the intensive use of fossil fuels for production of energy has seriously affected our biosphere. Hence, production of energy by sustainable green technologies is of current interest. In this regard, solar cells are promising to produce clean and sustainable energy. Moreover, solar cells can be used in portable electronic devices, home, garden, electric vehicles, etc.

A solar cell is a device that converts incident solar radiation into electricity by the photoelectric principle. A schematic diagram of a simple solar cell is shown in Figure 14.3 (Husain et al. 2018).

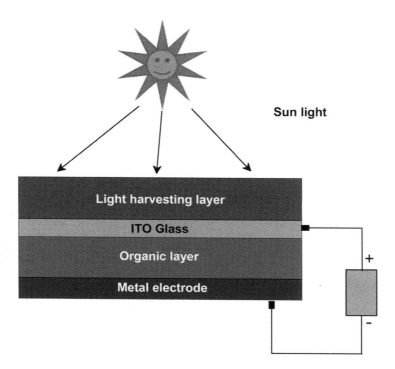

FIGURE 14.3 Schematic representation of a solar cell.

The main components of a solar cell are conductive glass substrate, an active layer, and a metal electrode. It is generally constructed in the following way:

(a) A polymer film is made in between the photo anode and cathode.
(b) A double-layer junction is developed between the donor and acceptor material. CPs and methano fullerene are used as donor and acceptor material, respectively.

The conversion of solar radiation into electricity occurs in the following steps: (a) The photoanode absorbs incident solar radiation, which excites the active layer from the ground state to the excited state, (b) excited electrons diffuse toward the interface between donor and acceptor material, (c) charge separation occurs through the interface, and (d) charge transport and charge collection take place at opposite electrodes.

Generally, the light harvesting capacity of a solar cell is highly dependent on the type of material used. Therefore, the choice of appropriate materials could significantly improve the efficiency of solar cells. In this regard, CPs are considered suitable light-absorbing materials owing to the following reasons.

(a) High light harvesting capacity—The CP has stable photoexcitation with UV-visible light, and thereby achieves high photon harvesting efficiency.
(b) They are lightweight materials (Awuzie 2017).
(c) They can be easily coated on flexible substrates. For example, polymer-based solar cells are constructed by simple screen/inkjet printing, coating, and doctor blading techniques (Krebs 2009).
(d) Relatively, monomers of CP are cheaper. Hence, they significantly reduce the cost of devices, especially in dye-sensitized solar cells (DSSCs). Usually in DSSCs, platinum (Pt) is used as a counter electrode, which is an extremely rare and costly metal. Therefore,

Advanced Conducting Polymers for Electrochemical Applications 237

replacement of Pt counter electrode with cheaper materials is highly desirable. Herein, CP nanostructures are also considered as one of the best alternatives for Pt counter electrodes (Wang et al. 2018).

(e) Improved photoconversion efficiency—The CP based solar cells exhibit greater photoconversion efficiencies than Pt electrode tailored DSSCs (Liu et al. 2014).

Though pure CP-based solar cells have several merits over the inorganic solar cells, there are some disadvantages such as low conversion efficiency and short lifetime. It was found that sunlight induces the photochemical reaction within a polymer chain, which causes chain scission (Mehmood et al. 2016). This reduces the charge transport property of conducting polymers, which results in a decrease of photoconversion efficiency. To eliminate these drawbacks, the following measures have been adopted.

(a) Structural modification/functionalization (Siddiqui et al. 2017)—The property of polymers can be altered through the functionalization of suitable functional groups. For instance, poly(1,5napthyridine-[3-hexylthiophene]), a functionalized thiophene polymer, has been synthesized and its application in DSSCs studied.

(b) Conducting-polymer-based nanocomposites—To achieve better performances, a number of binary nanocomposites such as CP-metal, CP-alloys, CP-metal oxide, CP-metal chalcogenide, CP-carbon materials, CP-C_3N_4, and CP-CP were prepared and their photoconversion efficiency determined. These composites showed improved light harvesting capacity and hole transporting properties than their pure polymers. Also, these nanocomposites exhibited greater or comparable photoconversion efficiency than Pt counter electrode tailored DSSCs (Figure 14.4). For example, PtM@PANI (M = Ni, FeNi) as counter electrodes were fabricated for DSSCs and their photoconversion efficiencies evaluated. It was found that PtNi@PANI and PtFeNi@PANI counter electrodes (Duan et al. 2018a) exhibited 8.52% and 8.39% photoconversion efficiency, which is greater than Pt electrodes. Recently, PPY/carbon black nanocomposites coated counter electrode was used in DSSCs. The PPY/carbon-coated electrode showed efficiency of 7.2%. The observed better efficiency of

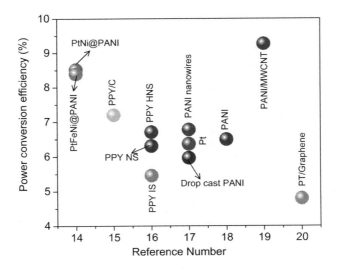

FIGURE 14.4 Comparison of photoconversion efficiency of conducting-polymer-based counter electrodes. C, carbon; HNS, hierarchical nanospheres; NS, nanospheres; IS, irregular sheets; MWCNT, multiwalled carbon nanotubes. (Data from Duan et al. 2018a,b, Peng et al. 2012, Chang et al. 2015, Wang et al. 2013a, Zhang et al. 2015a, Bora et al. 2015.)

PPY/carbon black composite is attributed to its good catalytic property and greater active surface area (Peng et al. 2012).

(c) Morphology—If the morphology of amorphous polymers changes, then their photocurrent efficiency will also be changed. For example, PPY irregular sheets, PPY hierarchical nanospheres, and PPY nanospheres were prepared by the chemical polymerization method and constructed as counter electrodes in DSSCs. The photoconversion efficiency was found to be of the following order (Chang et al. 2015): PPY hierarchical nanospheres > PPY nanospheres > PPY irregular sheets. From this result, it could be inferred that morphology has a key role in the performance of DSSCs.

(d) Processing in oxygen-free atmosphere—The photodegradation of polymer active layers can also be decreased by processing them in an oxygen-free atmosphere and at relatively low temperatures.

14.3.2 Batteries

Batteries are the most common power source for handheld devices to large-scale industrial applications. A battery consists of a cathode, anode, and electrolyte. They can be classified as two types. They are (a) primary battery or nonrechargeable battery, and (b) secondary or rechargeable battery (Landi et al. 2009). Among them, rechargeable batteries are used in many devices. However, conventional rechargeable batteries have some limitations like low capacity and short durability.

On the other hand, lithium-sulfur (Li-S) batteries are superior due to their relatively greater energy density, low weight, longer cycle life, and low environmental impact. However, to further lift their energy densities, further research is being performed worldwide. In addition, they have issues like the insulating nature of sulfur and safety concerns that need to be solved (Zhu et al. 2019). In this view, conductive polymers could eliminate the observed drawbacks in Li-S batteries. Because sulfur molecules interact with polymer chains through hydrogen bonding that can stabilize polysulfides, enhanced electrochemical performance of Li-S batteries is achieved. For example, PPY/S composite (Wang et al. 2006) provides a discharge capacity of 1280 mAh g^{-1} at a current density of 50 mA g^{-1}. These values were higher than that of the cell with the pure sulfur (1100 mAh g^{-1}). The observed improvement is attributed to (a) improved electrical conductivity of the PPY layer and (b) porous surface morphology of PPY that enhances the absorption of polysulfides.

Conductive polymers are also used as interlayers in Li-S batteries, since they possess rich heteroatoms that can provide effective binding energies for polysulfide trapping. For instance, PANI (Zhang et al. 2013a), PPY (Ma et al. 2015), and PT have been incorporated in sulfur cathodes to trap polysulfides. The main advantages of conductive polymers as interlayers are (a) they decrease internal resistance by providing additional electron pathways and (b) abundant heteroatoms act as strong binding sites for polysulfides. In order to improve performance of CP-tailored Li-S batteries, various binary and ternary nanocomposites were developed (Table 14.1).

14.3.3 Fuel Cells

A fuel cell is a device that directly converts chemical energy into electric energy without emission of greenhouse gases. They are classified (Wang et al. 2016) as direct methanol, proton exchange membrane, polymer electrolyte membrane, microbial, and alkaline fuel cells. A fuel cell contains a cathode, anode, electrolyte, and fuel, such as oxygen, hydrogen, or methanol. At the cathode, oxygen reduction occurs, whereas at the anode fuel oxidation takes place.

Generally, both the anode and cathode are made up of Pt/carbon material. As we know, Pt is expensive and carbon suffers from corrosive degradation. Moreover, the high cost and short life cycle also hinder the commercialization of Pt-based catalysts. Therefore it is necessary to find an alternative catalyst for electrode material with improved performance. Hence, chalcogenides, metals, metal oxides, polymer nanocomposites, carbon-based catalysts, and CP-based catalysts are used in fuel cells due to their good conductivity, high flexibility, and variable morphologies. For

TABLE 14.1
Comparison of CP-Based Binary and Ternary Composites Tailored Li-S Batteries

Material[a]	Performances Capacity (Number of Cycles, Current Density)	Reference
PT/S	830.2 mAh g^{-1} (80 cycles)	Wu et al. 2011
PPY/V$_2$O$_5$	271.8 mAh g^{-1} (1st cycle, 0.1 C)	Ren et al. 2012
S/PPY/graphene	641.5 mAh g^{-1} (40th cycle, 0.1 C)	Zhang et al. 2013b
PPY/TiO$_2$@S	1,150 mAh g^{-1} (100th cycle, 0.1 C)	Zhao et al. 2016
PEDOT:PSS	638 mAh g^{-1} (100 cycles, 0.2 C)	Li et al. 2015
C-PANI-S@PANI	835 mAh g^{-1} (100 cycles, 0.2 C)	Wang et al. 2013b
PANI nanowires/S	725 mAh g^{-1} (100 cycles)	Zhang et al. 2013a
PPY/S	823 mAh g^{-1} (10 cycles, 0.1 C)	Nakamura et al. 2015

[a] S, sulfur; PSS, poly(styrenesulfonate); PEDOT, poly(3,4-ethylenedioxythiophene); C, carbon.

example, PANI as support for metal catalysts has been reported for its several advantages, like high flexibility, high conductivity, controllable morphologies, and enhanced dispersibility, low cost, and high electrochemical activity. A PANI/MoS$_2$ hybrid also exhibited high hydrogen evolution reaction activity (Zhang et al. 2015b). In addition, iron phthalocyanine/polyaniline/carbon black as an efficient electrocatalyst supports the oxygen–reduction reaction (Yuan et al. 2011). Similarly, CoFe$_2$O$_4$/PANI-MWCT catalyst for oxygen–reduction reaction has also reported. PANI/MWNCT provides more active sites. Due to the synergistic effect of PANI, MWCNT, and CoFe$_2$O$_4$ nanoparticles, electrical conductivity and stability of the catalyst was significantly increased (Liu et al. 2016).

PANI-supported Pt catalysts were also reported in fuel cells. For instance, PANI/carbon black (CB)/Pt composites (Wu et al. 2005) have been reported as the best catalyst in fuel cells. This is because CB significantly improves the degree of polymerization, lowers the defect in PANI, provides more active sites, and quickens the charge transfer at the electrode/electrolyte interface. The performances of various CP-based materials in fuel cells are compared in Table 14.2.

TABLE 14.2
Comparison of CP-Based Material Tailored Fuel Cells

Material[a]	Power Density	Current Density	Reference
PANI/MoS$_2$	—	50 mAcm^{-2}	Zhang et al. 2015b
Phthalocyanine/PANI/CB	630.5 mW m^{-2}	0.60 mA cm^{-2}	Yuan et al. 2011
CoFe$_2$O$_4$/PANI/MWCNT	—	10 mA cm^{-2}	Liu et al. 2016
PANI/CB/Pt	—	16.2 mA cm^2	Wu et al. 2005
PPY/GF and PT/GF	1.22 and 0.8 W/m^2	4.1 and 1.9 A/m^2	Sumisha and Haribabu 2018
SS/PANI-W and SS/PPy-W	0.288 and 0.187 mW cm^{-2}	1.108 and 0.713 mA cm^{-2}	Sonawane et al. 2018
Ni-Co/SPAni	659.79 mWm^{-2}	1441.35 mA/m^2	Papiya et al. 2018
Ti$_4$O$_7$/PANI/Graphene	2073 mW/m^2	—	Li et al. 2019
PPY/N-CNT	—	400 mA/cm^2	Ozturk et al. 2018
PPY/Pt/CP	—	7.19 mA cm^{-2}	Carrillo et al. 2018

[a] CB, carbon black; MWCNT, multiwalled carbon nanotube; GF, graphitic felt; SS-W, stainless steel wool; SPAni, sulfonated polyaniline; N-CNT, nitrogen-doped carbon nanotubes; CP, carbon paste.

TABLE 14.3
Comparison of CP-Based Material Performance in Supercapacitors

Material[a]	Discharge Rate	Specific Capacitance (F g^{-1})	Cycling/Capacity Retention (Number of Cycles)	Reference
PANI	10 mV s^{-1}	503	85% (10000)	Hu and Chu 2001
PANI/Ni(OH)$_2$	—	55.5	79.5 (25000)	Zhang et al. 2015c
PANI/CNT	5 mV s^{-1}	440	93% (1000)	Otrokhov et al. 2014
PANI-MnO$_2$	5 A g^{-1}	715	96.5% (5000)	Prasad and Miura 2004
MoS$_2$/PANI/G	1 A g^{-1}	618	78% (2000)	Sha et al. 2016
PPY/CuO	1 A g^{-1}	553.7	90% (500)	Ma et al. 2013
PPY/MoS$_2$	2.5 A/cm^3	1275.5 F/cm^3	100% (3000)	Qian et al. 2015
PPY/cellulose	300 mA/cm^2	127	93% (5000)	Wang et al. 2015
Ni/PPY/MnO$_2$	2 A/g	350	91.3% (5000)	Chen et al. 2015
Fe^{3+}/PT	—	108.1	21.2% (1000)	Wu et al. 2018
PT and PFPT	2.5 mA cm^{-2}	260 and 110	—	Laforgue et al. 1999
PEDOT/WO$_3$	1.4 A g^{-1}	689	96% (260)	Zhuzhelskii et al. 2019

[a] CNT, carbon nanotube; G, graphene; PFPT, polyparafluorophenylthiophene; PEDOT, poly-3,4-ethylenedioxythiophene.

14.3.4 Supercapacitors

A supercapacitor (ultracapacitor) is used to store electrical energy. It contains electrodes and dielectric material to achieve high specific capacitances. This device has applications in electric vehicles, uninterruptible power supplies, etc. In order to achieve high capacitance, a high surface area electrode material is essential. In this regard, carbon, metal oxides and conducting polymers are used as electrode material (Meng et al. 2017). Among them, CPs showed high specific capacitance due to their redox behavior. For example, PANI stores charges via redox reaction by changing its various oxidations states.

However, due to swelling, shrinkage, and cracking of the polymer, PANI showed poor cycle stability. Furthermore, the degradation of PANI may also occur at high potentials due to overoxidation. These issues make it necessary to develop conducting-polymer-based composites. This strategy also allows composite materials with synergistic activity, enhanced structural stability, optimized porosity, and improved electric conductivity, which lead to extra charge storage via improved charge transportation and kinetic behavior. The performances of some CPs and their composites are given in Table 14.3.

14.3.5 Electrochemical Sensors

In medical diagnosis and pollution monitoring, several types of electrochemical sensors are used. Electrochemical sensors are devices that transform chemical information into electrical signals. Electrochemical sensors contain two components (Figure 14.5). They are the (a) receptor that transforms the chemical information into a form of energy and (b) transducer that transforms the energy, bearing chemical information, into an electrical signal (Lima et al. 2018).

Generally, a working electrode acts as a receptor in electrochemical sensing devices. Glassy carbon, carbon paste, nickel foam, and platinum electrodes are used as working electrodes. It has been found that these electrodes suffer in real sample detection due to their poor sensitivity, selectivity, interference, and stability. Hence, various kinds of nanomaterials are used to construct the best electrode for practical applications (Ahmad et al. 2017). Among them, conducting polymers such as PANI, PPY, and PEDOT have been widely utilized as electrochemical sensors for the detection of various molecules (Table 14.4) due to the following reasons.

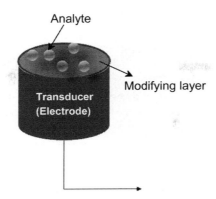

FIGURE 14.5 Pictorial representation of an electrochemical sensor.

TABLE 14.4
Analytes That Need to Be Determined

Analyte	Molecular Formula	Reason for Detection	Determination in
Ascorbic acid	$C_6H_8O_6$	Deficiency leads to scurvy	Fruit juices
Uric acid	$C_5H_4N_4O_3$	High level causes gout and kidney stones	Human urine, serum
Dopamine	$C_8H_{11}NO_2$	Parkinson's disease, nervous system problems	Blood
Cysteine	$C_3H_7NO_2S$	Deficiency causes disease	Human plasma
Nitrophenol	$C_6H_5NO_3$	Irritates eyes, skin, and respiratory tract	Soil and water
Hydrogen peroxide	H_2O_2	Extremely toxic to cellular life	*In vivo* and *in vitro*
Hydroquinone	$C_6H_6O_2$	Skin disorder	Tap water
Gallic acid	$C_7H_6O_5$	Has biological and pharmaceutical applications	Tea and plant extracts
Xanthine	$C_5H_4N_4O_2$	Xanthinuria, hyperuricemia, renal failure, etc.	Serum
Glucose	$C_6H_{12}O_6$	Low or high concentrations lead to health effects	Serum
Diclofenac	$C_{14}H_{11}Cl_2NO_2$	Pharmacological activity	Pharmaceutical preparations
Picric acid	$C_6H_3N_3O_7$	Causes anemia, cancer, cyanosis, and affects respiratory systems, liver malfunction	Soil and water
Hydroxylamine	H_3NO	Toxic to human, animals and plants.	Water
Folic acid	$C_{19}H_{19}N_7O_6$	Causes deficiency disease	Biological fluids
Formaldehyde	CH_2O	Carcinogenic	Air
Paraquat	$C_{12}H_{14}Cl_2N_2$	Toxic to human beings and animals	Ground and lake water
Aminobenzimidale	$C_7H_6N_2$	Pharmacological activity	Plant samples
Chlorpyrifos	$C_9H_{11}Cl_3NO_3PS$	Pharmacological activity	Tap water and Soil
Zinc ion	Zn^{2+}	Health effects	Water
Copper ion	Cu^{2+}	Health effects	Water
Ferrous ion	Fe^{2+}	Health effects	Tap water
Mercuric ion	Hg^{2+}	Acrodynia, Hunter-Russell syndrome, and Minamata disease	Tap water
Cadmium ion	Cd^{2+}	Chronic health effects	Tap water
Lead ion	Pd^{2+}	Chronic health effects	Tap water
Nitrate	NO_3^-	Health effects	Tap water

(a) Compatible with biological molecules—The conducting polymers modified themselves to bind biomolecules that need to be detected (Mulchandani and Wang 1996).

(b) Electrical conductivity and redox behavior—Conducting polymers are having greater electrical conductivity. For example, the conductivity of PANI, PT, a PPY are 1–100, 40–100, and 10–100 S cm^{-1}, respectively. Therefore, they are suitable to modify bare electrodes to attain high sensitivity. Also, they are redox active polymers that facilitate electron transfer from the analyte molecule into the electrode or vice versa. For example, PANI (Figure 14.6) can effectively mediate the reduction and oxidation of analytes (Ambrosi et al. 2008).

(c) Direct deposition of conducting polymer on the surface of the electrode—The success of a modified electrode depends on the uniform thickness of the modifying layer. Among different methods, electrochemical deposition is the best method. Due to their electroactive nature, conducting polymers can be easily deposited on bare electrodes by applying potential. It is also possible to control film thickness by this method (Gerard et al. 2002).

(d) The charge and nature of polymer chain—The doped conducting polymer chain has a positive charge and the dopant has a negative charge. Therefore, the polymer chain can attract and detect anionic molecules or ions more selectively, while the dopant attracts and detects cationic molecules. Also, aromatic groups interact with the aromatic moiety of analyte through π–π interactions. For example, PANI-doped with 5-sulfosalicylic acid–modified electrode (Chatterjee et al. 2013) was used to detect ammonia gas through coordination with sulfosalicylate molecule. Ammonia vapors absorb protons from emeraldine salt of PANI and form an emeraldine base as well as ammonium cations. Then, ammonium cations form coordination bonding with anionic dopant. The deprotonation of PANI nitrogen atoms results in a drop in electrical conductivity. This drop in conductivity is directly related to the concentration of ammonia vapors and thereby efficient ammonia detection was achieved.

(e) Nature of doping agent—The dopant in conducting polymers such as PANI has also played a significant role in their sensing behavior. PANI doped with ferrocene sulfonic acid (Fc-Fe^{3+}) as a sensor for H_2O_2 is the best example (Yang and Mu 2005). The Fe^{3+} ion in ferrocenesulfonic acid undergoes a redox reaction in a reversible manner. When H_2O_2 interacts with Fe^{3+}, a chemical reaction takes place first, which leads to the formation of an intermediate followed by the electrode reaction. The Fc-Fe^{2+} formed during the reaction process is immediately oxidized to Fc-Fe^{3+} at the electrode. It suggests that the Fc-Fe^{2+}/Fc-Fe^{3+} is involved in charge transfer, which enhances the rate of oxidation of H_2O_2 (Equations 14.1 to 14.3):

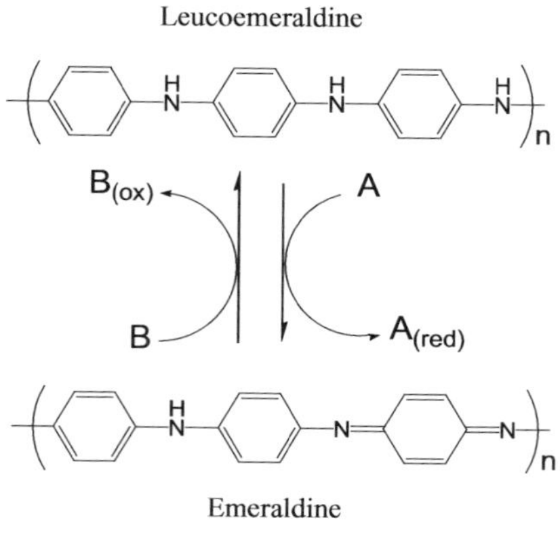

FIGURE 14.6 Electron transfer from analyte to PANI-modified electrode. A and B are analytes, A$_{(red)}$ is a reduced form of analyte A, and B$_{(ox)}$ is an oxidized form of analyte B.

Advanced Conducting Polymers for Electrochemical Applications

$$H_2O_2 + \underset{Fe^{3+}}{\text{Fc-SO}_3H} \longrightarrow \underset{Fe^{2+}}{\text{Fc-SO}_3H} + H\dot{O}_2 + H^+ \quad (14.1)$$

$$H\dot{O}_2 \longrightarrow O_2 + H^+ + e^{\ominus} \quad (14.2)$$

$$\underset{Fe^{2+}}{\text{Fc-SO}_3H} \longrightarrow \underset{Fe^{3+}}{\text{Fc-SO}_3H} + e^{\ominus} \quad (14.3)$$

Even though pure conducting polymers are widely used as the modifying layer for bare electrodes, they lack in sensitive and selective detection of many analytes. To improve upon this, the following methods were reported.

(a) Uses of derivatives of conducting polymer—Derivatives of PANI and PT have been used as sensor materials. For example, a PEDOT-modified glassy carbon electrode exhibits sensitive and selective detection of uric acid and ascorbic acid in milk with minimum interference (Motshakeri et al. 2018). The linear response range of this sensor is 6 to 100 µM for uric acid and 30 to 500 µM for ascorbic acid (Suresh et al. 2012).

(b) Fabrication of composites—It was found that a sensor composed of conducting polymers containing carbon, metal, semiconductors, or polymer (Figure 14.7) has enhanced sensitivity, high selectivity, low inference, and long-term stability. The synergistic effect, improved conductivity, and selective absorptivity are the major reasons for the observed improvement in nanocomposite-based sensors. For example, the electrochemical deposition of PEDOT/CNT film on a glassy carbon electrode showed high sensing ability toward dopamine, with a detection limit of 20 nM (Xu et al. 2013). This sensor is easy to

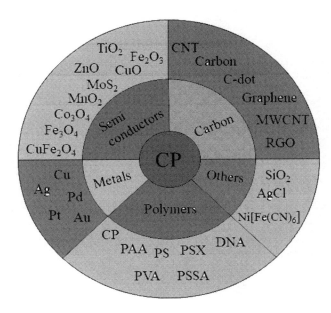

FIGURE 14.7 Conducting-polymer-based nanocomposites as electrochemical sensors. CP, conducting polymer; CNT, carbon nanotube; MWCNT, multiwalled carbon nanotube; RGO, reduced graphene oxide; PAA, poly(acrylic acid); PS, polysulfone; PSX, polysiloxane; DNA, deoxyribonucleic acid; PVA, polyvinyl alcohol; PSSA, poly(styrene sulfonic acid).

fabricate, very stable, and free of interference. The PANI/reduced graphene oxide nanocomposite modified glassy carbon electrode was also reported for selective detection of diclofenac (Ghanbari and Moludi 2016). The sensing performance of CP-based nanocomposite is given in Table 14.5.

(c) Immobilization of biomolecules—To construct sensitive biosensor, the enzyme is immobilized on conducting polymer modified working electrode. The reasons for choosing conducting polymers are (a) they are good electron transfer medium (Figure 14.8) and (b) they provide stable support for the enzyme through covalent linkages. For example, the PANI/PAA film has a carboxyl group that reacts with the amino group of glucose oxidase and forms amide linkages (Homma et al. 2014). Therefore PANI/PAA film acts as a stable support for glucose oxide through amide linkages resulting in good sensing behavior.

TABLE 14.5
Comparison of CP-Based Nanocomposites as Sensors

Sensor[a]	Analyte	Linear Range	Limit of Detection	Reference
PEDOT/CNT	Dopamine	0.1 to 20 μM	20 nM	Xu et al. 2013
ZnO/PANI/RGO	Dopamine	0.001–1 μM and 1–1000	0.8 nM	Ghanbari et al. 2016
AgCl@PANI	Dopamine	—	0.2 μM	Yan et al. 2008
Au@PANI	Dopamine	10–1700 μM	5 μM	Yang et al. 2012
Au/rGO/PANI	Serotonin	0.2–10.0 μmol L^{-1}	11.7 nmol L^{-1}	Xue et al. 2014
Au/SiO$_2$@PANI	Ascorbic acid	0.15 to 8 mM	3.775 μM	Weng et al. 2013
Cu/PANI	Ascorbic acid	0.005 to 3.5 mM	0.002 mM	Xi et al. 2010
P3PT/Pd, Pt	Dopamine	0.05–1 μM	9 nM	Atta and El-Kady 2010
PANI-PPY	Uric acid	2.5×10^{-6} to 8.5×10^{-5} M	1.0×0.10^{-6} M	Arslan 2008
NiHCF/PANI	H$_2$O$_2$	1×10^{-6} to 3×10^{-3} M	1.24×10^{-7} M	Wang et al. 2012
α-Fe$_2$O$_3$/PANI	Uric acid	0.01 to 5 μM	0.038 μM	Mahmoudian et al. 2019

[a] RGO, reduced graphene oxide; P3MT, poly(3-methylthiophene); NIHCF, nickel hexacynaoferrate.

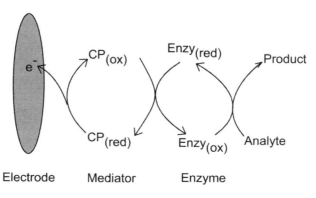

FIGURE 14.8 Electrochemical behavior of analyte at enzyme immobilized conducting polymer modified electrode.

14.4 CONCLUSIONS

In this book chapter, the significant role of conducting polymers in solar cells, batteries, fuel cells, supercapacitors, and electrochemical sensors was described. The following points were explored through an adequate literature survey.

(a) In solar cells, CP is used as electrode material. Photodegradation and low photoconversion efficiency were observed for pure CPs. Hence, strategies like functionalization, making nanocomposites, and controlled morphologies were reported. Especially CP-based nanocomposite exhibits superior performance in solar cells, even more than platinum-based counter electrodes.

(b) In Li-S batteries, CP was used as cathode material and as interlayer active material. Owing to the existence of heteroatoms, they effectively bind with polysulfides and enhance their efficiency of Li-S batteries. The battery performance was further improved by utilizing CP-based nanocomposites as electrode.

(c) In fuel cells, CP-based nanocomposites were effectively used as both cathode and anode materials. Because of their synergistic effect and fast charge transfer ability, nanocomposite-based electrodes act as effective catalysts for hydrogen evolution and oxygen reduction reactions.

(d) The role of CP-based nanocomposites is significant in supercapacitor devices. The fabrication of nanocomposites with suitable combinations gave high specific capacitance.

(e) Pure CP and their composites were vastly utilized as electrochemical sensors for clinically important biomolecules and pollutants. The major reasons for the application of CPs and improvements have been described.

ACKNOWLEDGMENTS

R. Suresh acknowledges the National Commission for Scientific and Technological Research (CONICYT), Santiago, Chile, for financial assistance in the form of a Post-Doctoral Fellowship (Fondecyt Project No: 3160499). The authors thank the support of the projects FONDECYT 1151296; and Center Optics and Photonics, Grant CONICYT-PFB-0824, CONICYT/FONDAP/15110019.

REFERENCES

Ahmad R, Tripathy N, Ahn MS, Bhat KS, Mahmoudi T, Wang Y, Yoo JY, Kwon DW, Yang HY, Hahn YB (2017) Highly efficient non-enzymatic glucose sensor based on CuO modified vertically-grown ZnO nanorods on electrode. *Sci. Rep.* 7:5715. doi:10.1038/s41598-017-06064-8

Ambrosi A, Morrin A, Smyth MR, Killard AJ (2008) The application of conducting polymer nanoparticle electrodes to the sensing of ascorbic acid. *Anal. Chim. Acta* 609:37–43. doi:10.1016/j.aca.2007.12.017

Arslan F (2008) An amperometric biosensor for uric acid determination prepared from uricase immobilized in polyaniline-polypyrrole film. *Sensors* 8:5492–5500. doi:10.3390/s8095492

Atta NF, El-Kady MF (2010) Novel poly(3-methylthiophene)/Pd, Pt nanoparticle sensor: synthesis, characterization and its application to the simultaneous analysis of dopamine and ascorbic acid in biological fluids. *Sens. Actuators B Chem.* 145:299–310. doi:10.1016/j.snb.2009.12.014

Awuzie CI (2017) Conducting polymers. *Mater. Today: Proc. Part E*, 4:5721–5726. doi:10.1016/j.matpr.2017.06.036

Bhandari S (2018) Polyaniline: structure and properties relationship, (Chapter 2), in Visakh PM, Della Pina C, Falletta E, eds., *Polyaniline Blends, Composites, and Nanocomposites*, pp. 23–60. Elsevier. doi:10.1016/B978-0-12-809551-5.00002-3

Bora C, Sarkar C, Mohan KJ, Dolui S (2015) Polythiophene/graphene composite as a highly efficient platinum-free counter electrode in dye-sensitized solar cells. *Electrochim. Acta* 157:225–231. doi:10.1016/j.electacta.2014.12.164

Carrillo I, Leo TJ, Santiago O, Accion F, Moreno-Gordaliza E, Raso MA (2018) Polypyrrole and platinum deposited onto carbon substrate to enhance direct methanol fuel cell electrodes behavior. *Int. J. Hydrogen Energ.* 43:16913–16921. doi:10.1016/j.ijhydene.2018.02.096

Chang LY, Li CT, Li YY, Lee CP, Yeh MH, Ho KC, Lin JJ (2015) Morphological influence of polypyrrole nanoparticles on the performance of dye–sensitized solar cells. *Electrochim. Acta* 155:263–271. doi:10.1016/j.electacta.2014.12.127

Chatterjee K, Dhara P, Ganguly S, Kargupta K, Banerjee D (2013) Morphology dependent ammonia sensing with 5-sulfosalicylic acid doped nanostructured polyaniline synthesized by several routes. *Sen. Actuators B Chem.* 181:544–550. doi:10.1016/j.snb.2013.02.042

Chen GF, Liu ZQ, Lin JM, Li N, Su YZ (2015) Hierarchical polypyrrole based composites for high performance asymmetric supercapacitors. *J. Power Sources* 283:484–493. doi:10.1016/j.jpowsour.2015.02.103

Das TK, Prusty S (2012) Review on conducting polymers and their applications. *Polym. Plast. Technol. Eng.* 51:1487–1500. doi:10.1080/03602559.2012.710697

Duan J, Duan Y, Zhao Y, Wang Y, Tang Q, He B (2018b) Bifunctional polyaniline electrode tailored hybridized solar cells for energy harvesting from sun and rain. *J. Energy Chem.* 27:742–747. doi:10.1016/j.jechem.2017.10.017

Duan J, Wang Y, Zhao Y, Yang X, Tang Q (2018a) Ternary hybrid PtM@polyaniline (M = Ni, FeNi) counter electrodes for dye-sensitized solar cells. *Electrochim. Acta* 291:114–123. doi:10.1016/j.electacta.2018.08.135

Gerard M, Chaubey A, Malhotra BD (2002) Application of conducting polymers to biosensors. *Biosens. Bioelectron.* 17:345–359. doi:10.1016/S0956-5663(01)00312-8

Ghanbari K, Moludi M (2016) Flower-like ZnO decorated polyaniline/reduced graphene oxide nanocomposites for simultaneous determination of dopamine and uric acid. *Anal. Biochem.* 512:91–102. doi:10.1016/j.ab.2016.08.014

Ghosh S, Maiyalagan T, Basu RN (2016) Nanostructured conducting polymers for energy applications: towards a sustainable platform. *Nanoscale* 8:6921–6947. doi:10.1039/C5NR08803H

Homma T, Sumita D, Kondo M, Kuwahara T, Shimomura M (2014) Amperometric glucose sensing with polyaniline/poly(acrylic acid) composite film bearing covalently-immobilized glucose oxidase: a novel method combining enzymatic glucose oxidation and cathodic O_2 reduction. *J. Electroanal. Chem.* 712:119–123. doi:10.1016/j.jelechem.2013.11.009

Hu C, Chu C (2001) Electrochemical impedance characterization of polyaniline coated graphite electrodes for electrochemical capacitors d effects of film coverage/thickness and anions. *J. Electroanal. Chem.* 503:105–116. doi:10.1016/S0022-0728(01)00385-0.

Husain AAF, Hasan WZW, Shafie S, Hamidon MN, Pandey SS (2018) A review of transparent solar photovoltaic technologies. *Renew. Sust. Energ. Rev.* 94:779–791. doi:10.1016/j.rser.2018.06.031

Jaymand M, Hatamzadeh M, Omidi Y (2015) Modification of polythiophene by the incorporation of processable polymeric chains: recent progress in synthesis and applications. *Prog. Polym. Sci.* 47:26–69. doi:10.1016/j.progpolymsci.2014.11.004

Krebs FC (2009) Fabrication and processing of polymer solar cells: a review of printing and coating techniques. *Sol. Energy Mater. Sol. Cells* 93:394–412. doi:10.1016/j.solmat.2008.10.004

Laforgue A, Simon P, Sarrazin C, Fauvarque JF (1999) Polythiophene-based supercapacitors. *J. Power Sources* 80:142–148. doi:10.1016/S0378-7753(98)00258-4

Landi BJ, Ganter MJ, Cress CD, DiLeo RA, Raffaelle RP (2009) Carbon nanotubes for lithium ion batteries. *Energy Environ. Sci.* 2:638–654. doi:10.1039/B904116H

Li Y, Yuan L, Li Z, Qi Y, Wu C, Liu J (2015) Improving the electrochemical performance of a lithium-sulfur battery with a conductive polymer-coated sulfur cathode. *RSC Adv.* 5:44160–44164. doi:10.1039/C5RA05481H

Li ZL, Yang SK, Song Y, Xu HY, Wang ZZ, Wang WK, Dang Z, Zhao YQ (2019) In-situ modified titanium suboxides with polyaniline/graphene as anode to enhance biovoltage production of microbial fuel cell. *Int. J. Hydrogen Energ.* 44:6862–6870. doi:10.1016/j.ijhydene.2018.12.106

Lima HRS, Silva JS, Farias EAO, Teixeira PRS, Eiras C, Nunes LCC (2018) Electrochemical sensors and biosensors for the analysis of antineoplastic drugs. *Biosens. Bioelectron.* 108:27–37. doi:10.1016/j.bios.2018.02.034

Liu Y, Li J, Li F, Li W, Yang H, Zhang X, Liu Y, Ma J (2016) A facile preparation of $CoFe_2O_4$ nanoparticles on polyaniline-functionalised carbon nanotubes as enhanced catalysts for the oxygen evolution reaction. *J. Mater. Chem. A* 4:4472–4478. doi:10.1039/C5TA10420C

Liu Y, Zhao J, Li Z, Mu C, Ma W, Hu H, Jiang K, Lin H, Ade H, Yan H (2014) Aggregation and morphology control enables multiple cases of high-efficiency polymer solar cells. *Nat. Commun.* 5:5293 (8 pages). doi:10.1038/ncomms6293

Ma G, Wen Z, Wang Q, Shen C, Peng P, Jin J, Wu X (2015) Enhanced performance of lithium sulfur battery with self-assembly polypyrrole nanotube film as the functional interlayer. *J. Power Sources* 273:511–516. doi:10.1016/j.jpowsour.2014.09.141

Ma GF, Peng H, Mu JJ, Huang HH, Zhou XZ, Lei ZQ (2013) In situ intercalative polymerization of pyrrole in graphene analogue of MoS$_2$ as advanced electrode material in supercapacitor. *J. Power Sources* 229:72–78. doi:10.1016/j.jpowsour.2012.11.088

Mahmoudian MR, Basirun WJ, Sookhakian M, Woi PM, Zalnezhad E, Hazarkhani H, Alias Y (2019) Synthesis and characterization of α-Fe$_2$O$_3$/polyaniline nanotube composite as electrochemical sensor for uric acid detection. *Adv. Powder Technol.* 30:384–392. doi:10.1016/j.apt.2018.11.015

Malinauskas A, Malinauskiene J, Ramanavicius A (2005) Conducting polymer-based nanostructurized materials: electrochemical aspects. *Nanotechnology* 16:R51–R62. doi:10.1088/0957-4484/16/10/R01

Mehmood U, Al-Ahmed A, Hussein IA (2016) Review on recent advances in polythiophene based photovoltaic devices. *Renew. Sust. Energ. Rev.* 57:550–561. doi:10.1016/j.rser.2015.12.177

Meng Q, Cai K, Chen Y, Chen L (2017) Research progress on conducting polymer based supercapacitor electrode materials. *Nano Energy* 36:268–285. doi:10.1016/j.nanoen.2017.04.040

Motshakeri M, Travas-Sejdic J, Phillips ARJ, Kilmartin PA (2018) Rapid electroanalysis of uric acid and ascorbic acid using a poly(3,4-ethylenedioxythiophene)-modified sensor with application to milk. *Electrochim. Acta* 265:184–193. doi:10.1016/j.electacta.2018.01.147

Mulchandani AK, Wang CL (1996) Bienzyme sensors based on poly(anilinomethylferrocene)-modified electrodes. *Electroanal* 8:414–419. doi:10.1002/elan.1140080503

Nakamura N, Yokoshima T, Nara H, Momma T, Osaka T (2015) Suppression of polysulfide dissolution by polypyrrole modification of sulfur-based cathodes in lithium secondary batteries. *J. Power Sources* 274:1263–1266. doi:10.1016/j.jpowsour.2014.10.192

Otrokhov G, Pankratov D, Shumakovich G, Khlupova M, Zeifman Y, Vasileva I, Morozova O, Yaropolov A (2014) Enzymatic synthesis of polyaniline/multi-walled carbon nanotube composite with core shell structure and its electrochemical characterization for supercapacitor application. *Electrochim. Acta* 123:151–157. doi:10.1016/j.electacta.2013.12.089

Ozturk A, Yurtcan AB (2018) Synthesis of polypyrrole (PPy) based porous N-doped carbon nanotubes (N-CNTs) as catalyst support for PEM fuel cells. *Int. J. Hydrogen Energ.* 43:18559–18571. doi:10.1016/j.ijhydene.2018.05.106

Papiya F, Pattanayak P, Kumar P, Kumar V, Kundu PP (2018) Development of highly efficient bimetallic nanocomposite cathode catalyst, composed of Ni:Co supported sulfonated polyaniline for application in microbial fuel cells. *Electrochim. Acta* 282:931–945. doi:10.1016/j.electacta.2018.07.024

Peng S, Tian L, Liang J, Mhaisalkar SG, Ramakrishna S (2012) Polypyrrole nanorod networks/carbon nanoparticles composite counter electrodes for high–efficiency dye–sensitized solar cells. *ACS Appl. Mater. Interfaces* 4:397–404. doi:10.1021/am201461c

Prasad KR, Miura N (2004) Polyaniline-MnO$_2$ composite electrode for high energy density electrochemical capacitor. *Electrochem. Solid State Lett* 7:A425. doi:10.1149/1.1805504.

Qian T, Zhou JQ, Xu N, Yang TZ, Shen XW, Liu XJ, Wu SS, Yan CL (2015) On-chip supercapacitors with ultrahigh volumetric performance based on electrochemically co-deposited CuO/polypyrrole nanosheet arrays. *Nanotechnology* 26:425402. doi:10.1088/0957-4484/26/42/425402

Ren X, Shi C, Zhang P, Jiang Y, Liu J, Zhang Q (2012) An investigation of V$_2$O$_5$/polypyrrole composite cathode materials for lithium-ion batteries synthesized by sol–gel. *Mater. Sci. Eng. B* 177:929–934. doi:10.1016/j.mseb.2012.04.013

Sha C, Lu B, Mao H, Cheng J, Pan X, Lu J, Ye Z (2016) 3D ternary nanocomposites of molybdenum disulfide/polyaniline/reduced graphene oxide aerogel for high performance supercapacitors. *Carbon* 99:26–34. doi:10.1016/j.carbon.2015.11.066

Siddiqui MN, Mansha M, Mehmood U, Ullah N, Al-Betar AF, Al-Saadi AA (2017) Synthesis and characterization of functionalized polythiophene for polymer-sensitized solar cell. *Dyes Pigments* 141:406–412. doi:10.1016/j.dyepig.2017.02.041

Sonawane JM, Patil SA, Ghosh PC, Adeloju SB (2018) Low-cost stainless-steel wool anodes modified with polyaniline and polypyrrole for high-performance microbial fuel cells. *J. Power Sources* 379:103–114. doi:10.1016/j.jpowsour.2018.01.001

Sumisha A, Haribabu K (2018) Modification of graphite felt using nano polypyrrole and polythiophene for microbial fuel cell applications-a comparative study. *Int. J. Hydrogen Energ.* 43:3308–3316. doi:10.1016/j.ijhydene.2017.12.175

Suresh R, Prabu R, Vijayaraj A, Giribabu K, Manigandan R, Stephen A, Narayanan V (2012) Poly(anthranilic acid) microspheres: synthesis, characterization and their electrocatalytic properties. *Bull. Korean Chem. Soc.* 33:1919–1924. doi:10.5012/bkcs.2012.33.6.1919

Vernitskaya TV, Efimov ON (1997) Polypyrrole: a conducting polymer; its synthesis, properties and applications. *Russ. Chem. Rev.* 66:443–457. doi:10.1070/RC1997v066n05ABEH000261

Wang G, Dong W, Yan C, Hou S, Zhang W (2018) Facile synthesis of hierarchical nanostructured polypyrrole and its application in the counter electrode of dye-sensitized solar cells. *Mater. Lett.* 214:158–161. doi:10.1016/j.matlet.2017.11.129

Wang H, Feng Q, Gong F, Li Y, Zhou G, Wang ZS (2013a) In situ growth of oriented polyaniline nanowires array for efficient cathode of Co(III)/Co(II) mediated dye sensitized solar cell. *J. Mater. Chem. A* 1:97–104. doi:10.1039/c2ta00705c

Wang H, Lin J, Shen ZX (2016) Polyaniline (PANi) based electrode materials for energy storage and conversion. *J. Sci.: Adv. Mater. Devices* 1:225–255. doi:10.1016/j.jsamd.2016.08.001

Wang J, Chen J, Konstantinov K, Zhao L, Ng SH, Wang GX, Guo ZP, Liu HK (2006) Sulphur-polypyrrole composite positive electrode materials for rechargeable lithium batteries. *Electrochim. Acta* 51:4634–4638. doi:10.1016/j.electacta.2005.12.046

Wang M, Wang W, Wang A, Yuan K, Miao L, Zhang X (2013b) A multi-core–shell structured composite cathode material with a conductive polymer network for Li–S batteries. *Chem. Commun.* 49:10263–10265. doi:10.1039/C3CC45412F

Wang Z, Sun S, Hao X, Ma X, Guan G, Zhang Z, Liu S (2012) A facile electrosynthesis method for the controllable preparation of electroactive nickel hexacyanoferrate/polyaniline hybrid films for H_2O_2 detection. *Sensor Actuat. B Chem.* 171–172:1073–1080. doi:10.1016/j.snb.2012.06.036

Wang ZH, Carlsson DO, Tammela P, Hua K, Zhang P, Nyholm L, Stromme M (2015) Surface modified nanocellulose fibers yield conducting polymer-based flexible supercapacitors with enhanced capacitances. *ACS Nano* 9:7563–7571. doi:10.1021/acsnano.5b02846

Weng CJ, Chen YL, Chien CM, Hsu SC, Jhuo YS, Yeh JM, Dai CF (2013) Preparation of gold decorated SiO_2@polyaniline core–shell microspheres and application as a sensor for ascorbic acid. *Electrochim. Acta* 95:162–169. doi:10.1016/j.electacta.2013.01.150

Wu F, Chen J, Chen R, Wu S, Li L, Chen S, Zhao T (2011) Sulfur/polythiophene with a core/shell structure: synthesis and electrochemical properties of the cathode for rechargeable lithium batteries. *J. Phys. Chem. C* 115:6057–6063. doi:10.1021/jp1114724

Wu G, Li L, Li JH, Xu BQ (2005) Polyaniline-carbon composite films as supports of Pt and PtRu particles for methanol electrooxidation, *Carbon* 43:2579–2587. doi:10.1016/j.carbon.2005.05.011

Wu K, Zhao J, Wu R, Ruan B, Liu H, Wu M (2018) The impact of Fe^{3+} doping on the flexible polythiophene electrodes for supercapacitors. *J. Electroanal. Chem.* 823:527–530. doi:10.1016/j.jelechem.2018.06.052

Xi L, Ren D, Luo J, Zhu Y (2010) Electrochemical analysis of ascorbic acid using copper nanoparticles/polyaniline modified glassy carbon electrode. *J. Electroanal. Chem.* 650:127–134. doi:10.1016/j.jelechem.2010.08.014

Xu G, Li B, Cui XT, Ling L, Luo (2013) X Electrodeposited conducting polymer PEDOT doped with pure carbonnanotubes for the detection of dopamine in the presence of ascorbic acid. *Sen. Actuators B Chem.* 188:405–410. doi:10.1016/j.snb.2013.07.038

Xue C, Wang X, Zhu W, Han Q, Zhu C, Hong J, Zhou X, Jiang H (2014) Electrochemical serotonin sensing interface based on double-layered membrane of reduced graphene oxide/polyaniline nanocomposites and molecularly imprinted polymers embedded with gold nanoparticles. *Sen. Actuators B Chem.* 196:57–63. doi:10.1016/j.snb.2014.01.100

Yan W, Feng X, Chen X, Li X, Zhu JJ (2008) A selective dopamine biosensor based on AgCl@polyaniline core–shell nanocomposites. *Bioelectrochemistry* 72:21–27. doi:10.1016/j.bioelechem.2007.07.003

Yang L, Liu S, Zhang Q, Li F (2012) Simultaneous electrochemical determination of dopamine and ascorbic acid using AuNPs@polyaniline core–shell nanocomposites modified electrode. *Talanta* 89:136–141. doi:10.1016/j.talanta.2011.12.002

Yang Y, Mu S (2005) Determination of hydrogen peroxide using amperometric sensor of polyaniline doped with ferrocenesulfonic acid. *Biosens. Bioelectron.* 21:74–78. doi:10.1016/j.bios.2004.08.049

Yuan Y, Ahmed J, Kim S (2011) Polyaniline/carbon black composite-supported iron phthalocyanine as an oxygen reduction catalyst for microbial fuel cells. *J. Power Sources* 196:1103–1106. doi:10.1016/j.jpowsour.2010.08.112

Zhang H, He B, Tang Q, Yu L (2015a) Bifacial dye-sensitized solar cells from covalent bonded polyaniline–multiwalled carbon nanotube complex counter electrodes. *J. Power Sources* 275:489–497. doi:10.1016/j.jpowsour.2014.11.046

Zhang JL, Shi L, Liu HD, Deng ZW, Huang LH, Mai WJ, Tan SZ, Cai X (2015b) Utilizing polyaniline to dominate the crystal phase of $Ni(OH)_2$ and its effect on the electrochemical property of polyaniline/$Ni(OH)_2$ composite. *J. Alloy Compd.* 651:126–134. doi:10.1016/j.jallcom.2015.08.090

Zhang K, Li J, Li Q, Fang J, Zhang Z, Lai Y (2013a) Improvement on electrochemical performance by electrodeposition of polyaniline nanowires at the top end of sulfur electrode. *Appl. Surf. Sci.* 285:900–906. doi:10.1016/j.apsusc.2013.09.010

Zhang N, Ma W, Wu T, Wang H, Han D, Niu L (2015c) Edge-rich MoS_2 naonosheets rooting into polyaniline nanofibers as effective catalyst for electrochemical hydrogen evolution. *Electrochim. Acta* 180:155–163. doi:10.1016/j.electacta.2015.08.108

Zhang Y, Zhao Y, Konarov A, Gosselink D, Soboleski HG, Chen P (2013b) A novel nano-sulfur/polypyrrole/graphene nanocomposite cathode with a dual-layered structure for lithium rechargeable batteries. *J. Power Sources* 241:517–521. doi:10.1016/j.jpowsour.2013.05.005

Zhao Y, Zhu W, Chen GZ, Cairns EJ (2016) Polypyrrole/TiO_2 nanotube arrays with coaxial heterogeneous structure as sulfur hosts for lithium sulfur batteries. *J. Power Sources* 327:447–456. doi:10.1016/j.jpowsour.2016.07.082

Zhu J, Zhu P, Yan C, Dong X, Zhang X (2019) Recent progress in polymer materials for advanced lithium-sulfur batteries. *Prog. Polym. Sci.* 90:118–163. doi:10.1016/j.progpolymsci.2018.12.002

Zhuzhelskii DV, Tolstopjatova EG, Eliseeva SN, Ivanov AV, Miao S, Kondratiev VV (2019) Electrochemical properties of PEDOT/WO_3 composite films for high performance supercapacitor application. *Electrochim. Acta* 299:182–190. doi:10.1016/j.electacta.2019.01.007

15 Electrochemical Studies for Biomedical Applications

Rajesh Parsanathan

CONTENTS

15.1 Introduction	251
15.2 General Working Principles of Electrochemical Sensors	252
15.3 Electrochemical Methods	252
15.3.1 Types of Voltammetry	252
15.4 Electrochemical Sensor and Circuit	254
15.4.1 Microelectrode	254
15.5 Biological Analytes and Electrochemical Methods	254
15.6 Gasotransmitter Electrochemical Sensors	255
15.6.1 Electrochemical Sensors for Carbon Monoxide (CO) Detection	255
15.6.2 Determination of Nitric Oxide (NO) Electrochemical Sensors	257
15.6.3 Hydrogen Sulfide (H_2S) Electrochemical Sensors	258
15.6.3.1 Ion-Sensitive Electrodes (ISEs) for Hydrogen Sulfide (H_2S)	258
15.6.3.2 Amperometric–Polarographic H_2S Sensors	259
15.7 Electrochemical Sensors for Nucleic Acids	259
15.7.1 Biomedical Applications of Electrochemical Sensors for Nucleic Acids	260
15.8 Redox Electrochemical Sensors	261
15.9 Electrochemical Immunoassays and Immunosensors	261
15.10 Paper-Based Electrochemical Sensors and Inkjet-Printed	261
15.11 Conclusions	262
Acknowledgments	262
References	262

15.1 INTRODUCTION

Analytical chemistry is the field where analytical techniques determine an analyte's concentration or characterize the chemical reactivity by measuring its potential, charge, or current (Siddiqui, AlOthman, and Rahman 2013). Electrochemistry is a branch of analytical chemistry that deals with the interaction of electrical and chemical effects; where the chemical changes are due to the passage of electrical current and the production of electrical energy by chemical reaction. Movement of electrons is an oxidation–reduction reaction, hence this field is named electrochemistry (Prabhulkar et al. 2012). Though there is much diversity in instrumentation, the majority of the electrochemical techniques and methods share numerous similar features (Kim et al. 2011). Because of its high sensitivity and precision, as well as a broad linear dynamic range with relatively economical instrumentation, these electrochemical methods (Figure 15.1) have become dominant and versatile analytical techniques in the field of biomedical applications (Nemiroski et al. 2014; Turner 2013). In traditional laboratories, this can be time-consuming and expensive. It is attractive to provide low-cost fabrication methods and point-of-care diagnostic tools with the capacity to detect and check various biological and chemical compounds. By combining well-established materials and fabrication methods, it is possible to produce electrochemical devices that meet the needs of many patients,

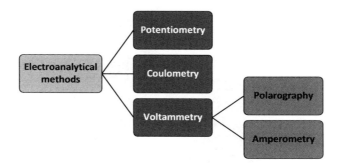

FIGURE 15.1 Schematic of classification of electroanalytical methods.

healthcare and medical professionals, and environmental specialists. Furthermore, continuing to innovate, by merging each of these favorable electrochemical sensing techniques and materials, is incredibly selective and provides accurate and repeatable quantitative results without expensive measurement equipment.

15.2 GENERAL WORKING PRINCIPLES OF ELECTROCHEMICAL SENSORS

The five critical interrelated and essential concepts of electrochemistry are as follows (Faulkner 1983):

(1) Analytes form at the electrode's surface determined by the electrode's potential.
(2) At a given time point, the concentration of analyte on the surface of the electrodes may not be equivalent to its concentration in bulk solution/biological matrix.
(3) An analyte may participate in any reaction apart from the oxidation–reduction reaction.
(4) The rate of current is a measure of the analyte's oxidation or reduction reaction at a given time point.
(5) Simultaneous control over both current and potential is not possible.

Hence, the analysis is based on electrochemical methods either within a medium or at the phase boundary and related to changes in the structure, chemical composition, or concentration of the compound being analyzed in any biological matrix qualitatively and quantitatively (Tripathi, Su, and Hwang 2018). Further, these methods are classified into five main groups: potentiometry, voltammetry, coulometry, conductometry, and dielectrometry.

15.3 ELECTROCHEMICAL METHODS

Working electrode types vary due to their variety of combinations, and hence, it is possible to change the parameters in the electrochemical experiments according to applications in the biomedical field. Electroanalytical techniques using different types of voltammetry and their parameters (advantages and limitations) are mentioned in Table 15.1.

15.3.1 Types of Voltammetry

(1) Sweep methods
  - Cyclic voltammetry
  - Linear sweep voltammetry
  - Rotating disk electrode

(2) Step and pulse methods
- Step voltammetry
    (a) Chronoamperometry
    (b) Chronocoulometry
    (c) Chronopotentiometry
- Pulse voltammetry
    (a) Normal pulse voltammetry
    (b) Differential pulse voltammetry
    (c) Square wave voltammetry

TABLE 15.1
Electroanalytical Techniques, Types of Voltammetry, and Parameters (Advantages and Limitations)

Electroanalytical Techniques	Parameters and Advantages	Limitations
Cyclic voltammetry (CV)	CV determines the nature of nernstian or non-nernstian behavior, and the rate/formation constants, formal potential, and diffusion coefficients.	Not suitable for quantitative analysis.
Linear sweep voltammetry (LSV)	Quantitative electrochemical analysis; the potential of the electrode varied at a constant rate during the reaction, and the resulting current is measured.	The bandpass limitations of the potentiostat and current measurer may affect the shape and characteristics of the LSV polarization curves.
Square wave voltammetry (SqWV)	SqWV has the advantages of a faster scan rate and increased sensitivity relative to differential pulse. Three current-potential plots are generated, namely reverse, forward, and difference current versus potential, thus resulting in higher sensitivity compared to CV technique.	Like the alternative current-voltammetry with sinusoidal alternating potential, it suffers from the fact that irreversible systems give low currents, especially at higher frequencies.
Chronoamperometry	Determination of diffusion coefficients and reaction kinetics investigation and its mechanisms.	Limited information about the identity of the electrolyzed species can be obtained from the ratio of the peak oxidation current versus the peak reduction current.
Chronopotentiometry	This technique determines higher concentration where a constant current is applied to the electrode, and the resulting potential change is plotted against time.	The measurement of the transition times negatively affected by the presence of adsorbed electroactive material or oxidation of the electrode surface. Hence, it is not commonly used for accurate measurements of diffusion coefficients.
Chronocoulometry	It is a modified version of chronoamperometry, which advances relatively more accurate measurement of the kinetic rate constant with detection of reactant adsorption on the surface of an electrode.	The signal increases over time instead of decreasing. There are better ways to do both of these with more modern techniques.

15.4 ELECTROCHEMICAL SENSOR AND CIRCUIT

The circuit that controls the potential of the working electrode and converts the signal current to a voltage is called a potentiostat (Fitzpatrick 2015). A simplified diagram of the sensor and potentiostat is illustrated in (Figure 15.2). The signal (current) from the working electrode (WE) is converted to a voltage by the operational amplifier U2 (Dobbelaere, Vereecken, and Detavernier 2017). This circuit also maintains the voltage of the working electrode at the bias potential, Vbias. The reference electrode (RE) potential is compared to the stable input voltage, Vbias (Sarkar et al. 2014). The op-amp U1 generates a voltage at the counter electrode (CE), which is sufficient to produce a current that is precisely equal and opposite to the working electrode current (Dryden and Wheeler 2015). At the same time, a constant voltage is maintained between the reference electrode and the working electrode.

15.4.1 MICROELECTRODE

The aforementioned technology can be modified to give sensors that will respond to the different gases of interest. This can be accomplished by using different working electrodes, bias potentials, and chemically selective filters. Table 15.2 shows which catalysts are used for each gas (Cretescu, Lutic, and Manea 2017).

15.5 BIOLOGICAL ANALYTES AND ELECTROCHEMICAL METHODS

The foremost important application of electrochemical detection in the biomedical field was for quantitative detection at a deficient concentration in real time in biological media (serum, plasma, blood, tissues, cell culture media, which includes all biological materials) without loss of sensitivity.

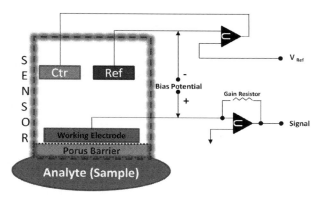

FIGURE 15.2 Schematic of electrochemical sensor and circuit.

TABLE 15.2
Biologically Important Gases and Frequently Used Electrocatalyst Metals

Gas Analyzed	Electrocatalyst Metals
CO	Platinum (Pt)
H_2S	Gold (Au)
H_2	Au
NO	Au
NO_2	Au
O_2	Au, Pt, Silver (Ag)
SO_2	Au

Electrochemical Studies for Biomedical Applications

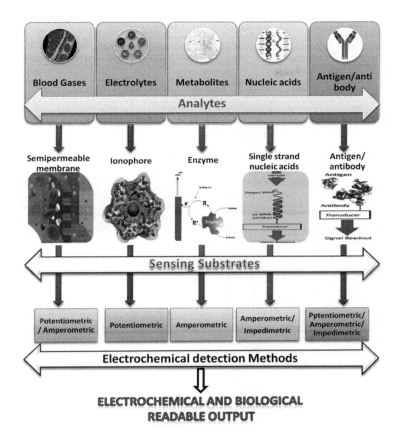

FIGURE 15.3 Electrochemical methods for determining biological analytes.

Electrochemical biosensing is one of the compelling methods to quantify concentrations of biological molecules (glucose, gasotransmitters [carbon monoxide, nitric oxide, hydrogen sulfide, carbon dioxide, and sulfur dioxide], ascorbic acid, uric acid, dopamine, hydrogen peroxides, nucleic acid, and proteins) in different matrices (biocompatibility) with the superior advantages of sensitivity, selectivity, low-detection limit, stability, long lifetime, wide linear range, fast response time, real-time, and reproducibility with simple operation (Krishnan et al. 2019). With these advantages, several types of research developed new electrochemical methods for determining biological molecules (Figure 15.3). Henceforth, it can provide vital information in a variety of applications related to chemistry, biology, and medicine. A few of these applications will be described in the following sections.

15.6 GASOTRANSMITTER ELECTROCHEMICAL SENSORS

Gases such as CO_2, NO_2, SO_2, O_2, CO, and H_2S are biologically important since those gases have a role in the pathophysiology (Chen et al. 2007; Mathew, Schlipalius, and Ebert 2011). Electrochemical gas sensors are a kind of gas detector that measure the levels of a target gas by either oxidizing or reducing the gas at an electrode and measuring the resulting current (Liu et al. 2012).

15.6.1 Electrochemical Sensors for Carbon Monoxide (CO) Detection

Carbon monoxide (CO) has high affinity to hemoglobin, which impedes oxygen delivery leading to intoxication and death (Blumenthal 2001). CO has many effects in biology due to its complex biochemical activities (Blumenthal 2001; Queiroga, Almeida, and Vieira 2012). Whether CO is derived from exogenous or endogenous sources, its cellular activity is related to its concentration and the concentration of molecular O_2, as well as to the availability of reduced transition metals such as Fe(II) (Queiroga, Almeida, and Vieira 2012). Finally, CO is involved in several cellular processes, acting as an anti-inflammatory, cytoprotective, maintenance of tissue homeostasis, and, in some particular cases, antiproliferative and vasodilator (Boczkowski, Poderoso, and Motterlini 2006; Motterlini and Otterbein 2010; Queiroga et al. 2011; Queiroga, Vercelli, and Vieira 2015; Queiroga, Almeida, and Vieira 2012).

Compared to other gasotransmitters, such as nitric oxide (NO) and hydrogen sulfide (H_2S), CO is a quite inert molecule. CO needs to be activated by coordination with low-valent metals or ions to react chemically. In biological systems, Fe^{2+} of reduced heme proteins are the main CO targets (Boczkowski, Poderoso, and Motterlini 2006). The best-described candidates are hemoglobin (erythrocytes), myoglobin (myocytes), and cytochrome c oxidase (mitochondrial complex IV) (Almeida, Figueiredo-Pereira, and Vieira 2015).

Cytochrome c oxygenase (COX), the final electron acceptor of the mitochondrial respiratory chain, was found to be the main mitochondrial target for CO at cytochrome a and a3 (Brown and Piantadosi 1990; Chance, Erecinska, and Wagner 1970). COX is involved in CO-induced cytotoxicity due to the CO capacity of inhibiting its activity and cell respiration (Chance, Erecinska, and Wagner 1970). The binding of CO to COX is highly dependent on oxygen levels because, under hyperbaric oxygen conditions, there is the dissociation of a cytochrome and a3–CO complex (Brown and Piantadosi 1990).

The carbon monoxide sensor consists of three electrodes immersed in a liquid electrolyte (a nonmetallic liquid that conducts electricity, usually through acids or dissolved salts) (Bai and Shi 2007). The three electrodes are the working electrode, the reference electrode, and the counter electrode (Guy and Walker 2016). The most important of these is the working electrode (WE) (Guy and Walker 2016). The working electrode is made of platinum, which is a catalytic metal to CO (it catalyzes the oxidation of CO to CO_2), backed by a gas-permeable but hydrophobic (waterproof) membrane (Cheng, Liu, and Logan 2006). The CO gas diffuses through the porous membrane and is electrochemically oxidized (see Equation 15.1) (Scott 1995).

Figaro electrochemical-type gas sensors are amperometric fuel cells with two electrodes. The essential components of two electrode gas sensors are a working (sensing) electrode, a counter electrode, and an ion conductor in between them. When toxic gas such as carbon monoxide (CO) comes in contact with the working electrode, oxidation of CO gas will occur on the working electrode through a chemical reaction with water molecules in the air:

$$CO + H_2O \rightarrow CO_2 + 2H^+ + 2e^- \tag{15.1}$$

Connecting the working electrode and the counter electrode through a short circuit will allow protons (H^+) generated on the working electrode to flow toward the counter electrode through the ion conductor. Also, generated electrons move to the counter electrode through the external wiring. A reaction with oxygen in the air will occur on the counter electrode:

$$(1/2)\,O_2 + 2H^+ + 2e^- \rightarrow H_2O \tag{15.2}$$

The overall reaction is shown in Equation 15.3. Figaro electrochemical-type gas sensors operate like a battery with gas being the active material for this overall battery reaction.

$$CO + (1/2)\,O_2 \rightarrow CO_2 \tag{15.3}$$

By measuring the current between the working electrode and the counter electrode, this electrochemical cell can be utilized as a gas sensor.

Electrochemical Studies for Biomedical Applications 257

The electrons involved in the electrochemical reaction flows from the working electrode through the external circuit, producing the output signal of the sensor (Grieshaber et al. 2008). In order for the reaction to take place, the thermodynamic potential of the working electrode is of critical importance (Duret-Thual 2014). The reference electrode provides a stable electrochemical potential in the electrolyte (Frank and James 1994). The reference electrode is protected from exposure to the CO gas so that its thermodynamic potential is always the same and remains constant (Frank and James 1994). Also, no current is allowed to flow through the reference electrode (this would change the thermodynamic potential) (Elgrishi et al. 2018). A counter electrode is provided to complete the circuit of the electrochemical cell (Elgrishi et al. 2018). The counter electrode functions solely as the second half-cell and allows electrons to enter or leave the electrolyte (Elgrishi et al. 2018).

The CO sensor is also equipped with a chemically selective filter. This filter removes potentially interfering gases before they reach the working electrode. With the filter operating correctly, the sensor will have minimal response to interfering gases (Helm, Jalukse, and Leito 2010).

15.6.2 Determination of Nitric Oxide (NO) Electrochemical Sensors

In 1987, nitric oxide (NO) was identified as being responsible for the physiological action of endothelium-derived relaxing factor (Ignarro et al. 1987). NO is now known to play a critical functional role in a variety of physiological systems. Within the vasculature, NO induces vasodilation, inhibits platelet aggregation, prevents neutrophil/platelet adhesion to endothelial cells, inhibits smooth muscle cell proliferation and migration, regulates programmed cell death (apoptosis), and maintains endothelial cell barrier function (Rosselli, Keller, and Dubey 1998; Majed and Khalil 2012).

Since NO plays a significant role in a variety of biological processes, where its physiological and pathological spatial and temporal concentrations are of extreme importance, measurement of NO is difficult due to its shorter half-life; further, the complexity arises by its high reactivity with superoxide, oxygen, thiols, and other biological components (Zielonka and Kalyanaraman 2010; Griendling et al. 2016). From the classical Griess methods, several techniques (chemiluminescence, paramagnetic resonance spectrometry, paramagnetic resonance imaging, spectrophotometry, and bioassay) have been developed for the measurement of NO from different sources including biological samples (Csonka et al. 2015; Samouilov et al. 2007). The sensitivity of the electrochemical (amperometric) detection of NO is the real-time sensitive method up to date. With the recent development of micro- and nanoelectrodes, the measurement of NO becomes more effortless in the biological systems with detection limits ranging from below 1 nM up to 100 µM (Zhang 2004).

The principles of determination of NO by electrochemical sensors (ISO-NOP) are almost the same: the sensor is immersed in a solution containing NO and positive potential of −800mV vs. Ag/AgCl reference electrode (Serpe and Zhang 2007). This first electrochemical NO sensor was based on a classical Clark electrode design (Zhang 2008). NO diffuses across the gas permeable (NO$^-$ selective membrane) and oxidizes at the surface of the working electrode producing a redox current (Zhang 2008). The NO cation formed due to the transfer of one electron to the electrode is as follows:

$$NO^- + e^- \rightarrow NO^+ \tag{15.4}$$

Further, NO cation is a relatively strong Lewis acid and in the presence of OH$^-$ irreversibly converts into nitrite:

$$NO^+ + OH^- \rightarrow HNO_2 \rightarrow H^+ + NO^{2-} \tag{15.5}$$

Nitrite can then be further oxidized into nitrate. The amount of NO oxidized is thus proportional to the current flow between the working and the reference electrodes, which is measured by a NO meter (Sarkar, Lai, and Lemay 2016). A typical redox current is relatively low for NO oxidation

(1–10 pA) (Zhang 2008). Hence, a sensitive amperometric-based electrode for NO and ultra-low noise amplification circuitry is needed; it was overcome by electronic devices ISO-NO and ISO-NO Mark II NO meters (Zhang 2008). These meters can measure as low as 0.1 pA.

Various other types of carbon fiber NO sensors, which utilize a variety of different coatings, have been described (Huffman and Venton 2009; Ozel, Wallace, and Andreescu 2011), including recently a coating of a 10 μM Pt disc electrode with a cross-linked sol–gel Langmuir-Blodgett film of siloxane polymer to render the electrode permselective for NO. Another report shows the addition of sol–gel films to a Pt electrode for NO detection (Matharu et al. 2007).

15.6.3 Hydrogen Sulfide (H_2S) Electrochemical Sensors

H_2S, which has evolved from a noxious gas to a physiological gasotransmitter, is deemed as the third endogenous gaseous mediator exhibiting properties similar to nitric oxide and carbon monoxide (Farrugia and Szurszewski 2014). However, much like CO and NO, the initial negative perception of H_2S has developed with the discovery that H_2S is generated enzymatically in animals under normal conditions (Xiao et al. 2018). A multitude of evidence supports that H_2S is widely involved in the roles of the physiological and pathological process, including hypertension, atherosclerosis, angiogenesis, myocardial infarction, Alzheimer's disease, Parkinson's disease, inflammation, and metabolic abnormalities (Wen, Wang, and Zhu 2018). In recent years, its cytoprotective properties have been recognized in multiple organs and tissues (Xiao et al. 2018). In particular, H_2S plays essential roles in combating oxidative species such as reactive oxygen species (ROS) and reactive nitrogen species (RNS), and protecting the body from oxidative stress (Di Meo et al. 2016; Shefa et al. 2018; Birben et al. 2012).

Due to the extensive attention in its biological function, *in vitro* and *in vivo* H_2S detection methods have evolved from simple colorimetric assays to the more recently used techniques, including fluorescence spectroscopy electrochemiluminescence (ECL), high-performance liquid chromatography, gas chromatography, electrochemistry gas sensors such as ion sensitive electrodes, and polarographic H_2S sensors (Richter 2004). Investigation of these processes requires a reliable and effective analytical system of low detection limit, fast response, and the possibility for *in situ* measurements (Everton et al. 2016; Shrivastava and Gupta 2011).

Similar to NO, under physiological conditions, the stability of H_2S and HS^- is influenced by inorganic and organic components that catalyze oxidation reactions, and this reactivity limits the practicality of most standard available analytical methods for real-time detection of H_2S (Olson 2012; Hartle and Pluth 2016). Though several analytical procedures are highly sensitive to detect micromolar to submicromolar levels, due to multiple steps and procedures, it is unclear if these methods also measure H_2S that existed as a persulfide or other labile species (Ono et al. 2014).

In the chemical industry, commonly employed H_2S quantitation methods are solid-electrolyte sensors (metal oxide-based sensor and solid polymer electrolyte (SPE)-Pt) (Chiou and Chou 2002). There were two well-established methods/devices used for the detection and quantification of H_2S in case of biological samples: ion-sensitive electrodes and amperometric (polarographic) H_2S sensors (Kraus, Doeller, and Zhang 2008).

15.6.3.1 Ion-Sensitive Electrodes (ISEs) for Hydrogen Sulfide (H_2S)

The ISE-based method is most easily obtained and commonly used to measure plasma H_2S with a detection range of 1–100 μM (Giep et al. 1996). ISE has a good sensitivity toward sulfide anions with a dynamic range of sulfide concentrations within a minute (Xu et al. 2016). The Ag/Ag_2S prepared by dipping a silver wire into a saturated solution of sodium sulfide. The slow reaction of silver wire with the S^{2-} ions, forming a thin dark layer of Ag_2S on the surface of the wire, acts as protection membrane preventing the interference of the sample (Allen and Larry 2001). Earlier, this method had been reported without detailed procedure, but Khan et al. later described the detailed procedure to the measurement of plasma H_2S (Khan and Malik 1971). Recent commercially available micro

ISE for H$_2$S detection requires a lesser volume of sample (10 µl) with a detection limit as low as 100 nM (Khan et al. 2015).

Careful inspection of the electrode surface is necessary; often, without a Ag$_2$S layer, it loses its selectivity due to Ag response to redox changes in the solution. Also, daily reconditioning and calibration of electrode surface is more laborious (Cheng and Da-Ming 2005). Further, it requires an alkaline environment that favors the equilibrium of S^{2-}, operating in a low concentration range, and maintaining the connection between the electrolyte bridge and the reference electrode (Brewer, Stoica, and Brown 2011).

15.6.3.2 Amperometric–Polarographic H$_2$S Sensors

A polarographic H$_2$S sensor (PHSS) has three essential features like other polarographic sensors (Doeller et al. 2005): the anode, cathode, and electrolyte (alkaline K$_3$Fe(CN)$_6$), which are protected from solution constituents by H$_2$S-permeable polymer membrane that permits diffusion of only free H$_2$S across the membrane and facilitates the interaction electrochemically with the polarized anode (Anani et al. 1990). H$_2$S was dissociated into HS$^-$ after crossing the permeable membrane and then chemically oxidized S$_0$, which simultaneously reduces ferricyanide to ferrocyanide, which donates electrons to the anode. The ferrocyanide electrochemically oxidized back to ferricyanide (100–200 mV) on the surface of the platinum electrode, producing a current proportional to sample H$_2$S concentration (Huang et al. 2007). The background current, which is subtracted from the signal current, dictates the PHSS lower limit of detection (Kraus, Doeller, and Zhang 2008). Compared to the ISE-based method, this method detects H$_2$S without an external electrical potential, with simple structure and good reproducibility, quick response time, and less containment of noble metals. The disadvantage of the liquid electrolyte sensors are that they leak easily, are prone to dry up, and have sizeable residual current due to impurities in the solution (Xu et al. 2016).

The applications of PHSS in biological samples have been studied on tissue homogenates, cultured and isolated cells, intact tissues and organs, and isolated mussel gill mitochondria (Wen, Wang, and Zhu 2018). The dynamic changes in H$_2$S concentration reported by the PHSS were not firmly seen with other techniques of H$_2$S measurement in biological samples (Vitvitsky and Banerjee 2015). The PHSS design is continually evolving to change based on measurement needs; the miniaturization allows *in situ* measurement of H$_2$S in blood, tissues, cells, and organelles (Doeller et al. 2005).

15.7 ELECTROCHEMICAL SENSORS FOR NUCLEIC ACIDS

Current sensing methods, such as lateral flow immunoassay, fluorescent microarray, polymerase chain reaction (PCR)-based methods, DNA microarrays, DNA sequencing technology, enzyme-linked immunosorbent assay (ELISA), require expensive reagents, high-precision instruments, and quantification methods to achieve highly sensitive detection (Gui and Patel 2011; Song et al. 2014). Additionally, most of the reactions cannot be monitored quantitatively in real-time. Thus, in the biomedical field, sensitive, specific, and fast analysis of nucleic acid is strongly needed (Vidic et al. 2019). Electrochemical sensor properties such as robust, simple, sensitivity, selectivity, and low cost for nucleic acid detection makes them more attractive than other available standard methods (Zhu et al. 2015). These methods are employed to detect selected DNA sequences or mutated genes associated with human disease (Drummond, Hill, and Barton 2003).

Currently, PCR-based and other target amplification strategies are most extensively used in practice. At the same time, such assays have limitations that can be overcome by alternative approaches (Gerasimova and Kolpashchikov 2014). There has been a recent explosion in the design of methods that amplify the signal produced by a nucleic acid target, without changing its copy number (Wang et al. 2015; Gerasimova and Kolpashchikov 2014). DNA-based electrochemical sensors exploit a range of different chemistries, but all take advantage of nanoscale interactions between the target in solution, the recognition layer, and a solid electrode surface (Drummond, Hill, and Barton 2003). Numerous approaches to electrochemical detection have been developed, including direct electrochemistry of DNA, electrochemistry at polymer-modified electrodes, electrochemistry of DNA-specific redox reporters, electrochemical amplifications with nanoparticles, and electrochemical devices based on DNA-mediated charge transport chemistry (Drummond, Hill, and Barton 2003).

15.7.1 Biomedical Applications of Electrochemical Sensors for Nucleic Acids

The testing of nucleic acids for various diseases is practiced mostly in clinical settings, mainly PCR-based, complex, and expensive. Alternative user-friendly options remain challenging in a clinical setup (Niemz, Ferguson, and Boyle 2011). Due to the promising nature of detection and diagnosis of various diseases such as metabolic disorders (cancer, diabetes, cardiovascular diseases) and infectious diseases (tuberculosis, hepatitis, dengue, and foodborne diseases like diarrhea, cholera, salmonellosis, etc.), electrochemical-based sensors for nucleic acids and aptamer in various communicable and noncommunicable diseases/disorders are listed in Table 15.3.

Ongoing efforts with DNA chips and other smart nanomaterials, however, are directed toward developing PCR-free DNA detection systems, which are yet to be fully commercialized (Du and Dong 2017). Tremendous progress has been noted in electrochemical DNA-based biosensors and arrays (Mok and Li 2008). Various types of electrodes immobilized with specific probes are at various stages of development, fabrication, and testing for the diagnosis of fatal diseases (Huang et al. 2017). Carbon nanotubes based on the electrochemical biosensor, surface plasmon resonance (SPR), quantum dot, and piezoelectric biosensors are promising candidates in molecular diagnosis (Tilmaciu and Morris 2015). The use of DNA biochip technology would eliminate the role of PCR in the future. New fabrication methods are emerging to develop fiber-optic DNA biosensors (Wei, Lillehoj, and Ho 2010).

The high stability and specificity of nucleic-acid-based biosensors can be promising candidates in the future for the clinical diagnostic market (Inan et al. 2017). However, the major bottleneck is

TABLE 15.3
Electrochemical-Based Sensors for Nucleic Acids and Aptamer Detection in the Diagnosis of Various Diseases

Target	Disease	Detection Type	Low Detection Limit or Range	Reference
IS*6110* gene	Tuberculosis	Labeled probe	0.01 ng/μl	Torres-Chavolla and Alocilja 2011
Mycobacterium tuberculosis	Tuberculosis	Labeled probe	0.065 ng/μL	Das et al. 2010
31-mer oligonucleotide sequence	Dengue	Label-free	$2.7 \times 10{-12}$ M	Deng and Toh 2013
miRNA-21	Cancer	Direct detection	100 aM	Hong et al. 2013
miRNA-21	Carcinoma	Indirect detection	0.006 pM	Meng et al. 2013
MCF-7 cells	Breast cancer	Sandwich type	~38 cells/mL	Yan et al. 2013
Thrombin	Inflammatory disorder	Labeled probe	$2.3 \times 10{-11}$ mol/L	Cui, Fan, and Lin 2013
Vasopressin	Traumatic injuries	Label-free	43 pM	He et al. 2013)
C-reactive protein	Cardiovascular disease	Label-free	100–500 pg/ml	Qureshi et al. 2010
IFN-γ	Tuberculosis	Direct	100 fM (RNA), 1–10 pM (DNA) buffer dependent	Min et al. 2008
Mucin 1	Breast cancer	Sandwich assay	<100 cells	Zhu et al. 2013

the requirement of strategic investments for commercialization of biosensor technology in the clinical field. These biosensors can be particularly useful in life sciences and medicine since in clinical practice, biosensors with high sensitivity and specificity can significantly enhance patient care, early diagnosis of diseases, and pathogen detection (Peña-Bahamonde et al. 2018). In this review, we will present the research conducted with antibodies, DNA molecules, and enzymes to develop biosensors that use graphene and its derivatives as scaffolds to produce capable biosensors able to detect and identify a variety of diseases, pathogens, and biomolecules linked to diseases (Peña-Bahamonde et al. 2018).

15.8 REDOX ELECTROCHEMICAL SENSORS

In the electrochemistry field, great attention was paid for the development of methods able to detect superoxide, hydrogen peroxide (H_2O_2), reactive oxygen species (ROS), and reactive nitrogen species (RNS) in biomedical research. ROS are involved in normal cellular homeostasis; during different pathological conditions as a consequence of imbalance, redox homeostasis leads to disease manifestation or progression. To detect the extracellular production of ROS, the choice of methods are high-performance liquid chromatography (HPLC), liquid chromatography–mass spectrometry (LC-MS), and mass spectrometry (Dikalov and Harrison 2014). Methods such as electron paramagnetic resonance, fluorescence microscopy, and genetic probes are a few bioanalytical approaches not based on electrochemistry that are routinely employed (Armbrecht and Dittrich 2017). Since available methods have their caveats, the electrochemical monitoring of these reactive species in the biological system will grow in importance and diffusion. Superoxide and H_2O_2 are present in deficient concentrations due to relatively shorter half-lives in reducing cellular environments. Hence, monitoring superoxide and H_2O_2 produced by living cells in real-time are thus rather challenging. Although plenty of electrochemical sensors for superoxide and hydrogen peroxide were reported in the scientific literature, few articles describe their use in biological systems (Prasad et al. 2015; Lvovich and Scheeline 1997; Flamm et al. 2015; Sadeghian et al. 2015; Cai et al. 2019; Ganesana, Erlichman, and Andreescu 2012; Balamurugan et al. 2018; Gao et al. 2018)

15.9 ELECTROCHEMICAL IMMUNOASSAYS AND IMMUNOSENSORS

Electrochemical immunosensors merge the inherent selectivity of the biological component and the sensitivity of electroanalytical methods. The electrochemical immunosensor, a type of biosensor, employs an antibody as a capture agent and quantitatively measures the electrical signal resulting from the binding event between the antibody and target molecule (i.e., the analyte) "sandwich immunoassay." It is a miniaturized favorable device for the detection of point-of-care testing (POCT). A few of these sensor devices are in the commercial stage and are consistently used in clinical, environmental, industrial, and agricultural applications (Piro and Reisberg 2017; Cho et al. 2018; Pollap and Kochana 2019; Tang et al. 2019; Contreras-Naranjo and Aguilar 2019). The principle behind them is that catalytic or binding of the biological analyte with sensor selectivity ultimately produces an electrical signal monitored by a transducer that is proportional to the analyte concentration.

15.10 PAPER-BASED ELECTROCHEMICAL SENSORS AND INKJET-PRINTED

A low-cost paper-based/inkjet-printed platform for electrochemical analyses can be produced that meets the needs of patients and the medical community, and it is also very likely that these devices will be considered disposable (Das et al. 2016). Previous studies demonstrated that inkjet-printed/paper-based electrochemical sensors are suitable for this application due to their enormous advantages including carefully selected materials and fabrication methods with incredible control over the material deposition process; in case of paper substrates, pump-free microfluidic devices can be set up due to their wicking property (Das et al. 2016; Cinti 2019; Gebretsadik et al. 2019; Loo et al. 2019; Felix, Baccaro, and Angnes 2018; Evans et al. 2018; Nyein et al. 2018). By merging factors of innovation with low-cost, sensitivity at the point-of-care sensors make the electrochemical technique more favorable.

15.11 CONCLUSIONS

Finally, it may be concluded that the development and research of electrochemical (bio)sensors or any analytical method is becoming a trendy scientific area at the intersection of the biological and engineering sciences and possibly will be extensively used shortly. The future possibility of development of simple, disposable, biocompatible, multiplexing, and low-cost implantable electrochemical sensors for biomedical and clinical applications are of great interest for their ability to monitor clinically relevant analytes such as blood gases (CO, NO, H_2S, CO_2, and SO_2), bodily electrolytes, macromolecules (protein, DNA, RNA and others), and metabolites in real-time. Furthermore, continuing to innovate, by merging each of these favorable electrochemical-sensing techniques and materials is incredibly selective and provides accurate and repeatable quantitative results without expensive measurement equipment. As an alternative to conventional methods, advance, electrochemical analytic methods might be promising for future needs of the biomedical field for many patients, healthcare and medical professionals, and environmental specialists.

ACKNOWLEDGMENTS

The author acknowledges Mr. William McLean for critical proofreading and excellent editing of this book chapter.

REFERENCES

Allen, J. B., and R. F. Larry. 2001. *Electrochemical Methods: Fundamentals and Applications*. 2nd ed. New York: Wiley, 1–736.

Almeida, A. S., C. Figueiredo-Pereira, and H. L. Vieira. 2015. Carbon monoxide and mitochondria-modulation of cell metabolism, redox response and cell death. *Front. Physiol.* 6:33.

Anani, A., Z. Mao, Ralph E. White, Srikrishna Srinivasan, and A. J. Appleby. 1990. Electrochemical production of hydrogen and sulfur by low-temperature decomposition of hydrogen sulfide in an aqueous alkaline solution. *J. Electrochem. Soc.* 137:2703–2709.

Armbrecht, L., and P. S. Dittrich. 2017. Recent advances in the analysis of single cells. *Anal. Chem.* 89 (1):2–21.

Bai, Hua, and Gaoquan Shi. 2007. Gas sensors based on conducting polymers. *Sensors (Basel, Switzerland)* 7 (3):267–307.

Balamurugan, M., P. Santharaman, T. Madasamy, et al. 2018. Recent trends in electrochemical biosensors of superoxide dismutases. *Biosens. Bioelectron.* 116:89–99.

Birben, E., U. M. Sahiner, C. Sackesen, S. Erzurum, and O. Kalayci. 2012. Oxidative stress and antioxidant defense. *World Allergy Organ. J.* 5 (1):9–19.

Blumenthal, I. 2001. Carbon monoxide poisoning. *J. R. Soc. Med.* 94 (6):270–272.

Boczkowski, J., J. J. Poderoso, and R. Motterlini. 2006. CO-metal interaction: Vital signaling from a lethal gas. *Trends Biochem. Sci.* 31 (11):614–621.

Brewer, P. J., D. Stoica, and R. J. Brown. 2011. Sensitivities of key parameters in the preparation of silver/silver chloride electrodes used in Harned cell measurements of pH. *Sensors (Basel)* 11 (8):8072–8084.

Brown, S. D., and C. A. Piantadosi. 1990. In vivo binding of carbon monoxide to cytochrome c oxidase in rat brain. *J. Appl. Physiol. (1985)* 68 (2):604–610.

Cai, X., H. Chen, Z. Wang, et al. 2019. 3D graphene-based foam induced by phytic acid: An effective enzyme-mimic catalyst for electrochemical detection of cell-released superoxide anion. *Biosens. Bioelectron.* 123:101–107.

Chance, B., M. Erecinska, and M. Wagner. 1970. Mitochondrial responses to carbon monoxide toxicity. *Ann. N. Y. Acad. Sci.* 174 (1):193–204.

Chen, T. M., J. Gokhale, S. Shofer, and W. G. Kuschner. 2007. Outdoor air pollution: nitrogen dioxide, sulfur dioxide, and carbon monoxide health effects. *Am. J. Med. Sci.* 333 (4):249–256.

Cheng, K. L., and Zhu Da-Ming. 2005. On calibration of pH meters. *Sensors (Basel)* 5 (4):209–219.
Cheng, Shaoan, Hong Liu, and Bruce E. Logan. 2006. Power densities using different cathode catalysts (Pt and CoTMPP) and polymer binders (Nafion and PTFE) in single chamber microbial fuel cells. *Environ. Sci. Tech.* 40 (1):364–369.
Chiou, Chiou-Yen, and Tse-Chuan Chou. 2002. Amperometric SO2 gas sensors based on solid polymer electrolytes. *Sens. Actuators, B: Chem.* 87:1–7.
Cho, I. H., J. Lee, J. Kim, et al. 2018. Current technologies of electrochemical immunosensors: Perspective on signal amplification. *Sensors (Basel)* 18 (1):207.
Cinti, S. 2019. Novel paper-based electroanalytical tools for food surveillance. *Anal. Bioanal. Chem.* 411 (19):4303–4311.
Contreras-Naranjo, J. E., and O. Aguilar. 2019. Suppressing non-specific binding of proteins onto electrode surfaces in the development of electrochemical immunosensors. *Biosensors (Basel)* 9 (1):15.
Cretescu, Igor, Doina Lutic, and Liliana Manea. 2017. *Electrochemical Sensors for Monitoring of Indoor and Outdoor Air Pollution, Electrochemical Sensors Technology.* IntechOpen, 65–84.
Csonka, C., T. Pali, P. Bencsik, A. Gorbe, P. Ferdinandy, and T. Csont. 2015. Measurement of NO in biological samples. *Br. J. Pharmacol.* 172 (6):1620–1632.
Cui, H. F., Fan, H., and Lin, Y. 2013. A new electrochemical aptasensor for protein detection based on target protein-induced strand displacement. *Adv. Mater. Res.* 753–755:2113–2116.
Das, Maumita, Gajjala Sumana, Rajamani Nagarajan, and Bansi Malhotra. 2010. Zirconia based nucleic acid sensor for Mycobacterium tuberculosis detection. *Appl. Phys. Lett.* 96:133703.
Das, S. R., Q. Nian, A. A. Cargill, et al. 2016. 3D nanostructured inkjet printed graphene via UV-pulsed laser irradiation enables paper-based electronics and electrochemical devices. *Nanoscale* 8 (35):15870–15879.
Deng, J., and C. S. Toh. 2013. Impedimetric DNA biosensor based on a nanoporous alumina membrane for the detection of the specific oligonucleotide sequence of dengue virus. *Sensors (Basel)* 13 (6):7774–7785.
Di Meo, S., T. T. Reed, P. Venditti, and V. M. Victor. 2016. Role of ROS and RNS sources in physiological and pathological conditions. *Oxid. Med. Cell Longev.* 2016:1245049.
Dikalov, S. I., and D. G. Harrison. 2014. Methods for detection of mitochondrial and cellular reactive oxygen species. *Antioxid. Redox Signal.* 20 (2):372–382.
Dobbelaere, Thomas, Philippe M. Vereecken, and Christophe Detavernier. 2017. A USB-controlled potentiostat/galvanostat for thin-film battery characterization. *HardwareX* 2:34–49.
Doeller, J. E., T. S. Isbell, G. Benavides, et al. 2005. Polarographic measurement of hydrogen sulfide production and consumption by mammalian tissues. *Anal. Biochem.* 341 (1):40–51.
Drummond, T. G., M. G. Hill, and J. K. Barton. 2003. Electrochemical DNA sensors. *Nat. Biotechnol.* 21 (10):1192–1199.
Dryden, M. D., and A. R. Wheeler. 2015. DStat: a versatile, open-source potentiostat for electroanalysis and integration. *PLoS One* 10 (10):e0140349.
Du, Yan, and Shaojun Dong. 2017. Nucleic acid biosensors: recent advances and perspectives. *Anal. Chem.* 89 (1):189–215.
Duret-Thual, C. 2014. 1 - Understanding corrosion: basic principles. In T. Liengen, D. Féron, R. Basséguy, and I. B. Beech, Eds., *Understanding Biocorrosion.* Oxford: Woodhead Publishing, 409–423.
Elgrishi, Noémie, Kelley J. Rountree, Brian D. McCarthy, Eric S. Rountree, Thomas T. Eisenhart, and Jillian L. Dempsey. 2018. A practical beginner's guide to cyclic voltammetry. *J. Chem. Educ.* 95 (2):197–206.
Evans, D., K. I. Papadimitriou, N. Vasilakis, et al. 2018. A novel microfluidic point-of-care biosensor system on printed circuit board for cytokine detection. *Sensors (Basel)* 18 (11):4011.
Everton, Sarah K., Matthias Hirsch, Petros Stravroulakis, Richard K. Leach, and Adam T. Clare. 2016. Review of in-situ process monitoring and in-situ metrology for metal additive manufacturing. *Mater. Design* 95:431–445.
Farrugia, G., and J. H. Szurszewski. 2014. Carbon monoxide, hydrogen sulfide, and nitric oxide as signaling molecules in the gastrointestinal tract. *Gastroenterology* 147 (2):303–313.
Faulkner, Larry R. 1983. Understanding electrochemistry: Some distinctive concepts. *J. Chem. Educ.* 60 (4):262.
Felix, F. S., A. L. B. Baccaro, and L. Angnes. 2018. Disposable voltammetric immunosensors integrated with microfluidic platforms for biomedical, agricultural and food analyses: A review. *Sensors (Basel)* 18 (12).
Fitzpatrick, Dennis. 2015. Chapter 4 - Glucose biosensors. In D. Fitzpatrick, Ed., *Implantable Electronic Medical Devices.* Oxford: Academic Press, 37–51.
Flamm, H., J. Kieninger, A. Weltin, and G. A. Urban. 2015. Superoxide microsensor integrated into a sensing cell culture flask microsystem using direct oxidation for cell culture application. *Biosens. Bioelectron.* 65:354–359.

Frank, J. A., and R. D. James. 1994. Factors affecting the accuracy of reference electrodes. *Mater. Perform.* 33 (11):14–17.

Ganesana, M., J. S. Erlichman, and S. Andreescu. 2012. Real-time monitoring of superoxide accumulation and antioxidant activity in a brain slice model using an electrochemical cytochrome c biosensor. *Free Radic. Biol. Med.* 53 (12):2240–2249.

Gao, L. X., C. Bian, Y. Wu, et al. 2018. Label-free electrochemical sensor to investigate the effect of tocopherol on generation of superoxide ions following UV irradiation. *J. Biol. Eng.* 12:17.

Gebretsadik, T., T. Belayneh, S. Gebremichael, W. Linert, M. Thomas, and T. Berhanu. 2019. Recent advances in and potential utilities of paper-based electrochemical sensors: beyond qualitative analysis. *Analyst* 144 (8):2467–2479.

Gerasimova, Y. V., and D. M. Kolpashchikov. 2014. Enzyme-assisted target recycling (EATR) for nucleic acid detection. *Chem. Soc. Rev.* 43 (17):6405–6438.

Giep, T. N., R. T. Hall, K. Harris, B. Barrick, and S. Smith. 1996. Evaluation of neonatal whole blood versus plasma glucose concentration by ion-selective electrode technology and comparison with two whole blood chromogen test strip methods. *J. Perinatol.* 16 (4):244–249.

Griendling, K. K., R. M. Touyz, J. L. Zweier, et al. 2016. Measurement of reactive oxygen species, reactive nitrogen species, and redox-dependent signaling in the cardiovascular system: a scientific statement from the American Heart Association. *Circ. Res.* 119 (5):e39–e75.

Grieshaber, D., R. MacKenzie, J. Voros, and E. Reimhult. 2008. Electrochemical biosensors - sensor principles and architectures. *Sensors (Basel)* 8 (3):1400–1458.

Gui, J., and I. R. Patel. 2011. Recent advances in molecular technologies and their application in pathogen detection in foods with particular reference to yersinia. *J. Pathog.* 2011:310135.

Guy, Owen J., and Kelly-Ann D. Walker. 2016. Chapter 4 - Graphene functionalization for biosensor applications. In S. E. Saddow, Ed., *Silicon Carbide Biotechnology (Second Edition)*, Elsevier, 85–141.

Hartle, M. D., and M. D. Pluth. 2016. A practical guide to working with H2S at the interface of chemistry and biology. *Chem. Soc. Rev.* 45 (22):6108–6117.

He, P., V. Oncescu, S. Lee, I. Choi, and D. Erickson. 2013. Label-free electrochemical monitoring of vasopressin in aptamer-based microfluidic biosensors. *Anal. Chim. Acta* 759:74–80.

Helm, I., L. Jalukse, and I. Leito. 2010. Measurement uncertainty estimation in amperometric sensors: a tutorial review. *Sensors (Basel)* 10 (5):4430–4455.

Hong, C. Y., X. Chen, T. Liu, et al. 2013. Ultrasensitive electrochemical detection of cancer-associated circulating microRNA in serum samples based on DNA concatamers. *Biosens. Bioelectron.* 50:132–136.

Huang, Tina H., Gail Salter, Sarah L. Kahn, and Yvonne M. Gindt. 2007. Redox titration of ferricyanide to ferrocyanide with ascorbic acid: illustrating the Nernst Equation and Beer–Lambert Law. *J. Chem. Educ.* 84 (9):1461.

Huang, Y., J. Xu, J. Liu, X. Wang, and B. Chen. 2017. Disease-related detection with electrochemical biosensors: a review. *Sensors (Basel)* 17 (10):2375.

Huffman, M. L., and B. J. Venton. 2009. Carbon-fiber microelectrodes for in vivo applications. *Analyst* 134 (1):18–24.

Ignarro, L. J., G. M. Buga, K. S. Wood, R. E. Byrns, and G. Chaudhuri. 1987. Endothelium-derived relaxing factor produced and released from artery and vein is nitric oxide. *Proc. Natl. Acad. Sci. USA* 84 (24):9265–9269.

Inan, H., M. Poyraz, F. Inci, et al. 2017. Photonic crystals: emerging biosensors and their promise for point-of-care applications. *Chem. Soc. Rev.* 46 (2):366–388.

Khan, Mohd Mahfooz, and Anees Uddin Malik. 1971. Studies on mixed ligand complexes: Part III. Mechanism of the reaction of heterocyclic amines with tristhiourea copper(I) chloride and iodides. *J. Electroanal. Chem. Interfacial Electrochem.* 29 (2):421–427.

Khan, S. N., F. Shaeib, T. Najafi, et al. 2015. Diffused intra-oocyte hydrogen peroxide activates myeloperoxidase and deteriorates oocyte quality. *PLoS One* 10 (7):e0132388.

Kim, D., S. Koseoglu, B. M. Manning, A. F. Meyer, and C. L. Haynes. 2011. Electroanalytical eavesdropping on single cell communication. *Anal. Chem.* 83 (19):7242–7249.

Kraus, David W., Jeannette E. Doeller, and Xueji Zhang. 2008. Chapter 8 - Electrochemical sensors for the determination of hydrogen sulfide production in biological samples. In X. Zhang, H. Ju, and J. Wang, Eds., *Electrochemical Sensors, Biosensors and Their Biomedical Applications*. San Diego: Academic Press, 213–235.

Krishnan, Siva Kumar, Eric Singh, Pragya Singh, Meyya Meyyappan, and Hari Singh Nalwa. 2019. A review on graphene-based nanocomposites for electrochemical and fluorescent biosensors. *RSC Advances* 9 (16):8778–8881.

Liu, X., S. Cheng, H. Liu, S. Hu, D. Zhang, and H. Ning. 2012. A survey on gas sensing technology. *Sensors (Basel)* 12 (7):9635–9665.

Loo, J. F. C., A. H. P. Ho, A. P. F. Turner, and W. C. Mak. 2019. Integrated printed microfluidic biosensors. *Trends Biotechnol.* 37 (10):1104–1120.

Lvovich, V., and A. Scheeline. 1997. Amperometric sensors for simultaneous superoxide and hydrogen peroxide detection. *Anal. Chem.* 69 (3):454–462.

Majed, B. H., and R. A. Khalil. 2012. Molecular mechanisms regulating the vascular prostacyclin pathways and their adaptation during pregnancy and in the newborn. *Pharmacol. Rev.* 64 (3):540–582.

Matharu, Zimple, G. Sumana, Sunil K. Arya, S. P. Singh, Vinay Gupta, and B. D. Malhotra. 2007. Polyaniline Langmuir–Blodgett film based cholesterol biosensor. *Langmuir* 23 (26):13188–13192.

Mathew, Neal D., David I. Schlipalius, and Paul R. Ebert. 2011. Sulfurous gases as biological messengers and toxins: comparative genetics of their metabolism in model organisms. *J. Toxicol.* 2011:394970.

Meng, X., Y. Zhou, Q. Liang, et al. 2013. Electrochemical determination of microRNA-21 based on bio bar code and hemin/G-quadruplet DNAenzyme. *Analyst* 138 (12):3409–3415.

Min, K., M. Cho, S. Y. Han, Y. B. Shim, J. Ku, and C. Ban. 2008. A simple and direct electrochemical detection of interferon-gamma using its RNA and DNA aptamers. *Biosens. Bioelectron.* 23 (12):1819–1824.

Mok, W., and Y. Li. 2008. Recent progress in nucleic acid aptamer-based biosensors and bioassays. *Sensors (Basel)* 8 (11):7050–7084.

Motterlini, R., and L. E. Otterbein. 2010. The therapeutic potential of carbon monoxide. *Nat. Rev. Drug Discov.* 9 (9):728–743.

Nemiroski, A., D. C. Christodouleas, J. W. Hennek, et al. 2014. Universal mobile electrochemical detector designed for use in resource-limited applications. *Proc. Natl. Acad. Sci. USA* 111 (33):11984–11989.

Niemz, A., T. M. Ferguson, and D. S. Boyle. 2011. Point-of-care nucleic acid testing for infectious diseases. *Trends Biotechnol.* 29 (5):240–250.

Nyein, H. Y. Y., L. C. Tai, Q. P. Ngo, et al. 2018. A wearable microfluidic sensing patch for dynamic sweat secretion analysis. *ACS Sens.* 3 (5):944–952.

Olson, K. R. 2012. A practical look at the chemistry and biology of hydrogen sulfide. *Antioxid. Redox Signal.* 17 (1):32–44.

Ono, K., T. Akaike, T. Sawa, et al. 2014. Redox chemistry and chemical biology of H2S, hydropersulfides, and derived species: implications of their possible biological activity and utility. *Free Radic. Biol. Med.* 77:82–94.

Ozel, R. E., K. N. Wallace, and S. Andreescu. 2011. Chitosan coated carbon fiber microelectrode for selective in vivo detection of neurotransmitters in live zebrafish embryos. *Anal. Chim. Acta* 695 (1–2):89–95.

Peña-Bahamonde, Janire, Hang N. Nguyen, Sofia K. Fanourakis, and Debora F. Rodrigues. 2018. Recent advances in graphene-based biosensor technology with applications in life sciences. *J. Nanobiotechnol.* 16 (1):75.

Piro, B., and S. Reisberg. 2017. Recent advances in electrochemical immunosensors. *Sensors (Basel)* 17 (4):794.

Pollap, A., and J. Kochana. 2019. Electrochemical immunosensors for antibiotic detection. *Biosensors (Basel)* 9 (2):61.

Prabhulkar, S., H. Tian, X. Wang, J. J. Zhu, and C. Z. Li. 2012. Engineered proteins: redox properties and their applications. *Antioxid. Redox Signal.* 17 (12):1796–1822.

Prasad, A., A. Kumar, M. Suzuki, et al. 2015. Detection of hydrogen peroxide in Photosystem II (PSII) using catalytic amperometric biosensor. *Front. Plant Sci.* 6:862.

Queiroga, C. S., A. S. Almeida, P. M. Alves, C. Brenner, and H. L. Vieira. 2011. Carbon monoxide prevents hepatic mitochondrial membrane permeabilization. *BMC Cell Biol.* 12:10.

Queiroga, C. S., A. Vercelli, and H. L. Vieira. 2015. Carbon monoxide and the CNS: challenges and achievements. *Br. J. Pharmacol.* 172 (6):1533–1545.

Queiroga, Cláudia S. F., Ana S. Almeida, and Helena L. A. Vieira. 2012. Carbon monoxide targeting mitochondria. *Biochem. Res. Int.* 2012:9.

Qureshi, A., Y. Gurbuz, S. Kallempudi, and J. H. Niazi. 2010. Label-free RNA aptamer-based capacitive biosensor for the detection of C-reactive protein. *Phys. Chem. Chem. Phys.* 12 (32):9176–9182.

Richter, Mark M. 2004. Electrochemiluminescence (ECL). *Chem. Rev.* 104 (6):3003–3036.

Rosselli, M., P. J. Keller, and R. K. Dubey. 1998. Role of nitric oxide in the biology, physiology and pathophysiology of reproduction. *Hum. Reprod. Update* 4 (1):3–24.

Sadeghian, R. B., S. Ostrovidov, S. Salehi, Han Jiuhui, Chen Mingwei, and A. Khademhosseini. 2015. An electrochemical biosensor based on gold microspheres and nanoporous gold for real-time detection of superoxide anion in skeletal muscle tissue. *Conf. Proc. IEEE Eng. Med. Biol. Soc.* 2015:7962–7965.

Samouilov, A., G. L. Caia, E. Kesselring, S. Petryakov, T. Wasowicz, and J. L. Zweier. 2007. Development of a hybrid EPR/NMR coimaging system. *Magn. Reson. Med.* 58 (1):156–166.

Sarkar, S., K. Mathwig, S. Kang, A. F. Nieuwenhuis, and S. G. Lemay. 2014. Redox cycling without reference electrodes. *Analyst* 139 (22):6052–6057.

Sarkar, Sahana, Stanley C. S. Lai, and Serge G. Lemay. 2016. Unconventional electrochemistry in micro-/nanofluidic systems. *Micromachines* 7 (5):81.

Scott, K. 1995. Membranes for electrochemical cells. In K. Scott, Ed., *Handbook of Industrial Membranes*. Amsterdam: Elsevier Science, 773–790.

Siddiqui, Masoom Raza, Zedd A. AlOthman, and Nafisur Rahman. 2013. Analytical techniques in pharmaceutical analysis: a review. *Arab. J. Chem.* 10:S1409–S1421.

Serpe, M. J., and X. Zhang. 2007. The principles, development and application of microelectrodes for the in vivo determination of nitric oxide. In A. C. Michael, L. Borland, Eds., *Electrochemical Methods for Neuroscience*. Boca Raton, FL: CRC Press/Taylor & Francis.

Shefa, U., M. S. Kim, N. Y. Jeong, and J. Jung. 2018. Antioxidant and cell-signaling functions of hydrogen sulfide in the central nervous system. *Oxid. Med. Cell. Longev.* 2018:1873962.

Shrivastava, Alankar, and Vipin Gupta. 2011. Methods for the determination of limit of detection and limit of quantitation of the analytical methods. *Chron. Young Sci.* 2 (1):21–25.

Song, Y., Y. Y. Huang, X. Liu, X. Zhang, M. Ferrari, and L. Qin. 2014. Point-of-care technologies for molecular diagnostics using a drop of blood. *Trends Biotechnol.* 32 (3):132–139.

Tang, Q., L. Zhang, X. Tan, L. Jiao, Q. Wei, and H. Li. 2019. Bioinspired synthesis of organic-inorganic hybrid nanoflowers for robust enzyme-free electrochemical immunoassay. *Biosens. Bioelectron.* 133:94–99.

Tilmaciu, C. M., and M. C. Morris. 2015. Carbon nanotube biosensors. *Front. Chem.* 3:59.

Torres-Chavolla, E., and E. C. Alocilja. 2011. Nanoparticle based DNA biosensor for tuberculosis detection using thermophilic helicase-dependent isothermal amplification. *Biosens. Bioelectron.* 26 (11):4614–4618.

Tripathi, A. M., W. N. Su, and B. J. Hwang. 2018. In situ analytical techniques for battery interface analysis. *Chem. Soc. Rev.* 47 (3):736–851.

Turner, A. P. 2013. Biosensors: sense and sensibility. *Chem. Soc. Rev.* 42 (8):3184–3196.

Vidic, J., P. Vizzini, M. Manzano, et al. 2019. Point-of-need DNA testing for detection of foodborne pathogenic bacteria. *Sensors (Basel)* 19 (5):1100.

Vitvitsky, V., and R. Banerjee. 2015. H2S analysis in biological samples using gas chromatography with sulfur chemiluminescence detection. *Methods Enzymol.* 554:111–123.

Wang, Yi, Yan Wang, Ai-Jing Ma, et al. 2015. Rapid and sensitive isothermal detection of nucleic-acid sequence by multiple cross displacement amplification. *Sci. Rep.* 5:11902.

Wei, F., P. B. Lillehoj, and C. M. Ho. 2010. DNA diagnostics: nanotechnology-enhanced electrochemical detection of nucleic acids. *Pediatr. Res.* 67 (5):458–468.

Wen, Y. D., H. Wang, and Y. Z. Zhu. 2018. The drug developments of hydrogen sulfide on cardiovascular disease. *Oxid. Med. Cell. Longev.* 2018:4010395.

Wen, Ya-Dan, Hong Wang, and Yi-Zhun Zhu. 2018. The drug developments of hydrogen sulfide on cardiovascular disease. *Oxid. Med. Cell. Longevity* 2018:21.

Xiao, Q., J. Ying, L. Xiang, and C. Zhang. 2018. The biologic effect of hydrogen sulfide and its function in various diseases. *Medicine (Baltimore)* 97 (44):e13065.

Xu, Tailin, Nikki Scafa, Li-Ping Xu, et al. 2016. Electrochemical hydrogen sulfide biosensors. *Analyst* 141 (4):1185–1195.

Yan, M., G. Sun, F. Liu, J. Lu, J. Yu, and X. Song. 2013. An aptasensor for sensitive detection of human breast cancer cells by using porous GO/Au composites and porous PtFe alloy as effective sensing platform and signal amplification labels. *Anal. Chim. Acta* 798:33–39.

Zhang, X. 2004. Real time and in vivo monitoring of nitric oxide by electrochemical sensors--from dream to reality. *Front. Biosci.* 9:3434–3446.

Zhang, Xueji. 2008. Chapter 1 - Nitric oxide (NO) electrochemical sensors. In X. Zhang, H. Ju, and J. Wang (Eds.), *Electrochemical Sensors, Biosensors and Their Biomedical Applications*. San Diego, CA: Academic Press, 1–29.

Zhu, C., G. Yang, H. Li, D. Du, and Y. Lin. 2015. Electrochemical sensors and biosensors based on nanomaterials and nanostructures. *Anal. Chem.* 87 (1):230–249.

Zhu, X., J. Yang, M. Liu, Y. Wu, Z. Shen, and G. Li. 2013. Sensitive detection of human breast cancer cells based on aptamer-cell-aptamer sandwich architecture. *Anal. Chim. Acta.* 764:59–63.

Zielonka, J., and B. Kalyanaraman. 2010. Hydroethidine- and MitoSOX-derived red fluorescence is not a reliable indicator of intracellular superoxide formation: another inconvenient truth. *Free Radic. Biol. Med.* 48 (8):983–1001.

Index

ABPE, *see* Applied bias photon to current efficiency
Accuracy parameter, 130
Acid blue 113 dye, 189
A-CNTs, *see* Amine functionalized carbon nanotubes
Advance oxidation processes (AOPs), 184, 207
Ag–Ni alloys, 38
Al–Al electrodes, 202
ALD, *see* Atomic layer deposition
Amine functionalized carbon nanotubes (A-CNTs), 172
Amperometric–polarographic H_2S sensors, *see* Polarographic H_2S sensor
Amperometry, 4
    experiments, 5
Analytes, significance of, 124
Anode, significance of, 4
Anodic coatings, 10
Anodic oxidation, *see* Direct oxidation
Anodic polarization, 29
AOPs, *see* Advance oxidation processes
Applied bias photon to current efficiency (ABPE), 108
Aqueous medium electrolytes, 73–74
Artificial photosynthesis, *see* Photoelectrochemical process
As-synthesized nanotubes, 55
Asymmetric supercapacitors, 64–65
Atomic layer deposition (ALD), 55

Batteries, 238
BDD, *see* Boron-doped diamond
BDM, *see* Bockris-Devanathan-Müller model
Binary metal sulfides, 141
Biocatalyst metabolic losses, 93
Biocatalysts, 92
Biomedical applications, 251–252
    biological analytes and electrochemical methods, 254–255
    electrochemical immunoassays and immunosensors, 261
    electrochemical methods, 252–253
    electrochemical sensor and circuit, 254
    electrochemical sensors and working principles, 252
    gasotransmitter electrochemical sensors, 255
        carbon monoxide detection and electrochemical sensors, 255–257
        hydrogen sulfide electrochemical sensors, 258–259
        nitric oxide electrochemical sensors determination, 257–258
    nucleic acids and electrochemical sensors, 259–261
    paper-based electrochemical sensors and inkjet-printed, 261–262
    redox electrochemical sensors, 261
Biomolecules, immobilization of, 244
Biosensor
    glucose, 15–16
    hydrogen peroxide (H_2O_2), 16
Bockris-Devanathan-Müller (BDM) model, 32
BODIPY, 218–221
Boron-doped diamond (BDD), 158, 160, 185

BP-S configuration, 197–198
Butler-Volmer equation, oxygen reduction reaction of, 92

Carbazole-dendrimer, 221, 222
Carbon monoxide detection, electrochemical sensors for, 255–257
Cathode, significance of, 4
Cathodic coatings, 10
Cathodic polarization, 29
CE, *see* Counter electrode
Central Electrochemical Research Institute (CECRI), 42
Chemical oxygen demand (COD), 88, 98
    removal, 199, 201, 202
Chemiresistive gas sensing materials, 121–122
    analytes and, 124
    chemiresistive sensors and, 122
    complexes, 127
    composites, 126–127
    device design and mechanism, 122–124
    future directions, 130–131
    metal organic frameworks (MOFs), 128
    metal oxide sensors, 126
    nanoparticles, 127–128
    parameters, 129–130
    polymers, 126
    sensing platform, 129
    types of, 125
Chronoamperometry, 253
Chronocoulometry, 253
Chronopotentiometry, 253
Clay cup modified MFC, 95
Cobalt electrodeposition, 36
COD, *see* Chemical oxygen demand
Complexes, 127
Composite coatings, 43
Composites, 126–127
    fabrication of, 243–244
Conducting polymers (CPs), 69, 140, 233–234
    electrochemical applications of
        batteries, 238
        electrochemical sensors, 240–244
        fuel cells, 238–239
        solar cells, 235–238
        supercapacitors, 240
    examples of
        polyaniline (PANI) and derivatives, 234–235, 239
        polypyrrole (PPY), 235
        polythiophene and derivatives, 235
Conductivity studies of organic compounds, as additives in polymer electrolytes, 221–223
Congo red, 188
CoNiFe ternary alloy, 39
Convection, 30
Conway model, 32
Copper–indium alloy, 38
Copper oxide, 139
Copper-oxide-based electrodes, 110
Coulombic efficiency, 88

267

Index

Counter electrode (CE), 254, 256, 257
COX, *see* Cytochrome *c* oxygenase
CPs, *see* Conducting polymers
CuI nanoplates, 141–142
CuNi dentric nanostructures, 143
Cyclic voltammetry, 253
Cyclic voltammetry oriented electrochemical reduction procedure, 56
Cytochrome *c* oxygenase (COX), 256

Dead time, 129
Decolorization, of textile dye, 186–188
Deposit thickness, 28
D-Glucose, 135
Differential scanning calorimetry (DSC), 224
    with organic molecules as additives in polymer electrolytes, 223–224
Diffusion, 30
Dimensionally stable anodes (DSA), 201
Direct deposition (DP), 27
Direct electrolysis, 17
Direct electrolytic decontamination, 157
Direct oxidation, 158–159
Dissolvable anodes, *see* Sacrificial anodes
DLVO theory, 194–195
DP, *see* Direct deposition
Drift, 129
DSA, *see* Dimensionally stable anodes
DSC, *see* Differential scanning calorimetry
DSSCs, *see* Dye-sensitized solar cells
Dye elimination, 187
Dye-sensitized solar cells (DSSCs), 54–56, 217–219
    organic materials as photosensitizers in, 220–228
    preparation of, 220
Dynamic range, 129

EAOPs, *see* Electrochemical advanced oxidation processes
EAPs, *see* Electroactive polymers
EDL, *see* Electrical double layer
EDLC, *see* Electric double-layer capacitor
Electrical double layer (EDL), 31–32, 194–195
Electric double-layer capacitor (EDLC), 64–66
    material advances for, 68–69
Electroactive polymers (EAPs), 186
Electrocatalysts, for wastewater treatment, 153–154
    decontamination advances, 161–163
    fundamentals of, 154
        electrocatalytic anode material categorization, 160
        electrocatalytic oxidation mechanism, 156–160
        electrocatalytic reactor configuration, 155–156
    superiorities and inferiorities, 163–164
Electrocatalytic advance oxidation processes, 154
Electrochemical advanced oxidation processes (EAOPs)
    anodes, advantages of, 183–184
    general setup of, 183
    of textile dyes from wastewater, 186
        acid blue 113 dye, 189
        Congo red, 188
        indigo carmine dye, 188–189
        methylene blue dye, 186–188
        rhodamine B dye, 188
        triphenylmethane dye, 189–190
        yellow 3 (DY3) dye, 189
    types of, 182, 184
    direct method, 184–185
    by indirect method, 185–186
Electrochemical immunoassays and immunosensors, 261
Electrochemical sensors, 240–244
Electrochemical supercapacitors, 63–65
    electric double-layer capacitors, 65–66
        material advances for, 68–69
    pseudocapacitors, 66–68
        advances in materials for, 69–72
    and reactions, advancements in electrolytes for, 72–73
        aqueous medium, 73–74
        ionic liquids, 74–75
        organic solvent medium, 74
Electrocoagulation, 193–194
    fundamentals of, 194
        colloidal stability, 194–195
        principle, 195–196
    future perspectives, 206–209
    of industrial wastewater, materials for, 200
        food industry, 200–201
        oil industry, 201–202
        pharmaceutical industry, 202–204
        pulp and paper industry, 202
        textile and tannery industry, 200
        wastewater containing heavy metals, 204, 206
    operating parameter effects on performance of, 196
        current density, 198
        electrode configuration, 197–198
        electrode material, 196
        initial pH, 196–197
        solution conductivity, 199–200
        temperature, 198–199
Electrode–electrolyte interface, 29
Electrode fouling, 136
Electrodeposition, 26
    fundamentals of, 27–32
    material tunable properties by parameters of, 34–35
        composition, 35–36
        morphology, 40–41
        structure, 37–40
    mechanism of, 29–31
    of metals and alloys, 5–10
    of metals, as sacrificial coatings, 10
    of organic materials, 10
    parameters affecting, 32
        agent addition, 33
        agitation effect, 34
        bath concentration effect, 33
        bath temperature effect, 33
        complexing agent, 33
        current density effect and distribution, 32
        hydrogen embrittlement, 34
        pH effect, 33
        plating time effect, 33–34
Electrodes, 4, 28, 86, 136; *see also individual entries*
Electrolysis, 17
Electrolyte, significance of, 86
Electrolyzer mode, 4
Electrophoretic deposition (EPD), 10–11
Electrospinning, 75
Enzymatic glucose sensor, 136–137
EPD, *see* Electrophoretic deposition

Index

Faradaic current, 28
Faradaic reaction, 66
Faraday's laws of electrolysis, 27–28, 33
FESEM, see Field emission scanning electron microscopy
FF, see Fill factor
Field emission scanning electron microscopy (FESEM), 170
Figaro electrochemical-type gas sensors, 256
Fill factor (FF), 219
Flexible supercapacitors, 68
Fluorinated tin oxide (FTO), 217–218
Food industry, 200–201
Fossil fuels, 105
FTO, see Fluorinated tin oxide
Fuel cells, 238–239

Gallium nitride (GaN), 110
Galvanostatic potential difference, 29
GaN, see Gallium nitride
Gasotransmitter electrochemical sensors, 255
    carbon monoxide detection and electrochemical sensors, 255–257
    hydrogen sulfide electrochemical sensors, 258–259
    nitric oxide electrochemical sensors determination, 257–258
Giant magneto resistance (GMR), 42
Gibbs free energy, 88–89
Glucose biosensors, 15–16
Glucose dehydrogenase, 137
Glucose oxidase, 136, 137
GMR, see Giant magneto resistance
Gouy-Chapman double layer, 31, 65
Grahame layer, 32
Graphitic carbon nitride, 111

Health, energy, and environment, electrochemistry application for
    biosensor, 15–16
    chemiresistive gas sensor, 12–14
    hydrogen generation, 11–12
    wastewater treatment, 17
Helmotz double layer, 31
History, of electrochemistry, 1–4
HRT, see Hydraulic retention time
Hybrid solar cells, 52–54
Hybrid supercapacitors, 67–68
Hydraulic retention time (HRT), 87
Hydrogen
    production, from biomass and water, 113–116
    and solar energy, 106
Hydrogen embrittlement, 34
Hydrogen peroxide (H_2O_2) biosensors, 16
Hydrogen sulfide electrochemical sensors, 258–259
Hydroxyl radicals, 156, 157, 182, 184

IHP, see Inner Helmholtz plane
Incident photon to current conversion efficiency (IPCE), 108
Indigo carmine dye, 188–189
Indirect electrolysis, 17
Indirect electrolytic decontamination, 158
Indirect oxidation, 159–160
Indium phosphide, 11
InGAN, 111
Inner Helmholtz plane (IHP), 65

Innovative material, in electrochemical glucose sensors, 135–136
    enzymatic glucose sensor and, 136–137
    nonenzymatic glucose sensors, 137–142
        alloying, 142–144
        doping, 145
        functionalization, 142
        morphology, 144
        nanocomposites, 145–146
Inorganic heterojunction and organic solar cells applications, 50–52
Intensiostatic/galvanostatic method, 4
Ionic liquids, 74–75
Ion-sensitive electrodes (ISEs), 258–259
IPCE, see Incident photon to current conversion efficiency
ISEs, see Ion-sensitive electrodes

Lewis basicity, 127
Limit of detection (LOD), 129
Linear response, 130
Linear sweep voltammetry, 253
LOD, see Limit of detection

Marcus model, 32
Mass transfer, 30
Materials synthesis, controlled electrochemical deposition for, 26–27; see also Electrodeposition
Material synthesis, by electrode position method
    electrophoretic deposition, 10–11
    metals and alloys electrodeposition, 5–10
    metals electrodeposition, as sacrificial coatings, 10
    organic materials electrodeposition, 10
Maxwell-Boltzmann statistic, 31
MDC, see Microbial desalination cell
MEC, see Microbial electrolysis cell
Metal azolate framework, 142
Metal ion doping, 145
Metal nanoparticles, 137–138
Metal organic frameworks (MOFs), 128
Metal oxides, significance of, 138–139
Metal oxide sensors, 126
Metals and alloys electrodeposition, 5–10
Metals electrodeposition, as sacrificial coatings, 10
Metal tellurides, 141
Methylene blue dye, 186–188
MFCs, see Microbial fuel cells
Microbial desalination cell (MDC), 97
Microbial electrolysis cell (MEC), 97
Microbial fuel cells (MFCs), 84–85
    activation losses, 92–93
    biocatalyst metabolic losses, 93
    components of, 86–87
    concentration losses, 93
    future directions, 100
    limitations, 97–98
        biochemical constraints, 98–99
        economic constraints, 99–100
        functional materials constraints, 99
    microbial conversion of organic material into electricity production prototypes, 94–97
    microbial conversion of organic matters into fuel, 97
    ohmic losses, 93
    parameters in treatment of, 87–88
    potential losses, 91–92

thermodynamics of, 89–91
Microelectrode, 254
Microorganisms, 87
Migration, 30
Mild steel corrosion and prevention, A-MWCNTs/tetra functional epoxy coatings for, 169–172
   experimental details, 172
      mild steel specimen surface preparation, 172–173
      sample preparation, 173
   results and discussion, 174–175
      electrochemical impedance study, 176
      morphology studies, 176–177
      thermal properties, 175–176
Mixed transition metal oxides (MTMOs), 69
MOFs, see Metal organic frameworks
Molybdenum disulfide, 113
MP-P configuration, 197, 198
MP-S configuration, 197, 198
MTMOs, see Mixed transition metal oxides
Multilayer coatings, 41–43
Multiwalled carbon nanotubes (MWCNTs), 170

N,N,N′N″-Tetrakis(2,3-epoxypropyl)-4,4-(1,4-phenylenedioxy) dianiline synthesis, 172
Nafion, 99
Nanobiosensors, electrochemical-based, 14
Nanocomposites, 145–146
Nanofiltration, 208
Nanoparticles, 127–128
NER, see Normalized energy recovery
Nernst equation, 90
Nernst-Monod model, 92–93
NHE, see Normal hydrogen electrode
Nickel–zinc alloy deposits, 35–36
Ni-Mn alloys, 37–38
Nitric oxide electrochemical sensors determination, 257–258
Nitride, in PEC, 110–112
Ni–Zn–Fe system, 36
Nondissolvable anodes, 26
Nonenzymatic biosensors, 16
Nonenzymatic glucose sensors, 137–142
   alloying, 142–144
   doping, 145
   functionalization, 142
   morphology, 144
   nanocomposites, 145–146
Non-faradaic current, 28
Nonpolarizable electrodes, 28
Normal hydrogen electrode (NHE), 90
Normalized energy recovery (NER), 88
Nucleic acids, electrochemical sensors for, 259–261

OER, see Oxygen evaluation reaction
Ohmic losses, 93
OHP, see Outer Helmholtz plane
Oil industry, 201–202
Organic electrolytes, 74
Organic materials, 217–219
   electrodeposition, 10
   photoelectrochemical measurements, 219–220
      solar cells preparation, 220
   as photosensitizers in DSSC, 220–228

Organic molecules, as additives in redox electrolyte in DSSCs, 220–221
Outer Helmholtz plane (OHP), 65
Oxidant generation mechanism, 183
Oxidation, significance of, 4, 26
Oxide electrodeposition, 36
Oxygen evaluation reaction (OER), 156, 160
Oxynitride, in PEC, 113

Palladium alloys, 38
PANI, see Polyaniline and derivatives
Paper-based electrochemical sensors and inkjet-printed, 261–262
Paper-cup-based integrated microbial fuel cell (PC-MFC), 95
PEC, see Photoelectrochemical process
PED, see Pulsed electrodeposition
PEDOT, see Poly ethylenedioxythiophene
Perovskite solar cells, 57–58
Pharmaceutical industry, 202–204
Phosphide, in PEC, 112
Photoelectrochemical (PEC) process, 106–107
   analysis, 225
   calculations in, 108–109
   experimental setup, 107
   hydrogen production from biomass and water and, 113–116
   mechanism of, 107–108
   quantum-dot-based photoelectrode and, 113
   water splitting
      metal oxide as photoelectrode in, 109–110
      nonmetal oxide as photoelectrode in, 110–113
Photovoltaic energy generation system, 49–50
   electrochemical synthesis
      for dye-sensitized solar cells, 54–56
      for hybrid solar cells, 52–54
      inorganic heterojunction and organic solar cells applications, 50–52
      in perovskite solar cells construction, 57–58
      for quantum dot-sensitized solar cells, 56–57
PHSS, see Polarographic H_2S sensor
Plant-intermitted MFC prototypes, 96
Point-of-care testing (POCT), 261
Polarizable electrodes, 28
Polarographic H_2S sensor (PHSS), 259
Polyaniline (PANI) and derivatives, 234–235, 239
Poly ethylenedioxythiophene (PEDOT), 56
Polymer electrolytes, 26
Polymers, significance of, 126
Polypyrrole (PPY), 126, 235
Polythiophene and derivatives, 235
Polyvinyl alcohol (PVA), 69
   removal of, 199
Positive electrode, 4
Potentiometry, 4
Potentiostatic setup, 4
Power, 87
Power output comparison, 98
PPY, see Polypyrrole
Pseudocapacitor, 64, 66–68
   advances in materials for, 69–72
Pt alloys, 142–143
Pulp and paper industry, 202
Pulsed electrodeposition (PED), 27
PVA, see Polyvinyl alcohol

Index

QDSSCs, *see* Quantum dot-sensitized solar cells
Quantum-dot-based photoelectrode, 113
Quantum dot-sensitized solar cells (QDSSCs), 56–57

RE, *see* Reference electrode
Reactive oxygen species (ROS), 261
Receptor function, 13–14
Recovery time, 129
Redox electrochemical sensors, 261
Redox-type electrolytes, 73
Reduced graphene oxide (rGO), 110
Reduction, significance of, 4, 26
Reference electrode (RE), 4, 26, 254, 257
Response time, 129
Reversability parameter, 129
rGO, *see* Reduced graphene oxide
Rhodamine B dye, 188
ROS, *see* Reactive oxygen species

Sacrificial anodes, 26
SAED, *see* Selected area electron diffraction
Scanning electron microscope (SEM), 226
    analysis, of organic-compound-doped polymer electrolytes, 225, 227–228
Selected area electron diffraction (SAED), 37
Selectivity
    parameter, 129–130
    poor, 136
Sensitivity
    parameter, 129
    poor, 136
Sensor response, 124
Sensors, definition of, 121
Separators, 86–87
Sodic–saline inoculum, electrical stress directed, 94
Solar cells, 235–238
Solar energy, 106, 217
Solar to electric energy conversion efficiency, 219
Solar to hydrogen (STH) efficiency, 107, 108
Specific energy consumption, 88
Spin trapping, 157
Square wave voltammetry, 253
Stability parameter, 129
Standard electrode potential, 29
Stern layer, 32, 65–66
STH, *see* Solar to hydrogen efficiency

Sulfide, in PEC, 113
Supercapacitor, 32, 64–65, 240
Symmetric supercapacitors, 64

Tantalum oxynitride (TaON), 113
TEM, *see* Transmission electron microscope
Ternary alloys, electrodeposition of, 36
Textile and tannery industry, 200
Textile dyes, 181–182; *see also* Electrochemical advanced oxidation processes
Titanium-dioxide-based electrodes, 109
TMOs, *see* Transition metal oxides
Transducer function, 13, 14
Transient current, *see* Non-faradaic current
Transition-metal-based alloys, 143
Transition metal oxides (TMOs), 69
Transition metal phosphides, 141
Transmission electron microscope (TEM), 37
Trasatti-Buzzanca model, 32
Trimethylamine, 128
Triphenylmethane dye, 189–190
Two-photon system, 12

Utility factor, 13, 14

Van der Waals (vdW) force, 194, 195
Vertical integrated MFC, 94–95
Voltammetry methods, 252–253

Wastewater treatment, 17; *see also* Electrocatalysts, for wastewater treatment
Water splitting over semiconductor photocatalysts, 12
WE, *see* Working electrode
Wire wool, 200
Working electrode (WE), 254, 256

X-ray diffraction (XRD) studies, 224

Yellow 3 (DY3) dye, 189

Zeta potential, 40
Zinc oxide, 139
Zinc-oxide-based electrodes, 109
Zn–Fe–Mo system, 36
Z-scheme photocatalysis, 12